Contemporary Clinical Neuroscience

Series Editors
Ralph Lydic
Helen A. Baghdoyan

For further volumes:
http://www.springer.com/series/7678

David W. McCandless

Kernicterus

 Humana Press

David W. McCandless
Department of Cell Biology & Anatomy
Rosalind Franklin University of
    Medicine & Science
Chicago Medical School
3333 Green Bay Road
North Chicago, IL 60064
USA
david.mccandless@rosalindfranklin.edu

ISBN 978-1-4419-6554-7   e-ISBN 978-1-4419-6555-4
DOI 10.1007/978-1-4419-6555-4
Springer New York Dordrecht Heidelberg London

Library of Congress Control Number: 2010930108

Humana Press is part of Springer Science+Business Media (www.springer.com)

This volume is dedicated to Jeffrey and Steven

Preface

Kernicterus (bilirubin encephalopathy) is a highly interesting example of metabolic encephalopathy. It fills all the characteristics of a metabolic encephalopathy in that it can develop rapidly, produces signature signs and symptoms, and is amenable to successful treatment. In the absence of treatment kernicterus can produce irreversible damage, devastating sequelae, and death.

The present volume will examine the history, biochemistry, and physiology of bilirubin, as well as its hepatic metabolism and renal excretion. Chapters will elaborate on bodily disposition of bilirubin and its neuropathology. Both early treatments and current therapy will be discussed in detail. Phototherapy will be presented, and its efficacy and influence on incidence thoroughly examined. New concepts relating to the way in which bilirubin is toxic, and damage to the acoustic system will be examined in detail. The promising treatment using gene therapy will be discussed. This therapy has been successfully used in Gunn rats.

When red blood cells break down, the resulting components include heme and globin. Bilirubin is contained in the heme moiety. The life span of red blood cells is about 120 days in adults, and about 90 days in the newborn. In utero, this significant amount of bilirubin is transferred across the placenta to be excreted by the maternal liver. At birth, the newborn liver is faced with increasing levels of bilirubin, and hepatic capacity for conjugation and excretion have yet to be "tested." It is not surprising that 20% (or more) of newborn children develop a transient physiological jaundice as the enzymatic machinery becomes functional.

Bilirubin travels in plasma bound to albumin. When taken up by the liver, bilirubin is transformed from lipid soluble to water soluble, facilitating excretion into the biliary system. When bound to albumin, or conjugated with glucuronic acid, bilirubin cannot enter the brain. In the "free" state, lipid soluble bilirubin can enter cerebral tissue, causing brain damage. Why bilirubin enters selective cerebral areas is a fascinating and unanswered question.

Many animal models of kernicterus have been used, but one stands out. A Wistar rat mutant described by C. K. Gunn had a genetic deficiency of the liver bilirubin-conjugating enzyme glucuronyl transferase. This defect correlates with the human inborn error, the Crigler–Najjar syndrome. This exciting finding permitted investigators to study a nearly perfect animal model of kernicterus. The Gunn rat model permitted studies on energy metabolism in neurologically-compromised animals, and defects in ATP metabolism were, not surprisingly, found. Early in vitro studies predicted these definitive results in that they showed conclusively that bilirubin uncoupled oxidative phosphorylation. Electron microscopy reinforced these findings, showing selective changes in cerebellar Purkinje cells and their mitochondria. More refined studies directly measuring ATP in nanogram layers of Purkinje cell-rich layers of the cerebellum of symptomatic Gunn rats showed similar effects on energy metabolites, but not in adjacent molecular and granular cell layers (Reynolds, S., and Blass, J. 1976).

Early neuropathological studies in humans noted the selective anatomical staining. The underlying cause of human kernicterus is frequently erythro-blastosis foetalis, or a deficiency of the liver bilirubin conjugation system. Human brains of kernicteric infants often show yellow staining in the basal ganglia, cerebellum, and brainstem. The focal staining is marked when the patient dies during the acute phase; if the patient dies later, the staining has largely vanished. Before phototherapy, many cerebral palsy patients were the result of hyperbilirubinemia.

One long-standing treatment for newborn jaundice is phototherapy. The benefits of light for ameliorating jaundice were serendipitously discovered by a pediatric nurse who noticed jaundiced newborns experienced fading of the yellow skin pigmentation when exposed to sunlight. Further studies on this important observation showed that phototherapy decreased serum unconjugated serum bilirubin levels. Widespread use of this therapy in the USA lagged behind its use in Europe. Bilirubin is structurally similar to the pigment phycobilin, which captures light energy, and so bilirubin is isomerized into compounds which are water soluble, not toxic, and easily excreted.

The wavelength of light used during phototherapy impacts the extent of bilirubin isomerization. Various studies have demonstrated that turquoise light (490 nm peak with 65 nm spectral width) is considered more effective than blue light (452 nm peak with 55 nm spectral width). Side effects to phototherapy are minor, and include epidermal cell death, ultraviolet light burn, and headaches and vertigo in hospital staff. Because of the limited side effects and strong benefits, phototherapy has continued as an effective treatment for hyperbilirubinemia in the newborn.

The present volume is designed to examine the biochemistry and physiology of bilirubin. Its formation, metabolism, transport, and excretion will be examined in detail. Moreover, the deposition of unconjugated bilirubin into highly specific newborn brain regions will be described in terms of mechanisms. Treatment modalities before and after bilirubin stains brain regions will be considered. The very important concept that kernicterus can cause subtle changes in mentation and behavior will be examined. This area of sequelae to mild degrees of bilirubin entry into brain tissue is critical and largely overlooked. Chapters detailing hyperbilirubinemia in older patients and examining albumin and blood transfusion treatment paradigms are included.

Kernicterus is an excellent example of a metabolic encephalopathy. It clearly produces a biochemical lesion early when the pigment first enters the brain. If initial staining can be minimized, or reversed, permanent damage can be reduced to a minimum. This takes strict attention to detail, and a keen ability for timely diagnosis.

The realization that the early diagnosis of any disorder which might have a bio-chemical (metabolic) basis has the potential for rapid reversal/improvement, if not outright cure, argues strongly for increased awareness by all health care workers. Kernicterus has a potential advantage in that hyperbilirubinemia usually appears in the newborn nursery when it can be recognized by nurses and by physicians. Another huge advantage for the study of kernicterus is the availability of the Gunn rat model. Hyperbilirubinemia and resultant kernicterus is quite another issue in Third World countries. These world health concerns are of such importance that they constitute the first chapter in this volume.

Chicago, Illinois   David W. McCandless

Acknowledgements

The author wishes to thank Dr. Joseph DiMario for many discussions regarding the content of this volume. Dr. Reinhard Oesterle was kind enough to translate many pages of early German studies on bilirubin and kernicterus. Charles Mininger provided important information regarding counseling for kernicteric patients. Laura Muncie, a neurophysiologist with many years experience with human auditory brainstem responses, provided highly valuable input for the ABR chapter. Dr. Michael Norenberg, Director of Neuropathology, Department of Pathology, University of Miami School of Medicine, Miami Florida graciously provided excellent slices of brains from kernicteric patients. One of these has been used on the cover of this volume.

Mrs. Cristina Gonzalez was tireless in obtaining both new and old references, helping to organize the book, and in preparing many tables. Mrs. Vilmary Friederichs was also essential in organizing the volume, and in facilitating its completion. My son, Jeffrey McCandless, NASA Ames Research Center, Palo Alto, California spent many hours preparing figures for this volume. Ms. Ann Avouris, Senior Editor at Springer Science and Business Media helped me in many ways. She can solve any problem, answer any question, and does so with incredible speed and accuracy. I have never known, nor even heard of any other publisher's editors performing in such an outstanding manner.

I am very fortunate to have worked with Dr. Steven Schenker early in my career. Steve has unbounded enthusiasm and motivation for clinically oriented research which is highly infectious. Words cannot express my gratitude to Steve.

Finally, my wife Sue, has provided many years of encouragement and support for these and other endeavors. Without her help, this volume would not have been possible. I cannot thank her enough.

Contents

Prologue: World Health Concerns

The enormous number of cases of hyperbilirubinemia worldwide can hardly be estimated. The World Health Organization estimates the number of people worldwide who have glucose-6-phosphate dehydrogenase (G-6-PDH) deficiency at 400 million. G-6-PDH deficiency is the disorder that contributes the largest number of jaundiced newborn infants. Endemic areas for G-6-PDH deficiency include sub-Saharan Africa, Asia, India, South America, and portions of the Middle East. These areas are also endemic for malaria. Rates of G-6-PDH deficiency include Nigeria, 28%, Cambodia, 15%, China, 8%, Myanmar, 11%, Thailand, 9.4%, Iran, 11.6%, etc. (Beutler, E., and Duparc, S. and the G-6-PDH working group 2007).

G-6-PDH deficiency results in severe hemolytic anemia, especially when triggered by some seemingly inoculus incident such as eating fava beans. Estimates place hyperbilirubinemia (greater than 15 mg/100 ml) at an incidence of 30% in G-6-PDH-deficient patients during a hemolytic crisis. Conversely, in severe neonatal jaundice, it has been shown that G-6-PDH is the most important factor in the hyperbilirubinemia (Bienzle, U., et al. 2008). In a similar vein, (Gibbs, W., et al. 2008), a total of 23 Jamaican infants with moderate/severe jaundice were examined for the presence of G-6-PDH. Results showed that the deficiency was present in 70% of the hyperbilirubinemic neonates. This was far greater than the 9.4% deficiency noted in nonmoderate/severely jaundiced newborn infants. As regards treatment in this group of neonates, phenobarbitol and phototherapy helped most, but still eight required exchange transfusion. Even with this treatment, two became kernicteric, and one of these died.

In a study of neonatal morbidity and mortality in newborn infants in a tertiary hospital in Ilesa Nigeria, 7,225 infants were included for study. The highest rate of indications for hospital admission was neonatal jaundice at 45%. In this study, jaundice was listed as the second highest cause of death in this group, at 14% (Owa, J., and Osinaike, A. 1998). Jaundice as a cause of death means the newborn infant died of kernicterus. The incidence of survivors with cerebral palsy is not mentioned. If this kind of morbidity/mortality occurs in a town with hospitals, etc., imagine what the mortality rate would be in rural areas in countries without any medical superstructure. In Zimbabwe, for example, several years ago there was only one neurologist in the country. The number of newborn dying each year of kernicterus

D.W. McCandless, *Kernicterus*, Contemporary Clinical Neuroscience,
DOI 10.1007/978-1-4419-6555-4_1, © Springer Science+Business Media, LLC 2011

must be in the thousands. We tend to become complacent given the low occurrence of kernicterus in developed countries.

One major problem regarding G-6-PDH deficiency is that in newborn infants, it cannot be diagnosed by clinical observation alone. Therefore, even in infants born under some level of medical supervision in third world countries, the mother/infant may be discharged before any "triggering" mechanism occurs. The result is the possibility for rapid onset of hemolytic anemia, and rapidly rising hyperbilirubinemia. This may leave the newborn brain damaged from unconjugated bilirubin staining of select brain areas.

One question is what medical treatment is available even if this rapid onset of hemolytic anemia occurs in a third world hospital setting? The answer would seem to be that such a setting would have few to none of the blue light photo therapy units deemed necessary to effectively reduce elevated serum bilirubin, as has been demonstrated in the USA. A recent paper from T. Hansen's group (De Carvalho, M., et al. 2007) suggests an alternative plan. They were able to demonstrate that seven ordinary day light fluorescent tubes placed close to the jaundiced infant effectively reduces unconjugated hyperbilirubinemia. This was achieved by placing the array of lights underneath a clear plexiglass bottomed crib. This treatment was compared to the conventional blue treatment.

Results showed that in 51 jaundiced newborn infants, decreases in serum bilirubin levels were greater in the two photo therapy methods than regular overhead lights (no photo therapy). The increased rate of bilirubin elimination by the two photo therapy was measured at 12 and 24 h after initiation of the photo therapy. The authors state that

"lack of access to expensive imported special blue lamps does not preclude delivery of effective photo therapy in developing countries."

As a side note, a student at Duke University has developed a simplified photo therapy device which uses blue lights at a reported cost of $500 per unit as compared to $4,000 for the commercially available unit. Prototypes are being tested in Tanzania.

In terms of world health concerns regarding hyperbilirubinemia and kernicterus, the major disorder producing morbidity and mortality is G-6-PDH deficiency. This genetic mutation is responsible for over 400 million cases of G-6-PDH deficiency. The number of newborn infants with the disorder, who are subsequently subjected to a precipitating event, which ends in kernicterus, is probably beyond estimation, but certainly runs to thousands of cases per annum. Many cases go unreported in developing countries due to political reasons and lack of knowledge about hyperbilirubinemia and kernicterus. Education and a closer attention to health issues is the first step to ameliorating this tragic and indefensible situation.

G-6-PDH deficiency was first described (Carson, P., et al. 1956) in patients undergoing treatment for malaria when primaquine was administered. The "primaquine sensitivity" was observed in ethnic groups, suggesting a genetic component. Later observations in several ethnic groups showed similar sensitivities to other precipitators such as fava beans. The disorder is one of severe hemolytic anemia accompanied with severe hyperbilirubinemia, sufficient to result in kernicterus and death. The World Health Organization estimates as many as 400 million people worldwide with G-6-PDH deficiency.

Glucose-6-phosphate dehydrogenase catalyzes the first (entry) step in the hexose monophosphate shunt (pentose phosphate pathway). This shunt, with the enzyme transketolase as the rate-limiting step, serves at least two functions. First, the shunt provides reducing power in the form of NADPH. This, for example, maintains glutathione in the reduced form, GSH. GSH in turn acts to detoxify H_2O_2, thereby protecting cells from the oxidizing effects of H_2O_2. Another function of the hexose monophosphate shunt is to produce ribose phosphate essential for the formation of the ribose used in RNA (see Fig. 1).

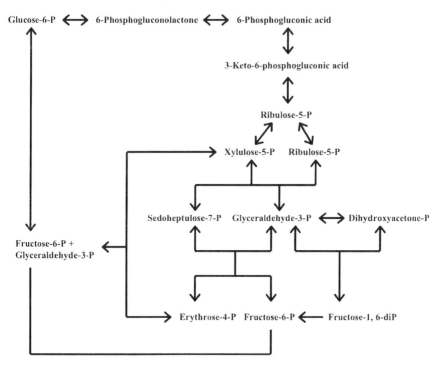

Fig. 1 Schematic representation of the hexose monophosphate shunt

As stated above, deficiencies of G-6-PDH are the most commonly inherited enzyme deficiencies. This genetic deficiency may affect as many as 25% of the population in endemic areas such as the Mediterranean, sub-Saharan Africa, and subtropical Asia. The inheritance of G-6-PDH deficiency is classified clinically as an x-linked recessive disorder. In spite of this, heterozygous females can develop symptoms. The genetic diversity of G-6-PDH deficiency is vast, with over 499 variants of the deficiency, each with unique biochemical features.

Many individuals who have G-6-PDH enzyme deficiency never show any significant symptoms. When symptomatic, the key feature is acute hemolytic anemia and accompanying jaundice. Several compounds may precipitate the symptoms of G-6PDH deficiency such as fava beans, primaquine, and toxins from infection. Primaquine-induced hemolytic anemia in blacks first led to the discovery of G-6-PDH deficiency. The association of fava bean ingestion and sickness (anemia) dates back as far as early Greece.

There is a relatively simple screening test for G-6-PDH deficiency based on fluorescence (Beutler, E., and Mitchel, M. 1968). Given the distribution of G-6-PDH deficiencies, as well as the thalassemias, it has been suggested that these two entities provide some protection from malaria (Luzzatto, L., Usanga, E., and Reddy, S. 1969). This has been borne out in females deficient in G-6-PDH, but not in males.

Clinically, once hemolytic anemia starts in a G-6-PDH deficient patient, within 2 days, fever starts, jaundice increases, and urine turns brown/black. There is a significant risk of renal tubular necrosis. Exchange transfusion can serve to remove damaged red blood cells (hemighost cells) which compromise microcirculation in the kidney (Chan, T., Chan, W., and Weed, R. 1982). see Fig. 2.

Fig. 2 Smear of G-6-PDH deficiency patient's blood. Note hemighost red blood cells

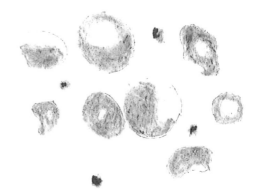

Several other illnesses, when present in G-6-PDH patients may precipitate a severe bout of hemolysis. Typhoid fever is one example (Chan, T., et al. 1971). This was demonstrated by transfusing G-6-PDH deficiency cells into non-G-6-PDH deficient patients just developing typhoid fever. Viral hepatitis is another disease which is worsened in patients with G-6-PDH deficiency.

It is clear that in G-6-PDH–prevalent areas in the world, there is a significantly higher incidence of neonatal jaundice and bilirubin encephalopathy/kernicterus. The jaundice usually starts to rise around the end of the first week, and many extend into the third or fourth postnatal weeks. A variety of factors may precipitate including various drugs and infections (Panizon, F. 1960; Yeung, C. 1992). Where possible, newborn infants should be screened for G-6-PDH deficiency, and bilirubin values monitored closely. When photo therapy is quickly and aggressively administered, requirements for exchange transfusions are reduced. Unfortunately, many areas, for example, rural sub-Saharan Africa, do not have access to G-6-PDH deficiency screening tests, no access to photo therapy, and little knowledge of the risks involved (see Table 1).

Table 1 Drugs which may precipitate hemolysis in G-6-PDH patients

Drugs with the potential to precipitate hemolysis in G-6-PD patients	
Furazolidone	Phenazopyridine
Glibenclanide	Phenylhydrazine
Isobutyl nitrate	Primaquine
Nalidixic acid	Sulfacetamide
Naphathalene	Sulfanilamide
Niridazole	Sulfapyridine
Nitro gurantoin	Urate oxidase

It is certain that if screening tests for G-6-PDH deficiency are available, and the means to implement photo therapy, and even transfusions, then the ravages of kernicterus (cerebral palsy) can probably be averted in most cases. It is imperative for these methods and education regarding newborn jaundice to be made available to areas which have many hundreds of cases of kernicterus per year, such as Africa and India.

The severity of the G-6-PDH deficiency in terms of enzyme inactivity may not always correlate with the degree of hemolytic anemia (Yoshida, A. 1973). Some variants of G-6-PDH deficiency have severe enzyme deficiencies like the Union and Markham varieties, but have little hemolytic activity. Other variations, such as the Alhambra and Triple varieties have a less pronounced loss of enzyme activity but have severe chronic hemolytic anemia.

The incidence of G-6-PDH deficiency in Egypt has been examined. In the first such study (Rageb, A., El-Alfi, O., and Abboud, M. 1966), 500 Egyptian men were treated for G-6-PDH deficiency, and the incidence was placed at slightly over 26%. In a later study (McCaffrey, R., and Awny, A. 1970), 650 males were tested for the deficiency. They were randomly selected from blood donors at a Cairo hospital. The tests administered were the ascorbate cyanide test and the methemaglobin reduction test for G-6PDH deficiency.

Results showed an excellent correlation between the two tests when both were used on the same patient. The incidence in this study was only 4.9% positive for the deficiency. This result is a striking difference from an earlier study showing a 26% incidence. The authors speculate that methods of testing may have produced false positives in the earlier study.

Another study examined the incidence, severity, and association with Gilberts syndrome in Sephardic Jews who were descendents of families who immigrated from Asia Minor (Kaplan, M., et al. 1997). The authors state that the worst consequence of G-6-PDH deficiency is neonatal hyperbilirubinemia, which can easily result in kernicterus and death. The exact cause of hyperbilirubinemia in these cases is uncertain, and could be related to a lack of glucuronyl transferase (glucuronosyl transferase) or increased bilirubin production due to massive hemolysis, or most likely by a combination of both. The rate of hemolysis is higher in Sephardic Jewish patients than others, but a precipitating agent is not always discernible. In G-6-PDH deficiency, the level of hyperbilirubinemia is not always related to the severity of hemolysis. Decreased bilirubin elimination is thought to be a factor in hyperbilirubinemic patients in this ethnic group. The possibility exists that the Gilberts disease (UDPGT1 deficiency) gene may be present in G-6-PDH deficiency thereby increasing the incidence for neonatal hyperbilirubinemia.

Results showed that of 371 patients in the study, 131 (35%) were found to be G-6-PDH deficient. Of these neonates, 23% developed hyperbilirubinemia, much higher than the control group. Hyperbilirubinemia was defined as values over 15 mg/100 ml. When the incidence of hyperbilirubinemia was compared to the combination of G-6-PDH and each of the three UDPGT1 genotypes, the incidence of elevated bilirubin levels increased. These data strongly suggest that a variant UDPGT1 promoter is playing a role in the neonatal hyperbilirubinemia associated with G-6-PDH deficiency.

The authors suggest that while G-6-PDH deficiency is thought to be precipitated by some compounding factor such as fever or drugs, these data suggest that in addition to G-6-PDH deficiency, hyperbilirubinemia requires an additional factor – a mutation in UDPGT1. The mutation by itself is innocuous, but in combination with G-6-PDH deficiency in newborn infants, may result in bilirubin values high enough to result in kernicterus. The authors note that not too many gene–gene interactions have been described, but that this type of phenomenon is present in Sephardic Jewish families.

In another study (Medina, M., et al. 1997), the molecular genetics of G-6-PDH deficiency was studied in Mexico. The incidence of G-6-PDH deficiency had been quoted as between 0.4 and 4.1% depending on the region analyzed. The highest rate occurred in the Mestizos area, while the lowest rate was from several Indian groups (Lisker, R., Cordova, M., and Zarate, G. 1969; Gonzalez-Quiroga, G., et al. 1994).

The results examining the molecular genetics of G-6-PDH deficiency showed that the G-6-PDH-A phenotype was common in Mexico. In addition, five different genotypes were seen among the patients studied. The authors note that their data also showed that in Mexico, G-6-PDH deficiency is heterogeneous at the DNA level.

Another study (Santhi, A., and Sachdeva, M. 2004) looked at the G-6-PDH deficiency incidence in Jats and Brahmins of Sampla, Haryana. These are two geographic areas in the northern portion of India. Results from this study were from 288 total individuals. Of these, 136 were Jats, and 152 were Brahmins. Screening for G-6-PDH deficiency was accomplished by the methemoglobin test. Results showed that the incidence for G-6-PDH deficiency in Jats was 11.5% and for Brahmins, 10.5%. This was a nonstatistically significant difference. Analysis by sex showed a statistically significant higher rate of the deficiency among males in each group. The authors note that these results are in general agreement with previous reports of a higher rate of incidence in northern and western portions of India (Bhasin, M., and Walter, H. 2001).

The incidence of G-6-PDH deficiency has been examined recently in Northeast Iran (Mohammadzadeh, A., et al. 2009). This was a prospective study to determine the prevalence of G-6-PDH deficiency in Mashhad city in Iran. Over a period of 8 months in 2006, all newborns were screened for G-6-PDH deficiency. Blood samples were tested by the fluorescent spot method.

Results showed a low percent (0.8%) of positive tests for G-6-PDH in a total of 2,570 neonates. This is compared to the WHO estimate of 2.9% of the world's population as being deficient in G-6-PDH (WHO working group 1989). Results from other areas of Iran ranged from 1.2% in Tehran to over 15% in the northern parts (Mazandaran) and southern portions (Shiraz) of the country. The overall incidence of G-6-PDH in Iran is estimated to be about 11.5%.

Another study looked at the prevalence of G-6-PDH deficiency in an area of the Brazilian Amazon which is endemic for malaria (Santana, M., et al. 2009). Two hundred patients' blood was tested for G-6-PDH deficiency, and 6 (3%) positive patients were identified. Testing for malaria disclosed an incidence of 1.5% in the 200, and half of the six G-6-PDH-deficient group.

The authors point out that in a hemolytic crisis, there may possibly be increased numbers of reticulocytes, which could lead to false negative values of G-6-PDH deficiency. The increased numbers of peripheral reticulocytes may themselves have normal levels of G-6-PDH. Chactacterization of the G-6-PDH isoenzymes showed high frequency of the phenotype A- G6PDH deficiency, verifying findings in Brazil.

The authors state that they observed a statistically significant protection in their patients with G-6-PDH deficiency against malaria. This also is in agreement with earlier reports of a protective action of the deficiency against malaria (Guindo, A., et al. 2007); Roth, E., et al. 1983). One outcome of the simultaneous occurrence of G-6-PDH deficiency and malaria (usually *Plasmodium vivax*) is a significantly higher level of hyperbilirubinemia requiring an increased frequency of blood transfusions. This study, the authors note, is of benefit because it shows the incidence of having both G-6-PDH deficiency and malaria (3%). This will serve to help health care workers to develop treatment policies for G-6-PDH deficiency in endemic malaria areas. These studies showing the decrease of hyperbilirubinemia in cases in which patients have both malaria and G-6-PDH deficiency is an alert to the need for vigilance in order to initiate therapy to lower bilirubin levels. This is especially true of afflicted newborn infants in order to prevent kernicterus.

Another review paper recently published (Beutler, E., Duparc, S., and the G-6-PDH deficiency working group 2007) the association of G-6-PDH deficiency and malaria, and malaria treatment regimes were studied. G-6-PDH deficiency is a hemolytic challenge in severe cases, and the possibility of augmented hemolysis from anti-malarial drugs should be avoided. The authors state that G-6-PDH deficiency in children has been shown to be associated with an approximate 50% reduction in the risk for malaria (Cappadoro, M., et al. 1998).

The present chapter focuses on the combination of G-6-PDH deficiency and malaria. In the presence of G-6-PDH deficiency, hemolytic anemia is the key symptom. Indeed, G-6-PDH deficiency was discovered based on investigations of primaquine-related hemolytic anemia. Decreased hemoglobin levels is an easily followed clinical sign of hemolytic anemia. The authors note that neonatal jaundice is the most catastrophic result of G-6-PDH deficiency, and can lead to severe kernicterus. A variety of drugs should be avoided in patients with G-6-PDH deficiency as they may precipitate hemolysis. Many of these drugs are listed in Table 1.

The authors state that it is difficult to predict the effects of anti-malarial drugs as regards hemolytic anemia in G-6-PDH-deficient patients. They suggest some sort of a classification which "ranked" the possibility for hemolytic anemia in G-6-PDH-deficient patients would be invaluable. While some animal (mouse) studies have been directed toward this problem, there are some differences in results as compared to humans. The authors suggest that human studies be utilized in early clinical trials, and to start by giving newly developed anti-malarial drugs to normal subjects and assessing hemolysis.

For patient screening for G-6-PDH deficiency, the fluorescence test (qualitative) is excellent. It, however, is only capable of reliably detecting G-6-PDH deficiency in males and homozygous females. In clinical trials (for example, chloroproquanil, dapsone, and artesunate being tested in Africa), G-6-PDH-deficient patients are included in the study. Patients were carefully monitored for the first few days, and if hemolysis begins, it would be noted. Patients were also followed up at home. If there is a rapid drop in hemoglobin, treatment is stopped, and the need for transfusion carefully watched. In terms of children, desferrioxamine has been used to reduce the time of hemolysis, thereby decreasing the frequency of the need for transfusions (al-Rimawi, H., et al. 1999).

As regards testing women for the possibility of G-6-PDH deficiency, the cyto-chemical assay vs. the fluorescent spot test is discussed (Peters, A., and Van Noorden, C. 2009). Results showed that the fluorescent spot test is widely used, and is cheap and easy to perform, but not reliable for testing women. It is concluded that the cytochemical assay is best for testing women for G-6-PDH deficiency. The cytochemical assay method is more expensive and difficult to perform, but its reliability is excellent. The authors suggest a simplified kit would be a welcomed development.

Another recent paper examines hyperbilirubinemia and G-6-PDH deficiency in the Zanjan province of Iran (Koosha, A., and Rafizadeh, B. 2007). In this study, 376 newborn infants were studied. G-6-PDH deficiency was found in eight neonates (2.1%). Interestingly, the etiology was not determined in slightly over 75% of the newborns. Of the eight newborns with G-6-PDH deficiency, seven were boys and one was a female. The maximum total bilirubin value in the G-6-PDH-deficient group was 24.9 mg/100 ml.Treatment consisted of photo therapy. No G-6-PDH-deficient newborn infant received an exchange transfusion.

The authors stress, as have others, that the incidence of G-6-PDH deficiency may vary greatly in different regions of the same country. This has been described above. This may be attributed to population variations. For example, the people of Shiraz and Tehran differ from the present study in ethnicity and the inclusion of immigrants from other provinces. The authors also stress that the devastating consequence of unchecked neonatal hyperbilirubinemia is severe kernicterus, which can result in death or serious neurological sequelae.

A recent paper explores criteria for initiating national screening for G-6-PDH deficiency in Canada (Leong, A. 2007). The author suggests that several criteria should be met before starting screening. These include that early treatment will decrease morbidity and mortality, a simple test is available, the disorder being screened is sufficiently serious to warrant testing, and knowledge of results can be communicated to patients/parents, and to health care workers.

The author notes that Singapore has had a G-6-PDH deficiency screening program in place since 1965. This has dramatically reduced the incidence of kernicterus such that there has only been one case in the last 20 years! The newborn infants who test positive are kept in the hospital for about 2 weeks after birth in order to protect them from stimuli which might precipitate a hemolytic crisis, and to educate the parents about future risks (Joseph, R., et al. 1999). In addition, asymptomatic G-6-PDH infants are identified and the parents also advised of risks. These newborn infants would be missed without testing, and be subjected to hemolytic crisis without knowledge of risk, and what to do if a crisis occurs.

In terms of the criteria for testing of newborns that there should be a high frequency of G-6-PDH deficiency, the exact numbers in Canada are unknown. No large epidemiological studies have been done in either Canada or the USA looking at the frequency of hyperbilirubinemia/G-6-PDH deficiency. It is however clear that in both the USA and Canada, rapid changes are occurring in the ethnic make up of the populations. Continued influx of peoples from all parts of the world will surely increase the incidence of G-6-PDH deficiency and associated hyperbilirubinemia. Vigilance as regards ethnicity of patients can serve to minimize the serious complications of G-6-PDH deficiency, namely hyperbilirubinemia and kernicterus. The risk, as always, is that people living away from cities might not be as aware of symptoms, and delay seeking medical advise until too late in the rapid course of rising serum bilirubin levels.

Another study emphasizes the correlation of neonatal jaundice with G-6-PDH deficiency and interaction with other risk factors (Yu, M., et al. 1992). In their report, a cohort of 333 G-6-PDH-deficient newborn infants were identified and compared to a group of 653 neonates who were G-6-PDH normal. Increased jaundice (over 15 mg/100 ml) was noted in male patients but not females, and was associated with G-6-PDH deficiency. Both hypoxia/asphyxia and maternal hepatitis B antigen were associated with an increased risk for jaundice (equal to or over 20 mg/100 ml) in G-6-PDH-deficient patients. Statistical analysis showed a significant additive synergistic effect on serum bilirubin levels. Untreated, this would be expected to result in an added morbidity and mortality rate due to the toxic effects of increased unconjugated bilirubin.

This chapter on world health concerns is placed as the prolog because of the enormous significance of these clinical disorders to human well-being. Untold thousands of children die from hyperbilirubinemia, and even more are stricken with the ravages of untreatable cerebral palsy each year. Compared to massive US federal spending on trivial items, very few dollars in expense could result in a virtual elimination of this devastating consequence of hyperbilirubinemia and kernicterus around the world.

History of Bilirubin

The history of bilirubin actually goes back many hundreds of years, when newborn infants were observed to be jaundiced. More recent accounts of jaundice in newborns seems to have begun in the late eighteenth century. One of those was written by Jean Baptiste Thimotee Baumes (Baumes, J. 1806). This description was published as a chapter in a book entitled *Traite de L'amaigrissemwnt des enfans*. The chapter relating to jaundice is entitled *Traite de L'ictere ou jaunisse des enfans de naissance.*

Before describing the Baumes chapter on jaundice and other historical landmarks relating to jaundice, a recent excellent review of the pioneers in the study of neonatal jaundice and kernicterus has been published (Hansen, T. 2002). Dr. Hansen has provided very interesting and thorough descriptions of early investigators of jaundice and kernicterus. In some cases, he has obtained information from still living grandchildren of these pioneers (Orth). Information in the Hansen manuscript cannot be found anywhere else, and the reader is referred to this outstanding paper on bilirubin history found in any other source, and the book in which resides a chapter on jaundice by Baumes is interesting in its own right. The second edition of the jaundice chapter was published in 1806, and is reproduced in a book of a collection of chapters on the maladies of infants. Other chapters in this book in which Baumes' paper on jaundice appears also include chapters on pneumonia, angina, gangrene, edema, etc.

The Baumes paper is 72 pages in length, and he is listed as a professor of pathology at the School of Medicine at Montpellier, and a member of many medical societies, so he must have been a much respected physician. The copy I have is the second edition, which Hansen says is not as clearly presented as the first edition.

Baumes work was based primarily on a description of ten newborn infants who exhibited jaundice. Baumes makes reference to the liver, so an association between the two was already known. Baumes also describes somnolence and poor feeding, symptoms of cerebral involvement in some of the patients he describes. In all cases, Baumes mentions meconium as having importance. He thought that a delay in passage of meconium might be associated with, or the cause of neonatal jaundice. The idea that meconium retention was prevalent at least as early as the 1850s, when Condie (Condie, D. 1853) stated that newborn jaundice was related to failure of free release of meconium.

D.W. McCandless, *Kernicterus*, Contemporary Clinical Neuroscience,
DOI 10.1007/978-1-4419-6555-4_2, © Springer Science+Business Media, LLC 2011

Baumes also believed that maternal breast milk, and especially colostrums were beneficial to the correction of neonatal jaundice. The very first observation of Baumes was of his own jaundiced daughter, and one wonders if this fueled his interest in neonatal jaundice. The second edition of Baumes work is in a book, "Meladies des Infants" and as stated above has other chapters, some dated as late as 1841. It would seem that advances in the jaundice field moved at a somewhat slower rate than today.

In his splendid review of the early pioneers in jaundice, Hansen states that Jaques F. E. Hervieux was a critic of Baumes. Hervieux states that Baumes' ideas on the involvement of delayed passage of meconium were "without doubt ingenious, but nevertheless only an intellectual theory." He also states that "Baumes supports his opinion with a very small number of observations, and then concludes with astonishing ease to the great majority of cases."

The first description of bilirubin staining of the brain of kernicteric newborn children is credited to Johannes Orth (Orth, J. 1875). Orth reported on the presence of yellow and red pigments and crystals in the organs of newborn infants. He presented data on 37 newborns, all of whom had evidence of pigment in their kidneys, and "most" others had small amounts of yellow pigment in all other organs. In one severe case who died only 2 days after birth, the brain was yellow, but more intense staining was noted in specific brain regions including the basal ganglia, hippocampus, cerebellum, and walls of the third and fourth ventricles. Examination of the brain microscopically by Orth, revealed that neurons in the basal ganglia contained yellow pigment whereas the surrounding glial components were not stained. This patient's skin was severely jaundiced, yet a pallor was detected, leading Orth to state that the jaundice might have had a hematologic cause. Orth was an assistant to the great pathologist Virchow in Berlin when these observations were made.

Christian Schmorl in Dresden was the first to coin the term "Kernicterus." Schmorl was chief of pathology at the University of Dresden, Medical Faculty. In the Schmorl paper (Schmorl, C. 1904), 120 jaundiced newborn infants who died were autopsied. Of these, most (114) were described as having brain icterus of a diffuse yellow nature with "fat bodies" observed in some cases. The others (six) had "core icterus" in which only particular parts of the brains were yellow. These were mostly in "central ganglia" – basal ganglia, and in the "elongated ganglia" – medulla oblongata. Microscopic study of these brains showed that the yellow pigment was selectively present in neurons in nuclei.

Schmorl also noted an absence of staining in the glial component of these nuclei. It is due to the more intense staining of nuclei in the brains of newborn infants dying from jaundice that Schmorl coined the term "kernicterus" (see Fig. 1). Schmorl correctly credited Orth with publishing similar findings on intra cerebral staining by the yellow pigment.

Schmorl may have also been the first to note that the yellow pigment in brain disappeared over time unless the tissue was preserved in formalin (see Schmorl, Fig. 1).

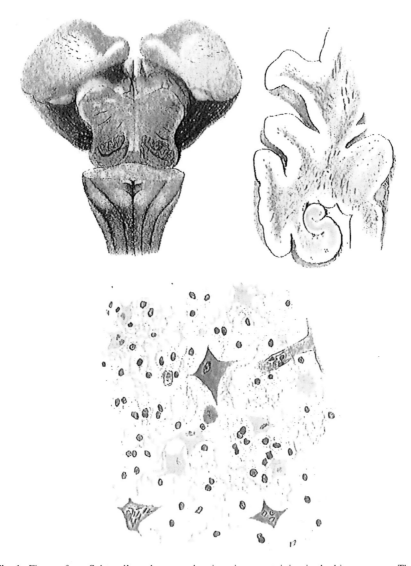

Fig. 1 Figures from Schmorl's early paper showing pigment staining in the hippocampus. These observations resulted in the term kernicterus

The concept evolved that neonatal jaundice could occur in families, and came to be called familial icterus gravis neonatorum. The understanding of the possible familial nature of jaundice was reported by several physicians, including Auden (Auden, G. 1905) and others. Interestingly, Hermann J. Pfannenstiel published a paper in 1908 (Pfannenstiel, H.J. 1908) describing a newborn case of familial icterus gravis neonatorium. This paper has been identified as the first of its kind, although several similar papers preceded it (see above). The outcome of this is that familial icterus gravis neonatorum has come to be called Pfannenstiel's disease in some quarters.

At this time, (early 1900s) many authors published cases of multiple examples of jaundice and kernicterus in single families. Arkwright (Arkwright, J.A. 1910), for example, reported a family which had 15 children, in which 14 were jaundiced and only 4 survived.

In another paper at about the same time, Rolleston, H.D. (1910), delivered a total of four newborns who became jaundiced, and three of these died. The mother of these three newborns had herself become jaundiced late in the pregnancy. The fourth pregnancy resulted in a normal newborn infant who never became jaundiced. The cause of the jaundice was stated to be "obscure."

In spite of an increasing number of published reports of neonatal jaundice, the link between hyperbilirubinemia and brain damage was not made until 1914 (Guthrie, L. 1914). This paper describes a newborn infant with familial icterus gravis neonatorum (erythroblastosis fetalis). In this paper, the neurological features of hypotonia and chorioatetosis were described in the patient at the age of 19 months. The suggestion was made that these symptoms may represent a manifestation of brain damage by bilirubin, and Guthrie is thus credited as the first to publish a description of what is now known as kernicterus.

In 1932, a key paper was published (Diamond, L., Blackfan, K., and Baty, J. 1932) which defined and combined under one heading the clinical conditions of erythroblastosis fetalis, newborn anemia, edema of the fetus, and icterus gravis neonatorum. The exact cause of these disease entities, now under one heading was unclear since the concept of blood incompatibilities was not well understood. This, nevertheless furthered knowledge of anemia and resultant hyperbilirubinemia and kernicterus. When the pathology of hemolytic disease of the newborn was ultimately described, the techniques for exchange transfusion in newborn infants were developed. This facilitated the removal and replacement of damaged red blood cells. It also reduced levels of unconjugated bilirubin, proving to be a satisfactory treatment for the reduction of serum bilirubin levels, and a dramatic reduction of kernicterus.

As a side note, in 1944, Blackfan, K., Diamond, L., and Leister, C. published an atlas of blood disorders in children. Leister was an artist of considerable talent, and produced a series of beautiful water color paintings of blood smears which constitute most of the figures in the atlas and are used with permission in this volume (Blackfan, K., Diamond, L., and Leister, C. 1944, see chapter 5).

In 1950, Vaughan, V., Allen, F., and Diamond, L. published an important paper looking at the relation of erythroblastosis and kernicterus, They also reviewed various aspects of kernicterus as it was known at that time. Earlier studies had indicated that a Rh-positive fetus born to a sensitized Rh-negative mother was at risk, and that the chances of recovery were inversely related to the degree of anemia. Frequency of kernicterus, however, was not related to other signs and symptoms.

The incidence of reported kernicterus in early studies had a wide range, from 55% to as low as 10%. In the study by Vaughn, Allen, and Diamond, an incidence over 4 years was 12%, with 4.9% of kernicteric newborns surviving. All survivors had some neurologic sequelae ranging from mild incoordination to severe motor disability and mental retardation.

The authors give a rather complete description of the clinical manifestations of kernicterus, which include developing jaundice on day 1, and by day 2–3 the jaundice becomes severe. This is accompanied by lethargy and poor feeding. Pathognomonic signs include a weakened or absent Moro response, and the presence of opisthotonic posturing. This includes rigidity and arm extension. It may also include brief high- pitched crying. Death in kernicterus is usually from respiratory failure, and seizures are not common, but do occur. Pulmonary hemorrhage is almost always associated with the end stages of kernicterus. Microscopically, the pathology consists of alveolar hemorrhage.

These authors found cerebral lesions in general agreement with those of early investigators. Specifically, lesions were found in the basal ganglia, hippocampus, cerebellum, olives, and medulla. These lesions were more diffuse in infants dying in 3 days or less, but were pronounced in newborns dying after 4–5 days. In infants surviving longer periods, the microscopic picture is one of neuronyl loss and glial proliferation. The location of the lesion is usually consistent with the signs and symptoms. Thus, newborn infants with motor involvement have lesions in the basal ganglia and cerebellum, whereas newborns with respiratory failure would show lesions in the medulla. (see Table 1 in the chapter "Neuropathology of Kernicterus", this volume).

The paper also describes the increase in incidence of kernicterus in newborns born prematurely, before 38 weeks of gestation, whether naturally or prematurely due to induction in sensitized women. The increased risk for kernicterus offsets the potential gains achieved in the prevention of stillbirths.

Kernicterus seemed to be a disease of neonatal life, reinforced by the findings that at birth, infants destined for kernicterus showed normal neurological reactions for a time before jaundice became severe. This indicates the brain was not damaged prior to birth. This odd course predates the findings of Schenker, who showed that unconjugated bilirubin easily crosses the placenta (Schenker, S. 1963). It is thought that death in cases of erythroblastosis fetalis is rarely from anemia, but that the greater the anemia, the higher the risk for developing kernicterus, the major cause of death in these cases. The authors downplay other hypotheses such as kernicterus acts to plug cerebral vessels leading to hypoxia, or that antibody transfer into the fetus at labor could be involved in the development of kernicterus.

Another milestone occurred when it was learned that the bilirubin in serum actually existed in two phases. The usual method for measuring bilirubin was by a spectrophotometric method in which bilirubin was coupled with diazotized sulfanilic acid-diazo reagent. It was afterward realized that there were two types of reactions occurring. One was termed direct, the other indirect – the reaction only occurred after alcohol was added to the reaction. The reason behind this peculiar phenomenon was at the time unclear.

Subsequent chromatography work showed that there was a slow moving fraction which gave the indirect diazo reaction, and a fast moving fraction which was the direct moiety. In 1956, three laboratories independently solved the mystery by showing that the direct moiety was formed by bilirubin glucuronide (Schmid, R. 1956; Talafant, E. 1956; Billing, B., Cole, P., and Lathe, G. 1956). Bilirubin glucuronide was a water-soluble form of bilirubin, and was easily excreted by the liver, as compared to the indirect form of bilirubin which was lipid soluble.

In another classic study, the protein binding of bilirubin was examined (Odell, G., 1959). Spectrophotometric studies showed that protein (albumin)-bound bilirubin was changed when organic anions such as salicylates and sulfonamides were injected into subjects. This was indicated by a drop in absorption at 460 mμ to values of from 420 μ to 440 mμ. The lower value is that which is associated with unbound unconjugated bilirubin.

The significance of this paper was to emphasize the competitive binding characteristics of certain organic ions such as sulfonamides and salicylates for albumin-binding sites. This acts to displace unconjugated bilirubin from albumin, facilitating its entry into brain structures, producing kernicterus. This also emphasizes the idea that simple measurement of total bilirubin levels may not be a completely accurate indication of kernicterus risk. A small dissociation constant for albumin-bound bilirubin would lead to a large percentage change in unbound bilirubin whenever there was a small increase in diffusible bilirubin.

Another very important paper was published in 1968 (Lucey, J., Ferreiro, M., and Hewitt, J. 1968). Dr Lucey's studies have centered on the prevention of hyperbilirubinemia and kernicterus in newborn infants using photo therapy. Prior to 1968, photo therapy had been used successfully in several countries in South America and Europe. There was some reluctance, however, for approval to be granted in the USA possibly due to uncertainty about potential undefined risks of photo therapy per se, and questions about the safety of the photo-degraded composition of products produced by the photo therapy.

Dr. Lucey and co-workers performed a controlled study in 111 newborn infants who were less than 2,500 g at birth. This was a randomized study in which newborn infants were assigned to either a photo therapy group, or to a control group. The assignments were made before 12 h of age. Results showed that the photo therapy treatment produced a statistically significant lowering of serum bilirubin between the fourth and sixth days. Furthermore, no correlation between the levels of albumin and the ability of photo therapy to lower serum bilirubin could be found.

No toxic effects were found which could be attributable to photo-degraded bilirubin products. Other authors had shown that photo-degraded products of photo therapy were rapidly excreted in bile and urine (Ostrow, J. 1967). This implies that the lipid-soluble nature of unconjugated bilirubin is rendered, by photo therapy, to water-soluble compounds which do not require conjugation in order to be excreted.

Since as many as 10% of preterm infants at this time had serum bilirubin levels over 20 mg/100 ml with significant risk for brain damage, the institution of photo therapy had the potential to reduce the risk of kernicterus to near zero. That is indeed what happened, and as stated in the chapter in this volume on photo therapy, the significance of this study, which demonstrated beyond doubt the safety of photo therapy, is immeasurable. Photo therapy has saved thousands of newborn infants from morbidity and mortality due to unconjugated bilirubin in the USA.

The study of Schenker, et. al. mentioned earlier, regarding the placental transfer of bilirubin was key in the understanding of why fetal hyperbilirubinemia was nonexistent, and answering in part why bilirubin levels tended to rise shortly after birth (Schenker, S., Dawber, N., and Schmid, R. 1964)

Advantage was taken of the availability of biosynthesized ^{14}C radiolabeled bilirubin. Radiolabeled unconjugated bilirubin, infused into fetal guinea pigs in vivo appeared in maternal bile as radiolabeled conjugated bilirubin. This excretion of fetal injected label began to appear in maternal bile within 15 min. Less than 2% of label appeared in fetal bile.

These results helped explain why fetal tissues are not stained in utero by unconjugated bilirubin. In regards to physiological and pathological hyperbilirubinemia, the results explain jaundice appearing within hours after birth. Since the placenta is a route for transfer of potentially toxic unconjugated bilirubin, the fetal liver is not significantly challenged as regards the enzymatic machinery (glucuronyl transferase) for conjugating bilirubin. At birth, the maternal mechanism for conjugating bilirubin is removed. This places a temporary but increasing load of bilirubin on the newborn liver. This had not been previously experienced by the liver, and a certain "lag" time exists in about 60% of newborn infants. Just a few days is enough to resolve the physiological unconjugated hyperbilirubinemia. In cases of hemolysis, the correction of hyperbilirubinemia may need clinical intervention.

Biochemistry and Physiology of Bilirubin

Early work on the biochemistry of bilirubin was conducted by Nobel Laureate Hans Fischer. Hemoglobin is an important constituent of the pyrrole group of molecules, and Fisher studied hemoglobin. Hemoglobin breaks into heme and globin in its first degradation step. The combination of heme and globin possesses the unique ability to loosely bind oxygen, enabling it to be released to body tissues. Heme bound to iron is termed hemin, which has the chemical formula: $C34\ H32\ O4\ N4\ FeCl$. When iron is removed from hemi, porphyrins are the result.

It was known that all porphyrins contain a pyrrole ring which consists of four carbon atoms, with the presence of a nitrogen atom which serves to close the pyrrole ring. Fischer was the first to show that the synthetic substance phorphin contained four pyrrole rings linked by four methane bridges ($=CH$) which serve to close the circlular structure. Subsequently, Fischer showed that all porphyrins have a porphin nucleus with four pyrrole rings in which the carbon atoms are linked by hydrogen. Fischer is credited with the discovery of porphyrin synthesis, and ultimately was successful in synthesizing over 130 isomers of porphyrin. Continued work on porphyrins led to the synthesis of a synthetic hemin which was indistinguishable from the naturally occurring hemin which is obtained from hemoglobin. (see Fig. 1).

Finally, Fischer turned his attention to the structures of biliverdin and bilirubin. Fischer showed that if one carbon atom was absent, the now opened ring resembled bile pigments. He subsequently elucidated the chemical structure of both biliverdin and bilirubin (Fischer, H., and Pleininger, H. 1942). (see Figs. 2 and 3).

Bilirubin is formed as a result of the breakdown of red blood cells. This process takes place in cells of the reticuloendothelial system. The system is widespread in the body, occurring in the liver, spleen, and bone marrow. In diseased states where there may be an increased production of bilirubin, or when the "regular" sites of lysis of red blood cells are not functional, several other locations may function to lyse the red blood cells. The heme moiety of hemoglobin results in at least 80% of unconjugated bilirubin. The smaller component of unconjugated bilirubin may come from a heme which has a very rapid turnover rate. Hepatic cytochromes may be the source (Schmid, R. et al. 1966).

As stated above, bile pigments are structures with four pyrrole rings linked by $–CH2$ or $=CH$ groups. These naturally synthesized pigments result from oxidative scission of ferroprotoporphyrin IX. This results in the IX alpha structure

D.W. McCandless, *Kernicterus*, Contemporary Clinical Neuroscience,
DOI 10.1007/978-1-4419-6555-4_3, © Springer Science+Business Media, LLC 2011

Fig. 1 Schematic
representation of porphoryn

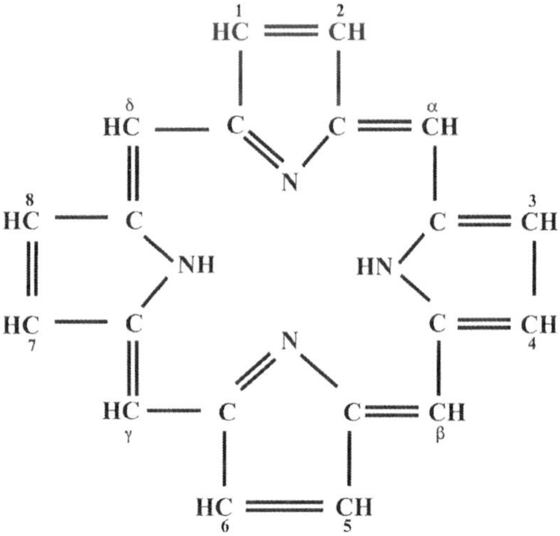

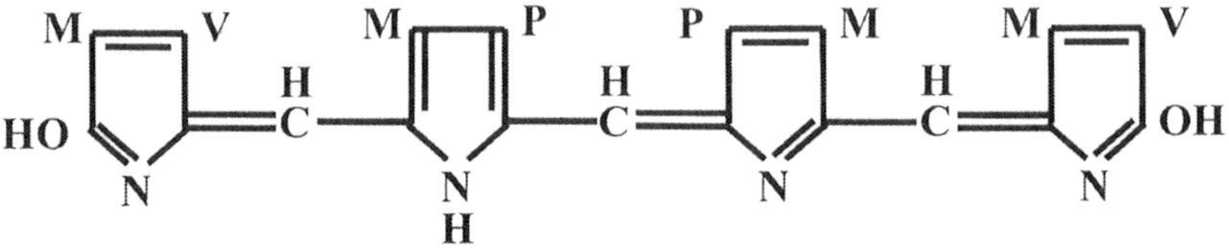

Fig. 2 Schematic representation of biliverdin

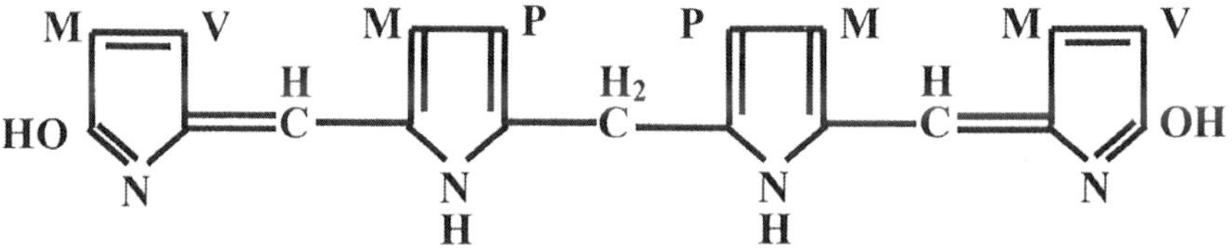

Fig. 3 Schematic representation of bilirubin

being present in the resultant bile pigments. The pigment biliverdin is very similar structurally to protoporphyrin. The central methane bridge of biliverdin is easily
reduced, transferring it into bilirubin.

The bilirubin molecule is stabilized by its presence of six hydrogen bonds. These
hydrogen bonds most likely represent the reason that this form of bilirubin is not
water soluble. This form of unconjugated bilirubin has at least three geometric isomers, and the conversion of bilirubin to these isomers may be the mechanism by
which photo therapy acts to produce forms which can be more readily eliminated.
Additionally, conjugation of bilirubin IX alpha to bilirubin glucuronide produces a
molecule which does not have hydrogen bonds, thereby creating a molecule which is

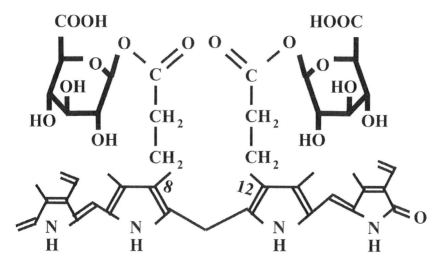

Fig. 4 Schematic representation of bilirubin glucuronide

water soluble, and can be readily excreted by liver cells into bile ducts (Grodsky, G., and Carbone, J. 1957). In aqueous solutions, bilirubin IX alpha can be cleaved and the halves may reform as isomers of bilirubin IX alpha, bilirubin III alpha, and bilirubin XIII alpha. Further bacterial or chemical reduction of bilirubin yields urobilins and urobilinogens. (see Fig. 4).

Conjugation of the lipid-soluble unconjugated bilirubin is a process of esterification of bilirubin's propionic acid with glucuronic acid. This may involve one or both of the propionic moieties. This results in a diminution of the ability of bilirubin to have strong hydrogen bonds, thereby rendering the now conjugated bilirubin water soluble. Thus, the conjugated bilirubin can be excreted in bile and urine, and the reabsorption in the digestive tract is impeded.

In normal bile, 80% of conjugated bilirubin is in the diglucuronide form, while the other 20% represents a single bilirubin/propionic bond – the monoglucuronic acid form. The enzyme catalyzing the conjugation of bilirubin is uridine diphospho-glucuronyl transferase (EC 2.4.1.17). This enzyme is located in the microsomal components of both rough and to a lesser extent, smooth endoplasmic reticulum. The enzyme uridine diphospho-glucuronyl transferase is specific for bilirubin. The interconversion of conjugated bilirubin between the monoglucuronide and diglucuronide forms is discussed in a later chapter in this volume.

The uptake and conjugation processes of metabolism of bilirubin have a greater capacity than does the excretion into bile. Therefore, the latter step is rate limiting in the entire process. When the conjugated bilirubin reaches the large intestine, it undergoes chemical changes due to bacterial action. The conjugated bilirubin is deconjugated and reduced to form several urobilins. The exact make up of these urobilins depends in part on the intestinal flora, and wide normal variation occurs. If antibiotics are administered, unconjugated bilirubin is directly excreted in feces.

The physiology of bilirubin is a highly complex process involving many steps, and anatomical sites in which problems can arise which lead to jaundice. Some aspects of this process have been known for over a century, while other details are still under investigation. Numerous clinical entities are associated with defects in the normal physiological process of the elimination of bilirubin. These syndromes/diseases are detailed in each chapter in this book. Before one can consider defective bilirubin metabolism, the normal physiological processes must be considered.

The liver is actually ideally suited anatomically for processing and eliminated bilirubin and other compounds. The hepatocyte is highly polarized in that one surface (sinusoidal surface) is in intimate contact with blood sinusoids. The sinusoids are modified capillaries. These channels are from 20 to 40 μ in diameter, and components destined for liver uptake are in direct contact with the hepatocyte. This surface consists of about 50% of the entire hepatocyte cytoplasmic membrane. (see Fig. 5).

Another surface, composing 15–20% of the entire cell forms, with other hepatocytes, the bile canaliculus. Tight junctions appear (Zonula occludens) as rows of contact points between adjacent cells. These tight junctions are at the margins of the bile canaliculi. The hepatic tight junctions may be an area of transport. The space of Disse is a very narrow (100 Å) which is positioned between the hepatocytes and the sinusoidal endothelial cell. It is most likely another area of transport between the sinusoid and the hepatocyte (Phillips, M., and Latham, P. 1982).

Hepatocyte intracellular organelles consist of the usual assortment of structures such as smooth and rough endoplasmic reticulum, Golgi apparatus, lysosomes, polysomes, coated/smooth vesicles, microtubules, microfilaments, and, remarkably, over 2,000 mitochondria per hepatocyte. The Golgi apparatus is usually located either near the nucleus or adjacent to the bile canaculus. Other inclusions found in hepatocytes include glycogen and lipids (triglycerides and lipoproteins).

The mitochondria are characteristic in that they are bound by a double membrane. The inner membrane is folded many times inward; the folds are called cristae. The cristae contain the elementary particles needed for the electron transport chain and other components and enzymes utilized in ATP production. (see Fig. 6). The mitochondrial matrix also contains DNA. A variety of abnormal inclusions are found in several disease states. Another type of cells seen in the liver in addition to hepatocytes are the Kupfer cells. These cells are phagocytes and may contain pigment granules, degenerating erythrocytes, and iron-containing granules.

The gross anatomy of the liver and its circulation are also ideally suited to its function. Most unique is the hepatic portal system. This portal system is one of two portal systems in the body, and includes all veins draining the digestive system from the diaphragm to the rectum, plus veins from the spleen, pancreas, and gall bladder. The portal vein enters the liver at the visceral surface carrying blood from the aforementioned viscera. The portal vein then branches many times to deliver blood to the hepatic sinusoids. From the sinusoids, blood passes through hepatic veins and enters the inferior vena cava. The significance of this system can hardly be overstated. The blood from the portal vein draining into hepatic sinusoids is actually entering into a

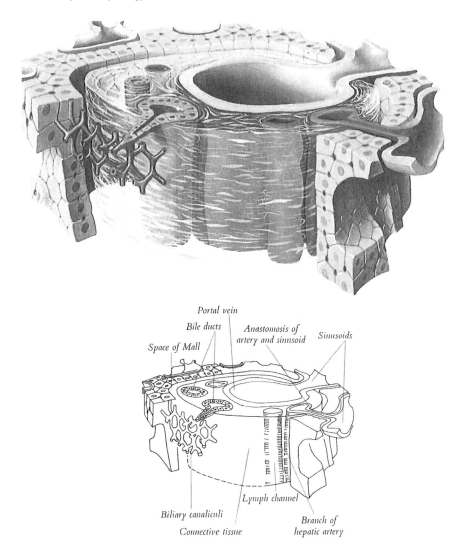

Fig. 5 Artist conception of hepatic portal area. With permission of E. Lily Co., Indianapolis, Ind

second capillary network before reaching the heart. The first network is that in the digestive tract and associated organs. Many substances are absorbed into the circulation in the first set of capillaries. The senoin set of capillaries in the liver allow potentially harmful substances and toxins to be absorbed and metabolized by the liver. Without the portal system, harmful chemicals would reach the heart and be distributed throughout the body, including the brain. In the adult, the portal vein has no valves which might impede flow into the liver.

The hepatic sinusoids are modified capillaries in that they are larger in diameter, and uniquely modified for transport of substances, including bilirubin, into the

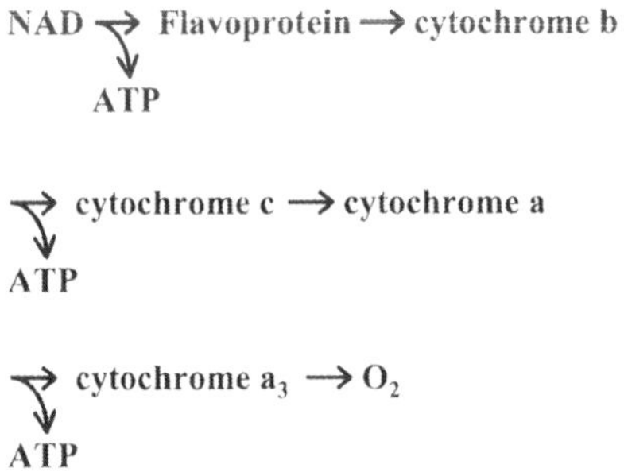

$$NAD \xrightarrow{} Flavoprotein \rightarrow cytochrome\ b$$
$$ATP$$

$$\xrightarrow{} cytochrome\ c \rightarrow cytochrome\ a$$
$$ATP$$

$$\xrightarrow{} cytochrome\ a_3 \rightarrow O_2$$
$$ATP$$

Fig. 6 Schematic representation of electron transport chain

hepatocyte. Oxygenated blood also enters the visceral surface of the liver along with the portal vein and hepatic artery, and these two blood sources mix in the sinusoids. Blood drained from the liver ultimately unites to form three large hepatic veins which drain into the inferior vena cava which is nestled in a fossa on the posterior surface of the liver.

A brief overview of bilirubin metabolism follows, and then the specific aspects will be examined. Chemically, bilirubin can be described as a compound composed of four pyrrole rings linked at the alpha positions by three methane groups (see Fig. 1). Bilirubin has six intramolecular hydrogen bonds which are believed to explain its insolubility at physiological pH. When the hydrogen bonds are not formed, bilirubin becomes water soluble and can be excreted.

Virchow (Virchow, R. 1847), 150 years ago, may have been the first to recognize that bilirubin (yellow pigment) resulted from red blood cell breakdown. Red blood cells destined for elimination are removed from circulation by cells of the reticuloendothelial system. These cells are components of liver, spleen, and bone marrow. As red blood cells are lysed, hemoglobin is quickly broken down into heme and globin. The vast majority (over 80%) of circulating bilirubin is the result of the breakdown of the heme component of hemoglobin into bilirubin and iron. Ten to twenty percent of bilirubin is derived from other sources. These sources are termed early labeled bilirubin, and appear to originate from a pool of heme compounds with a rapid rate of turnover.

The degradation of heme into bilirubin and iron is thought to be an enzymatic process involving a microsomal heme oxygenase system (Tenhunen, R., et al. 1968). This process takes place in liver, spleen, and bone marrow. The enzyme system requires both NADPH and 3 moles of oxygen to convert heme to bilirubin and carbon monoxide. (see Fig. 7).

Following formation of bilirubin by the reticuloendothelial system, it is transported to the liver in serum. Unconjugated bilirubin has a low water solubility, and therefore travels in serum bound to albumin. A one to one molar ratio results in tight binding of bilirubin to the albumin carrier. As the molar ratio rises, the tightness of the bond decreases such that some unconjugated bilirubin may not be bound. This in turn may allow pigment to enter tissue. Tissue deposition frequently occurs in high lipid areas, including adipose tissue and brain.

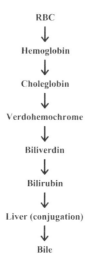

Fig. 7 Schematic of red blood cell breakdown

The distribution of bilirubin in vivo depends in part on a quite small percentage of bilirubin in serum which is not bound to albumin. This protein-free moiety is capable of entering tissue almost without limit. The ability of albumin to bind is dependent on the molar ratio of albumin and bilirubin as stated above, plus pH, and the presence of compounds which successfully compete with bilirubin for albumin-binding sites. Even the presence of free fatty acids in serum of infants from breast milk may serve to displace bilirubin from albumin-binding sites (Young, F., and Cheah, S. 1977). Anything which acts to decrease the albumin-binding capacity will increase the risk for kernicterus.

Even though there is a relatively tight binding of bilirubin to albumin in serum, it can readily cross biological membranes by carrier-mediated transport or sometimes diffusion. This is the mechanism by which bilirubin crosses the placenta, and also permits transport into liver and kidney. Conjugated bilirubin may distribute in tissues by a different mechanism than unconjugated bilirubin.

Hepatic uptake of unconjugated bilirubin occurs rapidly after formation. In a normal physiological situation, the unconjugated bilirubin concentrations in serum are 1 mg/100 ml or less. It seems that in hepatic sinusoids described above, bilirubin is dissociated from albumin-binding sites, crosses the space of Disse, and enters the hepatocyte. The uptake system has a capacity higher than that normally needed. Thus, hepatocyte uptake is not a rate-limiting step in bilirubin metabolism and subsequent excretion.

Once in the hepatocyte, bilirubin binds to cytosolic proteins such as ligandin and Z protein. These proteins are thought to act in intracellular binding/transport, and in the detoxification of the numerous compounds taken up by liver cells.

The process of intrahepatic conjugation is the key to the elimination of bilirubin. In this process, bilirubin is most frequently combined with glucuronic acid. This prevents hydrogen binding, rendering bilirubin less constrained, and thus bilirubin is more hydrophilic and easily excreted in bile. Other processes such as excretion in urine are also facilitated.

The formation of bilirubin conjugates is catalyzed by the enzyme UDP-glucuronyl transferase (E.C. 2.4.1.17). This reaction transfers glucuronic acid from uridine diphosphate glucuronic acid to unconjugated bilirubin, rendering the pigment conjugated and water soluble. The enzyme is thought to be specific for bilirubin (Lathe, G. 1972). Interestingly, UDP-glucuronyl transferase is, in addition to the liver, found in proximal tubule cells in the kidney and in the small intestine. In hepatocytes, conjugation most likely takes place in the microsomes.

A congenital absence of glucuronyl transferase is associated with hyperbilirubinemia of the unconjugated variety in newborn infants. This disorder is the Crigler–Najjar syndrome in newborn infants, and there is a perfect animal model, the Gunn rat. Both of these conditions are dealt with in separate chapters in this volume. If left untreated, the level of unconjugated bilirubin exceeds that which results in kernicterus.

Because of solubility differences between unconjugated and conjugated bilirubin, almost all of the bilirubin leaving the liver, and in the biliary system is conjugated. Conjugated bilirubin is excreted as a micellar complex with bile salts and cholesterol. Instestinal bacteria serve to deconjugate the conjugated bilirubin, and reduce it to form urobilins. Antibiotics can prevent this process, and bilirubin is then directly excreted in feces. Some urobilinogen may appear in urine, but most is reabsorbed in the small intestine and reexcreted by liver cells.

Renal excretion of conjugated bilirubin in any increased amount may occur in incidences of conjugated hyperbilirubinemia. The renal excretion of conjugated bilirubin depends on the level of pigment not bound to albumin. Unbound conjugated bilirubin is available for glomerular filtration, but this amount is less that 1%. As is the case for unconjugated bilirubin, compounds such as sulfisoxazole successfully compete with conjugated bilirubin for albumin-binding sites. This increases conjugated bilirubin's renal excretion. The renal excretion of conjugated bilirubin is not an efficient process as compared to the liver.

In a couple of papers (Dutton, G., and Storey, I. 1954; Storey, I., and Dutton, G. 1955), the nature of uridine-diphosphate glucuronic acid (UDP glucuronic acid) formation is described, and the above name of the compound is suggested. In the first paper, results showed that guinea pig liver extract was added to a liver suspension system which allowed for the conjugation of aminophenol, conjugation was increased. Boiling the entire mixture resulted in no conjugation taking place. These results suggested that liver extract contained a factor which led to the formation of glucuronides. The factor could not be demonstrated in kidney, muscle, or brain. Details of factors such as optimal pH, inhibition by sulfate, etc. were all examined. The authors state that the enzymatic activity seemed to be located in the cytoplasmic particulate material.

In the second paper, the same authors described studies in which the conjugating nucleotide was isolated, and degradation products examined. Results from this second series of experiments showed that the "factor" was a nucleotide composed of uridine-5-phosphate and glucuronic acid-1-phosphate, and that these two components are linked by a pyrophosphate bond. These findings serve to confirm the earlier paper's suggestion that the factor be called uridine-diphosphate-glucuronic acid.

The authors also point out that their studies showed that hydrolysis of the nucleotide with acid yielded uridine-5-phosphate, inorganic phosphate, and glucuronic acid. A formula for the action of the nucleotide in forming conjugates is suggested as follows UDP glucuronic acid +ROH>UDP + R-O-glucuronic acid, where ROH is the acceptor molecule.

In another paper (Strominger, J., et al. 1957), the enzymatic formation of uridine diphosphoglucuronic acid was elucidated. In these studies, it was established that the soluble fraction of liver homogenates contain an enzyme catalyzing the equation: UDPG+2DPN> UDPGA+ 2DPNH+ 2H.

The synthesis of UDPG had been proposed by members of the above group (Munch-Petersen, A., Kalckar, H., Cutulo, E., and Smith, E. 1953). The suggested reaction was uridine triphosphate + (alpha) glucose-1-phosphate <> UDPG+pyrophosphate.

Further results suggested that the oxidation reaction resulting in the formation of UDPGA is a one-way irreversible reaction with no intermediate compounds formed. The authors speculate that since UDPG is an alpha-glucoside, then probably UDPGA is also an alpha-glucoside. Several acceptors in the conjugation reaction are listed in this manuscript.

As discussed elsewhere in this volume (history) (Billing, B., Cole, P., and Lathe, G. 1956; Schmid, R. 1956; and Talafant, E. 1956), it became obvious that the so-called direct- reacting bilirubin was conjugated with glucuronic acid and was water soluble, and indirect-reacting bilirubin was unconjugated and lipid soluble. It is the unconjugated lipid soluble form of bilirubin which has the capacity to cross the blood–brain barrier, enter brain tissue, and produce bilirubin encephalopathy and kernicterus.

Briefly, the experimental design included preparing a liver homogenate, adding bilirubin, incubation in Warburg flasks, then estimating the production of direct and total bilirubin over time. Tissues in addition to liver were utilized.

Results showed that 15–30% of added bilirubin was converted to direct reacting bilirubin after 30-min incubation with liver homogenates. Agents which are exclusively conjugated with glucuronic acid (borneal) might be expected to inhibit the conjugation of bilirubin, and that was the result. Increasing amounts of borneal had a proportional amount of decrease in the conjugation of bilirubin.

Chromatographic results showed that only indirect (unconjugated) and direct bilirubin showed spots following incubation. This suggests that the conjugation observed was only with bilirubin. Further, as compared to other tissues, liver was the most efficient organ in producing direct bilirubin, but activity was present in kidney and brain. The enzyme beta-glucuronidase was effective in destroying the direct bilirubin. These results all point to bilirubin glucuronides being the form of "direct"-reacting acting bilirubin, and that bilirubin glucuronide probably represents the vast majority of conjugated bilirubin in vivo.

In another study (Lathe, G., and Walker, M. 1958), the defective formation of direct (conjugated)-reacting bilirubin was compared in adult animals, prematurely born humans, and in Gunn rats. Results showed that glucuronic acid seemed to be present in normal amounts, and that furthermore, addition of glucuronic acid to liver incubations failed to increase the formation of conjugated bilirubin. The authors speculate that the deficiency in the ability of liver to conjugate bilirubin lies in the deficiency of glucuronyl transferase.

The authors state that the jaundice and therefore defective conjugation of bilirubin seen in homozygous Gunn rats, leading to kernicterus, is permanent, whereas the jaundice in premature newborn infants is transitory. The initiation of conjugation of bilirubin in premature newborn infants is under control of developmental and/or environmental factors which were unclear, but suggested that treatment by stimulating glucuronyl transferase function might be a possibility.

In another report (Brown, A., and Zuelzer, W. 1957), the effect of age was examined in guinea pigs on the development of the ability of liver microsomes to conjugate bilirubin. Four-week fetal animals were compared to newborn guinea pigs as regards bilirubin conjugation ability. Results showed that liver microsomes had about a 0% ability to conjugate bilirubin at the fetal 4 week age, whereas by newborn day 15– 20, levels of conjugation by microsomes were close or at adult levels. This represented the "age" factor in the development of conjugating capacity.

While early studies suggested that bilirubin was conjugated almost exclusively with glucuronic acid, later studies (Isselbacher, K., and McCarthy, E. 1959) showed that radioactive sulfate, infused into rats with their bile ducts cannulated, excreted ^{35}S-labeled bilirubin. Results showed that the ^{35}S-labeled bilirubin migrated on paper chromatograms to the same spot as the major conjugated azobilirubin pigments. The amount of bilirubin conjugated with ^{35}S was on the order of 10%. There are other conjugating systems in human liver in addition to that involving glucuronic acid, so it was not surprising to find evidence of ancillary conjugating pathways for bilirubin in liver.

In a related paper (Schenker, S. 1963), the placental transfer and disposition of both unconjugated and conjugated bilirubin in guinea pigs was examined. In this study, pregnant guinea pigs were anesthetized, the bile duct cannulated, and the animal immersed in a warm (37°C) saline bath. The fetuses were delivered into the bath, and either radioactive (C14) unconjugated or radioactive conjugated bilirubin was injected into the fetus through the umbilical vein.

Results showed that when radio-labeled unconjugated bilirubin was injected into fetal guinea pigs still connected to their mother, one half of the label was excreted into maternal bile within 2 h. To determine if unconjugated bilirubin was dissociated from albumin when transferred across the placenta, I131albumin/C14 bilirubin was infused into fetuses. While the labeled bilirubin crossed, virtually none of the labeled albumin crossed the placenta. These data indicate that bilirubin was dissociated from albumin before it crossed the placenta.

Conversly, conjugated bilirubin was cleared from fetal circulation at a much slower rate than that of unconjugated bilirubin. In addition, results showed that conjugated bilirubin excretion into bile was similar to that of unconjugated bilirubin. This demonstrates that even if the fetal guinea pig was able to conjugate bilirubin, excretion into bile might be slow.

In another study examining albumin and bilirubin binding (Odell, G. 1959), it was demonstrated that certain organic anions were capable of uncoupling bilirubin from albumin-binding sites. Results from spectrophotometric studies of newborn human serum showed that protein-bound bilirubin is altered by organic anions. Thus, salicylates and sulfisoxazole, which have affinities for albumin, compete for binding and may displace bilirubin depending on molar ratios. The concentrations of serum organic anions necessary to affect this displacement of bilirubin from albumin-binding sites were within the range seen in clinical practice.

This study has special clinical relevance since, in terms of developing kernicterus, it is the free unbound to albumin unconjugated bilirubin which is key. In the presence of sulfasoxazole, kernicterus can readily develop in newborn infants at unconjugated bilirubin levels well below 20 mg/100 ml. Similiarily, knowledge of total bilirubin used widely today is less informative. The ability of sulfasoxazole to displace unconjugated bilirubin from binding sites should be common knowledge (Conney, A., and Burns, J. 1962).

Examining the cellular distribution of labeled bilirubin using differential centrifugation yielded interesting results (Brown, W., Grodsky, G., and Carbone, J. 1964). Results showed that after infusion of tritiated bilirubin, hepatic uptake was underway by 2 min. By 5 min, one-half of the infused tritiated bilirubin was in the liver, but only 2.5% was in bile. By 15 min, one-third of the label was in bile. At 30 min most of the label had been excreted, and was in the form of bile. Ultracentrifugation to look at subcellular localization showed most label in the supernant phase at 15 min after infusion. The fraction with the highest percent of label were the microsomes, lysosomes, mitochondria, and nuclei.

The authors note that the intracellular concentrations of labeled bilirubin relate more to the amount in serum than it is related to steps in uptake, conjugation, or excretion by the hepatocyte. The microsome level of labeled pigment was greatest as compared to other cell organelles. Since glucuronyl transferase is located in microsomes, this may indicate a localization at this conjugating site. There was also accumulation of bilirubin in lysosomes, and these organelles have been implicated as being involved in the final excretion step by electron microscopic observations. Bilirubin was also noted in mitochondria in this study.

In terms of bilirubin formation, it is the product of heme catabolism, but intermediate steps between heme and bilirubin were unclear. A study (Goldstein, G., and Lester, R. 1964) was done to examine the possible role of bilirubin in the heme catabolic steps. In this study, ^{14}C-biliverdin was administered to rats by intravenour infusion, and bile and urine were collected for scintillation spectroscopy.

Results showed that following infusion of ^{14}C-biliverdin, radioactivity was detectible within 30 min in bile. Most recovered radioactivity was as 14-C conjugated bilirubin. Only a small amount (1%) was recovered in urine. These results confirm that biliverdin is the precursor of bilirubin, and that the conjugating process is rapid.

The development of newborn transport into the liver and subsequent metabolism of bilirubin has been carefully studied (Gartner, L., et al. 1977). This study was done in Rhesus monkeys, and findings correlated in some measure with those from newborn humans. In these studies, pregnant Rhesus monkeys were obtained with established gestational age. Nineteen neonatal monkeys were untreated, and 14 additional monkeys were treated with phenobarbital I.M. daily 6 weeks prior to expected birth. After birth, the pattern of bilirubin in plasma was analyzed in the two groups of nontreated and treated neonates.

Results showed that total plasma bilirubin in untreated neonatal monkeys rose rapidly over the first 24 h. During the next 24 h, levels dropped, then rose again after 48 hours and were moderately elevated up to 96 h. This then, meant a biphasic characteristic pattern in which bilirubin rises immediately after birth, then falls. Later there is a second elevation, followed by a fall to about adult levels, which remain stable.

When pregnant Rhesus monkey mothers were treated daily with I.M. phenobarbitol, the offspring had unconjugated bilirubin levels lower by over 50% than untreated monkeys during the first 24 h. During the second phase (over 48 h), phenobarbitol seemed to have no appreciable effect on bilirubin levels. That is to say, the bilirubin levels did not rise during the second phase in treated monkeys.

When these results were compared to human newborn infants, results showed a similar biphasic pattern, but drawn out over a longer time frame. Thus, the phase 1 elevation of unconjugated bilirubin lasts until 120 h, after which it has declined. A second elevation occurs lasting about 11 days.

The authors note that phenobarbitol treatment increased glucuronyl transferase activity by nearly threefold. This increased enzyme activity was achieved without increasing the bilirubin load. These studies also suggest that the rate-limiting step was the enzymatic conjugation of bilirubin. In phenobarbitol-treated animals, the rate-limiting step had shifted to excretion by the hepatocyte.

The phase 2 increase was thought to be most likely the result of a continuing increasing bilirubin load, and a possible decrease in hepatocyte uptake of the pigment. That conjugation is not a factor in the phase 2 steps is indicated by the lack of rise in bilirubin in light of the enhanced conjugating capacity. Thus, phenobarbitol possibly does not alter bilirubin load, or enhance uptake.

Sulfisoxazole is not the only agent which can competitively displace bilirubin from its albumin-binding sites. Two papers appeared back to back (Shankaran, S., and Poland, R. 1977; Wennberg, R., Rasmussen, L., and Ahlfors, C. 1977) examining the ability of furosemide and two other diuretics to displace bilirubin from albumin. Since normally, bilirubin in serum is albumin bound, anything which might alter the binding capacity in newborn infants could act to facilitate entry of unconjugated bilirubin into tissue. Entry into brain produces risk of kernicterus.

Results from both studies indicated that furosemide was able to displace bilirubin from its binding to albumin as effectively as sulfisoxazole. Two other diuretics, ethacrynic acid and chlorothiazide, were equally effective in displacing bilirubin.

Furosemide is an effective diuretic, which was used to treat newborn infant conditions such as cardiac and renal failure, and respiratory problems such as edema.

These studies are important since they show that the normal physiologic situation in which bilirubin and albumin are bound, thereby preventing entry of unconjugated bilirubin into tissue, can be easily upset, to the detriment of the newborn, by drugs administered to treat serious newborn conditions. As unconjugated bilirubin enters tissue, serum levels drop, giving the impression of an improving hepatic condition, when in fact kernicterus may be occurring.

In terms of physiologic jaundice, hyperbilirubinemia sufficient to be detected visually (5–6 mg/100 ml) occurs in about 60% of normal full term neonates. The 95th percentile for newborn infants is 15–17 mg/100 ml peaking on the third to fourth days. The percentage of newborns reaching high levels is quite low (Billing, B. 1982).

Generally, physiological jaundice results when the generation of bilirubin exceeds the ability of the liver to eliminate it. Increased heme production due to high hemoglobin levels, shortened red blood cell life span, and brusing may all serve to elevate bilirubin levels. Glucuronyl transferase activity "matures" with age and with challenge, so prematurity may lead to inefficient conjugation capacity.

Conjugated bilirubin in the gut may undergo cleavage of the glucuronide component from bilirubin, thereby again creating unconjugated bilirubin. This unconjugated bilirubin can be reabsorbed into the circulation and act to increase the liver load. This process, enterohepatic circulation, is not uncommon.

Hemolysis is of concern since, when severe, significant hyperbilirubinemia can occur, with resultant kernicterus. Hemolysis due to blood incompatibilities is not as dangerous as it once was because of treatment modalities involving anti-D globulin administration to all Rh-negative mothers during pregnancy.

The enzyme glucose-6-phosphate dehydrogenase (G-6-PD) is commonly an enzyme which is deficient at birth. For an excellent review, see Kaplan and Hammerman (Kaplan, M., and Hammerman, C. 2005). Hyperbilirubinemia from G-6-PD deficiency can be sufficient to cause kernicterus, and may be responsible for up to 25% of all kernicterus cases. G-6-PD is a worldwide problem, with an estimate of hundreds of millions of people affected to some extent.

G-6-PD may be triggered by ingestion of the fava bean, and the result is massive hemolysis. In neonates, the onset of G-6-PD-associated hemolysis can be rapid and severe. In these cases, exchange transfusion is necessary. There are clearly defined genetic variations such that certain groups of people are affected differently. For example, Nigerian neonates seem more severely affected than are other ethnic groups.

In the normal physiologic sequence, biliverdin is reduced to bilirubin. The enzyme involved is biliverdin reductase. The gene which encodes biliverdin reductase has five exons and four introns, and does not have the TATA box. The role of alterations in this normal reduction of biliverdin to bilirubin is unclear.

Premature Birth

Definitions of premature birth vary. A definition of prematurity in which birth weight is less than 5.5 lb could include low birth weight, but not premature (in weeks of age). Definitions based on a gestational age of less than 39 weeks might not be accurate. It is relatively clear that morality rates are closely related to gestational age. For example, the mortality rate in a group of infants weighing between 2.2 and 3.3 lb was about 50%, and in infants less than 30 weeks, the mortality rate was 63%. Approximately, two thirds of all newborn mortality is linked to premature birth. The incidence of premature birth in the USA is 7%.

Causes of premature birth are unclear. Presumptive causes include precocious fetal endocrine activation, multiple pregnancies, maternal toxemia, infections, placental problems, premature membrane rupture, maternal malnutrition, etc. A significant number of premature births result from elective preterm delivery due to hypertensive disorders, trauma, intrauterine growth restrictions, etc. Increased prenatal education and better health care in underserved areas should result in a lower incidence of premature birth. However, the incidence rate for premature birth has remained more or less constant over the last 50 years, or has actually increased by some estimates. In the USA, the premature birth incidence reached 12.7% of all births in 2005.

When death occurs, careful pathologic examination frequently finds the cause of death. The exact cause of death is not premature birth, but can be attributed to conditions such as anoxia, hyaline membrane disease, septicemia, hyperbilirubinemia and accompanying kernicterus, etc. Before improved treatment modalities, kernicterus was found in from 2 to 20% of autopsied premature birth newborn infants. This wide range has to do with the extent of the autopsies. In many instances, brains were not examined due to parents' requests. Generally speaking, premature birth tends to increase the severity and reduce the clinical manifestations of newborn diseases such as intracranial hemorrhage, atelectasis, pneumonia, hyaline membrane disease, etc. These diseases and especially hyperbilirubinemia occur much more frequently among prematurely born infants than in full-term infants.

Care consists of preventing infection by safeguarding the newborn's food and objects in which he comes in contact. Reducing contact with people is also of key importance. A stable environment as regards temperature and humidity is important. Oxygen administration may be important in reducing mortality. In terms of feeding,

D.W. McCandless, *Kernicterus*, Contemporary Clinical Neuroscience, DOI 10.1007/978-1-4419-6555-4_4, © Springer Science+Business Media, LLC 2011

fatigue from nursing as well as aspiration of food should be avoided. The interval between feedings should be less than in full-term infants.

Another interesting statistic (Gilbert, W., Nesbitt, T., and Danielsen, B. 2003) is that for a premature birth in an infant weighing less than 700 g, neonatal costs are about a quarter million dollars, whereas a 7 lb infant has a medical expense of 1–2,000 dollars.

The role of infection in premature births has been reexamined and found to be involved in fully 25–40% of cases (Goldenberg, R., Hauth, J., and Andrews, W. 2000). The frequency of infection in premature birth is inversely related to gestational age, so the incidence would be expected to be higher in the earliest of premature infants. Some endotoxins released by microorganisms may in turn release prostaglandins. These may stimulate uterine contractions. Intrauterine infection can be a chronic process; organisms identifiable in the uterus include urea plasma urealyticum (Goldenberg, R., Hauth, J., and Andrews, W. 2000). Fetal infection can be linked to premature birth, and also to long-term problems such as cerebral palsy (Goldenberg, R., et al. 2008).

The condition of bacterial vaginosis is believed to be a key factor in premature births. Preterm delivery is a very important problem in obstetrics and for neonatalogists. A recent study aimed at assessing the association between bacterial vaginosis and premature birth (Hillier, S., et al. 1995). Bacterial vaginosis results from the replacement of the normal vaginal flora of lactobacillus with gardnerella vaginalis and mycoplasma hominis (Hillier, S., et al. 1993). The presence of bacterial vaginosis is associated with a number of conditions deleterious to full-term birth. These include premature rupture of fetal membranes, infection of the chorion and amnion, infection of amniotic fluid, and chorioamnionitis. This study was a large multicenter study, in which nearly 14,000 women were enrolled between 1984 and 1989.

Vaginal smears and pH were used to arrive at a diagnosis of bacterial vaginosis. Pregnancy outcome was evaluated. Estimate of gestational age was based on the last menstrual period and adjusted based on a pelvic exam. Preterm delivery was defined as birth before 37 weeks gestational age. Under 2,500 g was defined as low birth weight.

Results showed that of the 10,397 women followed through delivery, 504 infants (4.8%) were premature with low birth weights. Bacterial vaginosis was diagnosed in 1,645 women (16%). Characteristics associated with bacterial vaginosis included low socioeconomic class and unmarried. Low birth weight outcome data were restricted to women who did not use antibiotics during pregnancy to treat the vaginosis.

Results showed a statistically significant correlation between bacterial vaginosis and preterm delivery with low birth weight, and also with low birth weight alone. Women who received effective treatment for vaginosis had the same incidence of low birth weight/preterm delivery as control women without vaginosis. Delivery of a low-weight preterm infant in a previous pregnancy was the factor most strongly predictive of another preterm low birth weight offspring in this study. Having lost an earlier pregnancy for any reason was also associated with having a low birth weight baby.

The authors note that women with bacterial vaginosis were 40% more likely to deliver a premature low birth weight infant than were non-infected controls. Evidence in favor of a cause–effect relationship includes a biological gradient in which increased infection results in increased risk for an abnormal outcome, and consistency in outcome regardless of the infection's location. Previous studies (5–10 in Hilliar-choose 1) have also reported an increase in premature births in women with bacterial vaginosis. This has been shown in both case control and cohort studies in several countries. The highest risk appeared in cases in which the bacterial vaginosis was present early in pregnancy.

Biochemical vaginosis is a common genital infection in pregnancy. The exact mechanism by which bacterial vaginosis causes premature births is not known. The supposition is that it causes infection of the upper genital tract, thereby causing an amniotic fluid infection, and infection of the chorion and amnion. The presence of the normal vaginal flora bacterial lactobacilli may act to protect against premature birth.

Another study concerned with bacterial vaginosis and premature birth was undertaken to look at the question of the possible reduction of preterm delivery in cases treated with antibiotics. The study also looked at whether bacterial vaginosis further increases the incidence of preterm delivery in women already at risk (Hauth, J., et al. 1995).

About 624 women were randomly assigned to a group having active treatment and a second group receiving placebo. All were treated for bacteriurea. Metronidazole and erythromycin were used to treat bacterial vaginosis. Following treatment, bacterial vaginosis disappeared in 70% of the antibacterial-treated group, whereas the infection disappeared in only 18% of the placebo group. As a total group, of the 624 women studied, 29% delivered babies before 37 weeks gestation. Of these, 26% of the antibacterial-treated women gave birth to preterm infants, whereas 36% of placebo-treated women had premature offspring. This was significant at the 0.01 level of confidence. Another outcome of this study was that women who had bacterial vaginosis had a higher incidence of preterm infants whether they were treated or not.

The authors note that the reduction in premature birth in antibacterial-treated bacterial vaginosis women argues in favor of using the treatment in these cases. It also lowered premature delivery in women who had previous preterm births. The authors could not rule out the possibility that some other upper genital tract infections were primarily involved, and that the bacterial vaginosis merely served as a marker of other infectious processes.

Some clinical tests have been developed to indicate an increased risk for premature birth. Fetal fibronectin is one such biomarkes of increased risk (Lu, G., et al. 2001). Ultrasound of the cervix shows that a short cervix (less than 25 mm) is associated with premature birth (Leitich, H., et al. 1999). Finally, screening for bacterial infection as described above, is a predictor of an increased risk for premature birth.

In terms of prevention of premature birth, several interventions have been developed in order to maintain pregnancy whenever possible (Goldenberg, R., and

Rouse, D. 1998). Overall, due to these interventions, mortality among infants weighing from 1,000 to 1,500 g has decreased from 50% in 1960 to 5% in 1998, and from 95% in infants weighing 00–1,000 g in 1960 to 20% in 1998. About one half of all neonatal deaths occur in newborns weighing less than 1,000 g.

A variety of interventions have been utilized in order to reduce premature births. Many interventions are aimed at reducing so-called spontaneous preterm births. These include those which follow spontaneous labor and/or spontaneous rupture of membranes. While these two occurrences may seem separate, evidence shows that risk factors for the two are similar, and the distinction between them is clouded.

One "intervention" is that of good prenatal care. This means frequent visits to the obstetrician and attending classes regarding pregnancy and birthing. This is not, however, a clear cause/effect relation since those who generally have good prenatal care are less likely to have bacterial vaginosis. It has been shown that women at higher risk have less prenatal care. Preterm care needs to include patient education, home visits, and nutritional counseling.

A structurally (anatomically) weak cervix is seen in from 1/200 to 1/1000 women. This is viewed as a possible condition for second trimester birth. A simple treatment is to place a couple circumferential stitches (cerclage) in the cervix. In one study, this seemed to result in a statistically significant reduction in preterm births, but the correction was only about 5% (Grant, A. 1989).

Physiological studies show that progesterone levels decline before birth. Some studies therefore have been directed toward maintaining progesterone levels by injecting hydroxyprogesterone caproate into women at risk for premature birth (Johnson, J., et al. 1975). Analysis of the data indicate that progesterone supplementation produced a significant reduction in premature births. Because weekly injections were required and only a small percentage of women were helped, use of this treatment modality has fallen from favor. Attempts to reduce preterm births by limiting alcohol intake, smoking, and drug usage have all been attempted, but these compounds are more frequently associated with low birth weight rather than prematurity, and compliance is difficult.

Another area of intervention has centered around nutrition (Carmichael, S., and Abrams, B. 1997). These nutritional interventions have involved counseling, and protein, caloric, and vitamin supplementation. Results show that caloric supplementation may result in a small increase in birth weight. This is especially true in areas of famine. While birth weight is up, gestation may not be changed.

The efficiency of vitamin and mineral supplementation has not been well studied. Anemias were once thought to be associated with premature birth, but this relationship has not held up. Low zinc levels are associated with low birth weights, and may also decrease preterm birth (Goldenberg, R. 1995).

Another approach has been to try to educate women on signs of preterm labor, then to try to stop or delay this with tocolytic drugs. Instruction consists of an increased awareness of signs of labor such as vaginal discharge, pelvic pressure, and low back pain. Ultimately, this seemed not to result in a reduction of preterm births (Main, D., et al. 1985). Drugs such as the toxolytic class are effective in slowing birth for a couple days, but what is needed is a couple weeks delay or even more. The toxolytic class is beneficial in that it allows for the administration of other drugs such as corticosteroids which act to reduce neonatal respiratory distress syndrome. Bed rest and hydration have not been shown to be beneficial, although widely used.

In the case of preterm rupture of membranes, the current practice is to adapt a careful watch–and-wait approach. The risk for fetal infection has been lessened by more powerful antibiotics. Prophylactic treatment with antibiotics may act to prolong the period between preterm rupture of fetal membranes and delivery.

A recent study (Iannucci, T., Tomich, P., and Gianopoulos, J. 1996) has examined the outcomes of extremely low birth weight newborns (weights of 500–800 g). In this study, 111 neonates with extremely low birth weights were grouped according to the reason for premature delivery: group 1 were those with idiopathic preterm labor, group 2 were those patients with premature rupture of membranes, resulting in preterm delivery, group 3 were those intentionally delivered because of non-reassuring maternal or fetal status. All mothers in groups 1–3 were in their first pregnancy. Group 4 were all patients with multiple gestations.

Variables used for comparison included parity, age, history of preterm labor, prenatal care, treatment with corticosteroids, and the incidence of chorioamnionitis. The investigators found no significant difference between the groups as regards the above variables.

The study showed an increased survival rate as the gestational age increased. The overall survival rate was 71.5%. As regards outcomes, intraventricular hemorrhage, hyaline membrane disease, and fetal sepsis were evaluated. Results on outcomes showed no difference between groups as regards the incidence of these various outcomes. The four groups were also compared as to the number of days required in the neonatal intensive care unit, and again, no significant differences between the groups were noted.

The authors comment that previously it was thought that differences in outcomes might occur as a result of the cause of extremely low birth weight in preterm infants. This study showed no such differences. The authors had expected that infants in groups 3 and 4 would have had a shorter length of time on a respirator, and a shorter length of stay in the intensive care unit. The long-term morbidity of these extremely low birth weight infants awaits further study.

An older study (Miller, C., and Reed, H. 1958) attempted to correlate serum bilirubin levels with respiratory function in premature infants. In this study, infants born at the medical center and weighing between 1,000 and 2,500 g were included in the project. Infants were grouped according to their respiratory rates. Group 1 maintained a respiratory rate of about 40/min. Group 2 breathed more rapidly at first, then leveled off at 40/min. Group 3 had significantly increased rates of respiratory rates 1 h after birth. Previous studies had shown that all infants with cyanosis, and all who die during the first week of life came from group 3.

Bilirubin in the serum was determined on days 1, 2, 4, 6, and 8. The focus was on indirect bilirubin levels. Rh studies were performed if indirect bilirubin in serum exceeded 12 mg/100 ml. Blood cultures and tests for congenital syphilis were routinely performed.

Results showed bilirubin was elevated in the serum from 69 infants during the study period. Several infants showed signs of ABO incompatibility, but were not excluded from the study. About 12 of the 69 (17%) weighed less than 1,750 g, whereas the remaining (83%) weighed between 1,750 and 2,500 g. Categorization into the three respiratory groups was about equal. Regarding serum unconjugated bilirubin levels, there was a clear difference between the two weight groups. Infants weighing less than 1,750 g had a peak concentration of unconjugated bilirubin on days 5 and 6, with the mean concentration of 11.9 mg/100 ml. In infants weighing over 1,750 g, the peak bilirubin levels were reached on the third day, and the average indirect level was 7.2 mg/100 ml. The statistical significance of this difference was less than 0.05. The highest single value of unconjugated bilirubin in groups 1, 2, and 3 was 11.5 mg/100 ml, 13.4 mg/100 ml, and 26.5 mg/100 ml, respectively. The mean bilirubin level in group 3 was 12 mg/100 ml.

In this study, only five infants had unconjugated bilirubin levels over 14 mg/100 ml. All five of these were in group 3. Nine infants had respiratory distress such that oxygen was required. Of these, three had bilirubin levels greater than 14 mg/100 ml. Of the 60 infants not requiring oxygen therapy, only 2 had bilirubin levels in excess of 14 mg/100 ml.

Transfusion was performed on two infants. One had bilirubin levels of 19.8 mg/100 ml, the other had bilirubin levels of 26.5 mg/100 ml. Criteria for transfusion was coupled to levels of unconjugated bilirubin and to its rate of rise.

The authors point out that high bilirubin can be deleterious to newborn infants, especially when concentrations exceed 18–20 mg/100 ml. There would appear to be no good way to predict which newborn baby will develop high levels of serum bilirubin. Blood incompatibilities certainly raise suspicions, but not all such newborns have dangerously high indirect bilirubin levels. The rapidity of elevation of serum bilirubin levels should be noted with alarm.

The authors suggest that the grouping of infants according to their respiratory rates would be beneficial. In this study, most infants in group 1 and 2 had bilirubin levels less than 14 mg/100 ml. Two infants in group 1 had an ABO incompatibility and levels over 14 mg/100 ml. This would serve as a convenient easy to obtain criteria for suspicion of impending unexplained hyperbilirubinemia. The exact reason for the correlation between immediate after birth respiration rate and hyperbilirubinemia is not clear, but hypoxia could play a role.

The actual life span of newborn red blood cells has been determined by radio isotope labeling of the cells using radioactive [51]Cr. Results from these studies in red blood cells of the placenta after placental blood transfusions show no consistent difference between full-term and premature newborns. The ^{51}Cr half-life in newborn infants shows that in these infants, the life span is normal. This contrasts results in premature birth infants which show a decreased survival rate of red blood cells. This is in keeping with results showing a similar decrease in red blood cell life span in fetuses in utero (Kaplan, E. 1961).

These studies point out that increased hemolysis in full-term newborn infants due to shortened red blood cell life span may not be a factor. However, the results do show that premature infants may carry over from in utero results, a decreased red blood cell life span which would result in increased bilirubin production as compared to full-term babies. This could be predicted to further challenge the already overstressed hepatic conjugating system, and thereby contribute to hyperbilirubinemia.

In another study (Dancis, J., et al. 1958), N15 was administered to three premature infants, and hemoglobin and N15 was plotted against age. The total hemoglobin and total N15 located in hemoglobin were also measured. Based on the data from the three premature infants (average weight 1,780 g), a pattern of radioactive N15 could be discerned.

In the first phase, the N15 concentration in hemoglobin rose rapidly indicating active synthesis of newly formed red blood cells. This period was the initial 6–10 days after birth. Phase 2, from postnatal days 10–50, N15 concentration in hemoglobin rose more slowly, as the total circulating N15 remained about constant. During this phase, total hemoglobin increased. This then would be a neonatal period in which there was an increase in synthetic activity. Phase 4, corresponding to neonatal days 80–130, represent the last phase to be studied. At this time, the hemoglobin N15 levels continued to decrease, while total hemoglobin was rising. The total N15 circulating began to decrease. During phase 4 there was rapid synthesis of new red blood cells, and those red blood cells which were tagged with the label N15 were at the end of their life span.

The authors suggest that these data conclude that the anemia resulting from prematurity represents a decrease in the rate of synthesis of red blood cells. They suggest that this phenomenon is "magnified" by the increase in blood volume and rapid growth of the newborn. They state that these results show a normal life span. One of the infants studied had an abrupt drop in total circulating N15 indicating a possible hemolytic event. There were no clinical signs or symptoms of this hemolysis, and would not have been discerned were it not for this study.

In a recent review of jaundice in low birth weight newborns, pathobiology, etc., were described (Watchko, J., and Maisels, M. 2003). The authors state that premature and non-premature newborns develop jaundice by similar mechanisms. These include decreased conjugation (glucuronyl transferase) capacity, possible decrease in red blood cell life span, and increased enterohepatic circulation of bilirubin. Hyperbilirubinemia has been shown to be more prevalent, severe, and last longer than that seen in full-term infants. Premature infants may have a slower response to unconjugated bilirubin than full-term infants. The frequent delay in feeding in premature infants may result in delay in intestinal flora important in elimination of bilirubin, and contribute to enterohepatic circulation of bilirubin.

The authors state that in spite of hyperbilirubinemia, in most if not all premature infants, the development of kernicterus seems to be decreasing. Other investigators have suggested that mild neurological involvement could be occurring at unconjugated bilirubin levels as low as 10–14 mg/100 ml (Kim, H., et al. 1980).

The authors note that kernicterus is a pathological diagnosis made at autopsy. Some have defined actual kernicterus as occurring based on histological damage to neurons rather than just the presence of yellow staining on gross examination. This is because the yellow staining might be a feature of a terminal event rather than kernicterus per se.

Infants born prematurely have always been considered to develop ker-nicterus more rapidly than full-term infants. Total serum bilirubin levels above 20–24 mg/100 ml was defined as the level above which kernicterus might develop. Below 18–20 mg/100 ml total serum bilirubin, the development of overt kernicterus was unlikely. Even adverse neurodevelopmental sequelae were not associated with hyperbilirubinemia.

Some studies purported to show no correlation between kernicterus and several factors such as low birth weight, low serum albumin, asphyxia, sepsis, and total serum bilirubin. The authors of this review suggest that other factors should be looked for which might act to precipitate kernicterus. Cited as an example were the studies showing that preterm newborns who had IV catheters flushed with saline containing benzyl alcohol enjoyed a drop to 0% incidence when the ben-zyl alcohol was withdrawn from the saline flush. Speculation was that the benzyl alcohol affects membrane fluidity thereby increasing passage of bilirubin into the brain.

The authors of this review cite papers showing no correlation between total serum bilirubin level and later neurodevelopmental symptoms in very low birth weight infants. The authors cite studies by Graziani, I. (1992), who was unable to correlate the development of cerebral palsy or developmental delays to hyperbilirubinemia at levels up to 22.5 mg/100 ml. Several similar studies are quoted in this review article, all failing to show a correlation between total serum bilirubin and neurological or neurodevelopmental sequelae. The authors also cite an NICHHD phototherapy trial which showed no difference between phototherapy and control (non-phototherapy treated) in terms of cerebral palsy occurrence, ataxia, hypotonia, or low IQ. In the NICHHD study, exchange transfusions were mandated when bilirubin levels reached 10 mg/100 ml, so high levels greater than 20 mg/100 ml were not seen in this study. Surviving newborn infants in the NICHHD study were reexamined at age 6 and found to be similar to control children as regards the Wechsler verbal and IQ tests.

The authors conclude that overt kernicterus is currently a rare event in the USA Neuronal injury is much more complex than simply a cause/effect relation between bilirubin levels and the brain. The decrease in incidence may be due to improved care and aggressive use of phototherapy. Further studies are needed to clarify subtle differences between cases.

In a very interesting and recent paper, there was an examination of conditions surrounding the development of kernicterus in preterm infants (Okumura, A., et al. 2009). A series of eight preterm infants were studied. Six of the eight were aged 26 weeks or less at birth and five weighed less than 1,000 g. All eight had athetoid cerebral palsy by the end of the first year. Bilirubin levels ranged as high as 17.4 mg/100 ml total levels. The low was 7.1 mg/100 ml total. None were above the 20 mg/100 ml level which is usually a cause of some concern. All received phototherapy.

During infancy (from discharge–12 months), seven of eight had abnormal intensities as displayed by MRI in the globi pallidi on T2 weighted images. Acoustic brainstem responses (ABRs) (abnormal interwave separation) were also positive, showing bilateral abnormal responses again in seven of eight newborns. Evidence of cerebral palsy became apparent at about 6 months of age, and gradually grew worse. Findings ultimately showed athetoid movements, inability to sit, rigidity, etc.

The authors note that both MRI and ABR abnormalities were present shortly after birth, in the presence of elevated unconjugated bilirubin levels, but not levels as high as 20 mg/100 ml.

The authors further emphasize that the diagnosis of kernicterus should be made on tangible evidence such as MRI and ABR. These findings are characteristic and common in most preterm newborn infants with kernicterus. The authors stress that a lack of acute neurological symptoms frequently occurs in preterm infants with kernicterus. The authors state that all too often, decisions are made based on total serum bilirubin levels, which are clearly insufficient. They suggest measurement of the bilirubin/albumin ratio as a much better predictor of the possibility of bilirubin crossing the blood–brain barrier. Changes in behavior and ABR can and do occur with moderate unconjugated hyperbilirubinemia.

The suggestion is that many cases of kernicterus causing cerebral palsy go undiagnosed. Recognition of kernicterus as a causative agent becomes very important as aggressive treatment protocols become more common.

In a similar vein, (Ahlfors, C. 2010), the awareness of unbound unconjugated bilirubin (see the chapter "Hyperbilirubinemia Revisited," this volume) levels plus total bilirubin levels will serve to improve treatment. The goal (as stated by Okumura, A., et al. 2001) is to better predict the potential threat of kernicterus in the preterm infant. This facilitates aggressive treatment and lowers the chances of cerebral palsy developing many months after a newborn jaundice spike. Ahlfors repeats the concept that total bilirubin levels, which are frequently used in pediatric settings, are not adequate.

Erythroblastosis Fetalis

Hemolytic disease of the newborn (HDN) is a collective phrase which includes several acquired hemolytic anemias, of which erythroblastosis fetalis is one. The term erythroblastosis fetalis, applies to Rh isoimmunization resulting in hemolysis, but the term focuses on aspects of the disease which are not always specific or present in each case. This chapter will therefore refer to HDN to describe red blood cell hemolytic anemias. In a very general and obvious way, hemolytic anemias in the newborn can be deleterious because they represent an increase in bilirubin formation at a time when the newborn liver may not be enzymatically equipped to conjugate the pigment, facilitating its excretion.

Table 1 represents an outline of various conditions which can fall under the general heading of acquired hemolytic anemias. These all have features in common. The clinical onset is usually rapid. Nausea, vomiting, fever, and abdominal discomfort are frequent initial symptoms. Jaundice is also a key early symptom. Hepatosplenomegaly may be present within 24–48 h. The anemia can result in cardiac signs such as tachycardia and heart failure. Blood smears usually show normocytic and normochromic anemia. Polychromatophils and reticulocytes are seen. Serum bilirubin is almost always elevated, sometimes over 20 mg/100 ml. (see Figs. 1, 2, and 3).

Acquired hemolytic anemias due to toxins are usually due to the direct lytic action of agents such as lead, potassium dichlorate, benzol, phenol, and toxins generated by various bacteria. Hemolytic anemias may be associated with auto agglutinins, and may occur in conjunction with leukemia, infectious mononucleosis, and pneumonia.

In these cases, the patient forms antibodies against his own tissues. Most laboratory data are similar in all of these classifications of hemolytic anemia. The anemias associated with auto-agglutinins are usually acute and self-limiting. Hemolytic anemias may result from transfusion of erythrocytes into a patient who has iso-agglutinins for the infused erythrocytes, or if transferred material has isoimmune bodies for the patient's erythrocytes.

Hemolytic disease of the newborn (erythroblastosis fetalis) refers to several subtypes which can affect the fetus or newborn, resulting in hyperbilirubinemia, which may be severe. These hemolytic diseases can be associated with isoimmunization. This hemolytic disease is usually a result of an isohemagglutinin

D.W. McCandless, *Kernicterus*, Contemporary Clinical Neuroscience,
DOI 10.1007/978-1-4419-6555-4_5, © Springer Science+Business Media, LLC 2011

Table 1 Some sources of
acquired hemolytic anemias

Acquired hemolytic anemias

(A) Toxic
    (1) Lytic action of drugs/chemicals
    (2) Sensitive people- to sulfa drugs

(B) Idiosyncratic

(C) Infections

(D) Antibodies vrs wbc
    (1) Isoimmunization
        Rh Isoimmunization-erythroblastosis
        A or B Isoimmunization
    (2) Transfusion Reaction
    (3) Auto aglutining -Leukemia

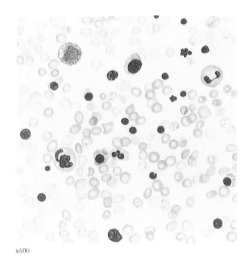

x600

Fig. 1 Blood smear from a patient with erythroblastosis. x600 Note nucleated RBCs. Water color painting by C. Leister. With permission from Blackfan, K., Diamond, L., and Leister, C. (1944)

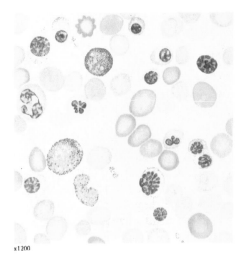

x1200

Fig. 2 Blood smear from a patient with erythroblastosis. x1200 Note nucleated RBCs. Water color painting by C. Leister. With permission from Blackfan, K., Diamond, L., and Leister, C. (1944)

A B
C D

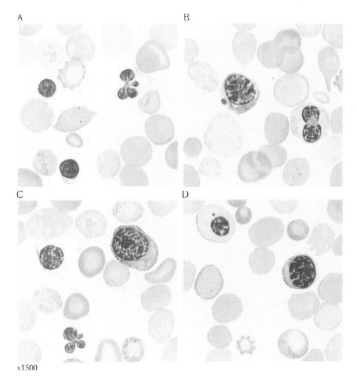

x1500

Fig. 3 Blood smear from a patient with erythroblastosis. x1500 Note nucleated RBCs. Water color painting by C. Leister. With permission from Blackfan, K., Diamond, L., and Leister, C. (1944)

in maternal serum. Several variations in incompatibility can result in hemolysis. Similar incompatibilities in the ABO system may result in hemolytic disease of the newborn.

Pathologic alterations are similar to all acquired hemolytic anemias. However, occurrence of these incompatibilities in the fetal/newborn period presents unique problems. The disorder runs a gamut of mild to severe signs and symptoms. In the severe expression, hepatosplenomegally with extramedullary erythropoiesis may be present. Nucleated red blood cells may be seen on blood smears, and cardiac involvement is common. Pathologic changes in infants who are severely affected even at birth include jaundice. Kernicterus is a feature most characteristic in newborn children who die in the first day or two after birth. The kernicterus involves the usual sites, including the cerebellum, basal ganglia, olives, hippocampus, etc.

Other factors, such as prematurity, hypoxia, and efficiency of neonatal care, all may influence the outcome. Clinically, some infants actually show no symptoms. Jaundice is a key feature, becoming apparent several hours after birth. This delay is due to the ability of unconjugated bilirubin to cross the placenta in utero, and be conjugated and excreted by the maternal liver. Jaundice may increase at an alarming rate, equaling as much as 1 mg/100 ml/hour.

When kernicterus develops, the symptoms include lethargy, opisthotonic posturing, rigidity, loss of the moro response, and a high-pitched cry. Most kernicteric infants not treated die within the first week. Aggressive treatment of hyperbilirubinemia in a proper setting with blue light therapy and possible transfusion if needed, may greatly reduce the incidence of kernicterus.

Early studies of HDN (erythroblastosis fetalis) examined the incidence of kernicterus and the nature of the pigment (Claireaux, A., Cole, P., and Lathe, G. 1953). It was found that kernicterus occurs at a rate of 1/1000 live births in a case survey from two hospitals. Kernicterus was found in about 9% of neonatal deaths. The greatest mortality rate occurred on neonatal day 3. Prematurity was a negative factor for kernicterus development, as one-quarter of total cases involved prematurity. These data were, of course, before blue light treatment. The yellow pigment was extracted from two brains, and identified as bilirubin. A bilirubin-retaining lipid was identified as the brain tissue.

One question regarding newborn anemia and hyperbilirubinemia centers on the the relative contributions of the maturation of the conjugating system as compared to the life span of newborn erythrocytes. In one study (Hollingsworth, J. 1955), radio activelabled chromium (^{51}Cr) was used to label newborn erythrocytes and subsequently determine lifespan.

In this study, blood was collected from the umbilical vein and labeled with ^{51}Cr. After washing in sterile saline, the labeled newborn blood was injected into normal adult volunteers and activity over time was assessed.

Results showed that the biologic half-life of normal adults was 24–36 days using this ^{51}Cr technique. This correlated well with results from other studies using different methods for measuring rbc half life. The survival rates of newborn labeled red blood cells in this ^{51}Cr study ranged from 15–23 days.

The rate of destruction of newborn red blood cells varied somewhat between subjects. The reason for the short life span was not clear, but there was no correlation between hemolysis and corpuscular volume or hemoglobin levels. One postulation was that red blood cells formed under relatively hypoxic conditions had shortened life spans when compared to those formed under normal oxygenation conditions following birth. The shortened life span of newborn red blood cells clearly influences the serum levels of bilirubin, at a time when conjugation and excretion of the pigment is at a critical period.

A later review article (Altman, K. 1959) examined various pertinent aspects of red blood cell physiology. As pointed out, red blood cells are unique among cells of the body in that they function in the absence of a nucleus, and in the absence of cytoplasmic organelles. The metabolism of the erythrocyte is almost exclusively based on glycolysis. There is little synthetic or degradative activity, and almost no oxygen consumption. Even though the rbc has no nucleus, it carries genetic information, and like other cells, the rbc is subject to genetic regulation.

The age of the rbc determines in part its ability to function. For example, acetyl cholinesterase activity decreases with increasing rbc age. This rate of decline differs from other declining enzyme activities such as glucose-6-phosphate dehydrogenase and glyceraldehyde-3-phosphate dehydrogenase. The rate of destruction of these two key enzymes may be increased when erythrocyte aging is accelerated in conditions such as hemolytic disease of the newborn. In this case, the stromal compartment may have an increased importance in rbc metabolism, especially if the rbc surface is damaged. If for example, the surface of the rbc becomes coated with isoimmune antibodies, the enzyme–stromal–lipoprotein complexes may be altered such that these complexes bring about a decrease in activity of key stromal enzymes.

Alteration of rbc surface enzymes by bacteria or viruses may lead to a shortened life span due to damage. The exact amount of damage which can be done to a rbc, as described above, and still have the cell survive is not clear. It would appear that stromal enzymatic changes seem to be the primary cause of rbc destruction (Gardos, G., and Straub, F. 1957).

The main metabolic pathway in red blood cells is that of glycolysis and also the hexose monophosphate shunt. All enzymes associated with these metabolic pathways are measurable in erythrocytes. Most ATP pathways and cycles are functionally inactive in red blood cells. This is a necessary milieu in order to maintain glycolytic activity. The so-called Rapoport–Luebering cycle is a cycle in which excess ATP may be utilized in a non-productive cycle aimed at reducing high energy when in excess in order to protect glycolysis (Rapoport, S., and Luebering, J. 1950; Rapoport, S., and Nieradt, C. 1951).

In terms of hemolytic anemias, those induced by drugs in people sensitive to certain drugs are of interest. If a patient is not exposed to a precipitating drug or agent, then no hemolytic consequences are noted. However, enzymatic determination in these individuals show levels of glucose-related enzymes such as glucose-6-phosphate dehydrogenase are 10–25% that of normal persons. This then leads to an increased sensitivity to drugs capable of producing hemolysis (Beutler, E. 2006).

There are many examples of enzymes present in quantities which are only just enough to fulfill normal or resting demands. Whenever increased capacity is required, the pathway may be threatened. The cerebellum has been investigated for example, and found to have such enzymatic restrictions (Reynolds & Blass).

In the case of drug-sensitive red blood cells, a so-called GSH sensitivity test has been developed. The test is based on the decreased levels of GSH in sensitive red blood cells of GSH to about 10% of normal levels. Both methemoglobin and Heinz body formation are features of this sensitivity to drugs. These acute hemolytic anemias produced by drug sensitivity can be associated with marked hyperbilirubinemia.

Pigment has been detected in significant amounts in the amniotic fluid during pregnancy in cases in which HDN (erythroblastosis fetalis) has been suspected. This was first described in 1953 (Bevis, D. 1953). Proof of the characteristics of the amniotic fluid pigment, and its identification as bilirubin was shown at a later date (Brazie, J., Ibbott, F., and Bowes, W. 1966).

In this study, amniotic fluid was withdrawn by paracentesis and kept in the dark. Spectral scans were made using a Carey spectrophotometer. The pigment was also extracted from the amniotic fluid using chloroform. The pigment was also diazotized and subjected to reverse phase kieselguhr column chromatography.

Results showed a spectral scan peak absorbance of amniotic fluid pigment of about 450 mμ. When the amniotic fluid was extracted with chloroform, the peak bilirubin absorption was at 450–453 mμ. Diazotization of the chloroform layer produced an absorption peak shift to 560 mμ, similar to the shift shown by pure bilirubin when dissolved in chloroform.

Finally, column chromatography (reverse phase kieselguhr), a single pigment band was at the top of the column, and was similar to that shown by pure bilirubin. As a control, serum taken from a patient with both direct and indirect hyper bilirubinemia showed three bands corresponding to unconjugated bilirubin and both mono- and diglucuronide unconjugated bilirubin. The position of bilirubin extracted by chloroform was identical to that shown by pure bilirubin extracted in chloroform, or pure bilirubin added to normal amniotic fluid.

The authors state that the yellow pigment found in amniotic fluid in cases of erythroblastosis fetalis has been presumed to be bilirubin, in some measure due to absorption spectra. The authors further state that their data provide additional evidence through diazotization absorption peak shifting, and results from the reverse phase kieselghur column chromatography data. The authors had tested 35 amniotic fluid samples from erythroblastosis fetalis cases, and all show similar patterns. These studies serve to confirm that the pigment present in amniotic fluid in erythro blastosis fetalis cases is beyond doubt, bilirubin.

In a large study over 6 years, a total of 742 infants under 1 week of age were admitted to a hospital with hemolytic disease of the newborn, and/or with indirect hyperbilirubinemia greater than 15 mg/100 ml (Hyman, C., et al. 1969). The children were observed for signs and symptoms such as extent of jaundice, muscle tone, moro reflex, nature of crying, and general neurological status. Of these patients, 405 of were available for follow-up study up to 4 years of age, and are the basis for this report.

Results from this study showed that of the 405 patients studied, 396 had HDN, with 348 due to Rh incompatability, and 48 had ABO problems. Nine patients were hyperbilirubinemic due to other causes. There was a slight favor for males –218 vs. 187 females, a statistic previously shown. Ten percent were premature. Of the original 742 patients, 49 died in the neonatal period and 405 were followed up to 4 years. As regards bilirubin levels, 119 patients had high bilirubin levels (greater than 20 mg/100 ml) and 286 patients had elevated levels, but under 20 mg/100 ml.

When CNS signs and symptoms were evaluated over the 4-year period, 85% were found to have no evidence of CNS of malfunction by either history or by physical examination. About 59 patients (15%) had one or more of the following symptoms: hearing loss, athetosis, seizures, strabismus, nystagmus, psychosis, and a variety of other minimal disorders.

Of the hearing loss group, 29% of the 59 CNS-positive patients showed this symptom, and 65% of these had bilirubin levels classed as high, which strongly suggested a cause/effect relationship. Only 5% had athetosis, and all had high bilirubin levels (all over 30 mg/100 ml). Nineteen percent of CNS-positive patients showed strabismus and 5% of these patients had high bilirubin levels. Thirty-six percent of CNS-positive patients had evidence of seizure activity. Of this group, only 12% had high bilirubin levels. This suggests possible CNS effects at lower than 20 mg/100 ml. One third of the CNS-positive group had minimal brain dysfunction. This category included symptoms such as clumsiness, tremors, learning and behavioral disorders, etc.

Interestingly, some patients with low bilirubin levels were severely affected. One for example with peak bilirubin levels of 16.4 mg/100 ml had grand mal seizures and moderate retardation. This patient was a full-term infant with no other complications. In EEG studies of 41 of the 59 CNS positive, no consistent pattern emerged and no significant finding indicative of bilirubin encephalopathy was noted.

In terms of psychological evaluation, 75% of patients were of normal intelligence, while four patients were high in intelligence, and the rest had low mental capacity. Perceptual problems occurred in 39% of patients, and involved the auditory pathway.

Of six infants exposed to bilirubin levels greater than 30 mg/100 ml, one half had classical kernicterus. These patients had athetosis, mental retardation, and hearing loss. Of equal or greater interest is that four patients of the 59 with low bilirubin levels (less than 20 mg/100 ml) had significant neurological disorders. This emphasizes the concept that one should not be complacent about any given patient with low to moderate hyperbilirubinemia. Another interesting statistic is that of the 376 infants with only mild or no CNS abnormalities on initial neurologic exam, fully 12% were noted to have significant CNS abnormalities on follow up. The authors state that this reflects the limitations of neonatal neurologic exams in predicting later sequalae.

Early investigators interested in newborn hyperbilirubinemia due to HDN frequently only worried about and measured total bilirubin levels. Even now in some clinical settings total bilirubin levels are used in decision making in jaundiced neonates. The importance of measuring both direct and indirect bilirubin in jaundiced newborns is illustrated by the following study (Jirsova, V., Jirsa, M., and Janovsky, M. 1958).

In this paper, two cases are described in which newborn infants with hyperbilirubin emia were found to have a significant percent of the total bilirubin consist of the direct (conjugated) kind. In the first case, the infant, born at home, had a high titer of Rh-negative antibodies. He was admitted 10 h after birth with cyanosis, hypotonia, and palpable liver and spleen. Total bilirubin was 23.8 mg/100 ml, 4.9 mg/100 ml was direct. A transfusion was initiated, but bilirubin had climbed to a total of 37 mg/100 ml, 13.0 being direct. One day later total bilirubin reached 37.2 mg/100 ml; the direct was 24 mg/100 ml of the total. By the eighth day the infant seemed moribund, apathetic, irritable, had depressed respiration, etc. Thiocapryllic acid was administered and improvement noted. By day 16, total bilirubin was back down to 10 mg/100 ml, six of that being direct., and by discharge, the patient was neurologically normal except for gazing spells.

In the second case, the precipitating cause was again an Rh-negative incompatibility. Two previous children had severe jaundice. By the third day, total bilirubin was almost 40 mg/100 ml, the direct component was about 19 mg/100 ml. Even at this level of hyperbilirubinemia, kernicterus did not develop. Exchange transfusion was done, but the infant did not do well. Hepatosplenomegaly developed, and by the third week, thiocapryllic acid therapy was initiated, and by the seventh week, improvement was sufficient to permit discharge.

The literature at this time rarely mentioned direct-reacting bilirubin when speaking about hyperbilirubinemia. The usual opinion was that direct-reacting bilirubin is negligible, and only total need be assayed. As stated above, in many clinical pediatric settings, this trend is still maintained. The authors state that since elimination of bilirubin involves first conjugation with glucuronic acid of indirect bilirubin, then second, the excretion of the direct-reacting bilirubin into bile, both forms should measured.

Based on the two cases reported in this paper, two types of hyperbilirubinemia may coexist. The first is the more common in which there is a lowered enzymatic capacity of the liver to conjugate bilirubin. Coupled with increased, as detailed earlier in this chapter, the resultant hyperbilirubinemia may be expected to be largely of the indirect type. However, as indicated by the two cases, which were severe from birth, there may well be liver damage with a negative alteration in both components of bilirubin in excretion–conjugation and excretion from the liver. Treatment of the two types of elevated bilirubin levels is different.

This division of hyperbilirubinemia into two types dictates treatment regimes. For example, if the hyperbilirubinemia is 90% or more indirect reacting, then this would suggest phototherapy or transfusion. By contrast, in cases of HDN in which both indirect and direct bilirubin levels are high, then in addition to blue light/transfusion therapy, therapy aimed at supporting liver function would be appropriate. The second group of cases is in danger of liver damage or complete hepatic failure, possibly necessitating liver transplantation. It is the first group of patients which is in danger of kernicterus.

The Gunn Rat Model

In 1938, an important paper appeared in *The Journal of Heredity*, which described for the first time a genetic mutation in Wistar rats resulting in neonatal jaundice (Gunn, C. 1938). The significance of this can hardly be overstated, as nearly all features of this murine model of hyperbilirubinemia correlate exactly with those of the human disorder known as the Crigler–Najjar syndrome. This has permitted extensive and meaningful data to be derived from animal studies. Because of this, knowledge of hyperbilirubinemia has been advanced at a rapid rate. This chapter will detail with some of the pertinent findings from the Gunn rat model.

In the paper by Dr. C.H. Gunn, the description included no less than 249 jaundiced newborn Wistar rats. Observations confirmed that none of the jaundiced pups ever reverted to the non-jaundiced state (see Fig. 1). Furthermore, most appeared normal in color when born to non-jaundiced mothers, but by 12 h of age were becoming noticeably yellow. Some born to jaundiced mothers appeared jaundiced at birth.

Breeding experiments showed that a cross of homozygous normal rats (JJ) with homozygous jaundiced rats (jj) yielded only heterozygous non-jaundiced rats. Conversely, when both parents were homozygous jaundiced, all offspring were similarly jaundiced. Non-jaundiced heterozygous rats when bred to homozygous jaundiced rats yielded a nearly 50% jaundiced number of offspring.

Developmentally, jaundiced newborn rats lagged behind non-jaundiced littermates. Furthermore, the jaundiced rats showed the presence of neurological difficulties, including ataxia, and a partial paralysis of the lower extremities. There was a marked difference in size and weight in the jaundiced littermates as compared to those that were non-jaundiced.

About 50% of heterozygotes showed an increased fragile nature of erythrocytes, and a degree of reticulocytosis. A couple of treatment regimes were tried in jaundiced rats, including splenectomy, treatment with liver extract, and hypertonic and antihemolysis agents. None of these treatment regimes was effective. Splenectomy in the human counterpart may offer some relief. The author comments on the similarity of the hereditary condition in Wistar rats to that seen in humans, and called at that time heredity acholuric jaundice.

Later studies by Johnson (Johnson, L., et al. 1959), examining a large number of Gunn rats reconfirmed Gunn's early results. Thus, Johnson showed that

D.W. McCandless, *Kernicterus*, Contemporary Clinical Neuroscience,
DOI 10.1007/978-1-4419-6555-4_6, © Springer Science+Business Media, LLC 2011

Fig. 1 Adult homozygous Gunn rat. Source: National Bioresource Project, Institute of Laboratory Animals, Kyoto University See appendix for source

homozygous newborn Gunn rats were not jaundiced, but were becoming yellow as early as 6 h of age. She further demonstrated that hyperbilirubinemia peaked around 2 weeks of age, then dropped by 4 weeks. In their study, kernicterus death and high bilirubin levels correlate well. Most deaths from kernicterus occurred in Gunn rats with bilirubin levels above 10 mg/100 ml. Another outcome of early work of Johnson was the observation that the severity and incidence of kernicterus could be increased by administering sulfonamides to homozygous newborns. This produced a drop in indirect bilirubin in serum, and an increase in the staining of brain nuclei. The bilirubin is effectively displaced from albumin-binding sites by sulfonamides, thereby permitting entry into brain tissue by unbound unconjugated bilirubin.

The sulfonamide treatment increases the observable severity of neurological manifestations and death from kernicterus. This, and further studies on binding of bilirubin to albumin have been described by Odell (Odell, JCI 38:823).

In a paper published in 1961 (Blanc, W), the characteristics of kernicterus in Gunn's strain of Wistor rats were described. Blanc emphasized the potential benefit of a rat model of human kernicterus, and stated that five criteria should be met. These are (1) evidence of CNS symptomology in jaundiced animals, (2) bilirubin staining of nuclei in brains of affected individuals, (3) persistence of yellow stain following formaldehyde fixation, (4) presence of yellow pigment in neurons, and (5) microscopic evidence of degeneration/damage to nerve cells. The author continues by saying that all five criteria are met in Gunn rats.

Results showed bilaterally symmetrical lesions in the hippocampus, basal ganglia, colliculi, and other areas (see Table 1). These areas showed severe neuronal

Table 1 Cerebral areas
stained in Kernicteric Gunn
rats

Hippocampus
Thalamus
Globus pallidus
Putamen
Caudate
Brain stem nuclei
Substantia nigra
Red nucleus
Subthalamus
Superior/Inferior colliculus
Cerebellar floccus parafloccus
Dentate nucleus

Adapted from Blanc, W. (1961)

damage, and extensive ganglion cell loss. Bilirubin is seen microscopically in ganglion cells fixed in formalin. In older Gunn rats, pigment is not seen but there is atrophy of above-mentioned brain areas and focal gliosis.

Spectrophotometric studies were performed on jaundiced ganglion cells visualized on smears of fresh brain tissue in Gunn rat brain. Absorption maximum peaks occurred around 410 mμ. Ultrafiltered diffusible bilirubin has a peak at about 420 mμ, and peaks in erythroblastosis fetalis brain are about 405–415 mμ. There is correlation between the amount of pigment at 405–415 mμ and the severity of symptoms.

The author states that the above biochemical, clinical, and neuropathological findings are certainly in keeping with the generation of kernicterus in both Gunn's rats and in jaundiced newborn infants.

The effects of bilirubin on ultrastructure in the brains of Gunn rats have been described (Schutta, H., Johnson, L., and Neville, H. 1970). In this study, jaundiced Gunn rats and non-jaundiced littermates from 1 day old to 18 months of age were studied. Brains were fixed by perfusion with 4% glutaraldehyde; staining was with lead citrate and uranyl acetate. Areas examined included the cerebral cortex, cerebellar Purkinje cells, colliculi, cerebellar nuclei, basal ganglia, and hippocampus. Emphasis was placed on mitochondrial structure because of many previous studies showing changes in the biochemistry of mitochondria in jaundiced Gunn rats. Animals other than Gunn rats have also been used as kernicterus models. These include animals infused with unconjugated bilirubin (see Figs. 2, and 3).

Results from electron microscopy study of Gunn rat Purkinje cells showed a normal appearance until day 3 when some mitochondria seemed enlarged. This was the first structural change seen with electron microscopy. By the fifth day, many Purkinje cells had intracytoplasmic inclusions and large oddly shaped mitochondria. The cristae of these mitochondria were tightly packed. Purkinje cells with cytoplasmic inclusions always had altered mitochondria from 8 days and up, and some mitochondria also contained granule-filled vesicles.

Some mitochondria showed lamellar inclusions, and a few mitochondria had an altered shape to the cristae. By 2 months of age, altered appearing mitochondria

Fig. 2 Coronal section of puppy brain with kernicterus. Permission from Rozdilsky, B. (1961)

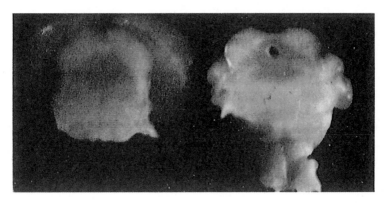

Fig. 3 Coronal slices through brainstem of a puppy with kernicterus. Permission from Rozdilsky, B. (1961)

were still to be seen in Purkinje cells, which were returning to normal in pro-file. Even by 18 months, some altered Purkinje cells could be seen. Pretreatment of sections with amylase eliminated the appearance of granules in Purkinje cells, suggesting they at least had a glycogen component.

Other brain areas such as the colliculi, hippocampus, basal ganglia, and cere-bral cortex at 1–1.5 weeks also had mitochondria identical to those seen in the cerebellum. Cytoplasmic inclusions were also seen. At 2 months of age, neurons with glycogen-like granules were noted. Mitochondrial changes were not seen in astropcytes or oligodendrocytes.

These observations demonstrated that mitochondria are damaged in vivo by bilirubin in neuronal cells. Further, the accumulation of glycogen in enlarged neu-ronal mitochondria suggest altered function. Such changes as were seen could result from a direct effect of bilirubin on surrounding cytoplasm, and mitochon-drial changes were secondary. This second possible mode of bilirubin toxicity seems unlikely, and is not supported by any direct evidence.

The authors favor a "cytoplasmic first" hypothesis based on the supposition that all mitochondria are similar, and if it were purely a mitochondrial disorder, then all mitochondria would be affected regardless of brain region. (This is questionable reasoning given other examples of metabolic encephalopathy which show highly localized lesions – Wernicke's disease and Leigh's disease being two fine examples). The authors quote a newborn guinea pig (Diamond, I., and Schmid, R. 1966b) study in which oxidative phosphorylation was not altered when unconjugated bilirubin was injected. Problems associated with the Diamond/Schmid study have been detailed elsewhere (see the chapter "Bilirubin and Energy Metabolism," this volume).

In another electron microscopy study, the effect of hyperbilirubinemia on the substantia nigra in Gunn rats was studied. Gunn rats and suitable controls from 2 weeks of age to 13 months were examined. Light microscopic observations showed the presence of periodic acid Schiff (PAS) positive inclusions, suggestive of glycogen granules in 2-week-old, and again in 6–8-week-old Gunn rats. PAS staining was only slightly positive in adult Gunn rats.

In 2-week-old Gunn rats, electron microscopy showed the presence of spheroid-shaped and enlarged mitochondria in the substantia nigra. The size of mitochondria in homozygous Gunn rats was nearly triple than those in control rats. Osmiophilic granules were also present in some mitochondria. Endoplasmic reticulum in proximity to granule-containing mitochondria also contained similar granules. At 6 weeks, many swollen granule-containing mitochondria were seen in cells of the substantia nigra. Granules in the endoplasmic reticulum were more numerous than before. Ribosomes were also present in areas in which enlarged mitochondria were found, and all were in proximity to the nucleus (Batty, H., and Millhouse, O. 1976) (see Figs. 4 and 5).

By 12 weeks and older, few mitochondria were enlarged, and only a few had granules. Osmophilic granules appeared in the endoplasmic reticulum. Neurons in adult Gunn rats were normal in appearance as compared to controls.

The authors suggest that the granules seen in mitochondria and endoplasmic reticulum are glycogen granules. This conclusion is based on the size being consistent to proven glycogen granules, the positive PAS reaction, and to the observations of others (Schutta, H., et al.) that these granules are not apparent in amylase-treated sections. These membranous intramitochondrial bodies have also been seen in feline GM gangliosidosis (Farrell, D., et al. 1973), and late infantile lipidoses (Gonatas, G., and Gonatas, J. 1965), among others.

Regarding mitochondria, the authors again suggest that the enlargement was either due to an increase in the matrix compartment, or to expansion of the intracrystal space. The suggestion is made that these mitochondrial changes represent an attempted compensation to decreased energy production due to the toxic effect of unconjugated bilirubin. The enlargement of mitochondria in 6–8 week Gunn rats is thought to be due to the presence of osmophilic granules identified as glycogen and located in the intracrystal space. Intramitochondrial glycogen may be found in a number of normal and pathological conditions. The accumulation of mitochondrial glycogen is viewed as reflecting a compensatory mechanism to altered

Fig. 4 Electron micrograph
showing mitochondrial
disruption and the presence of
300 Angstrom granules in
homozygous Gunn rats.
Permission from Batty, H.,
and Millhouse, O. (1976)

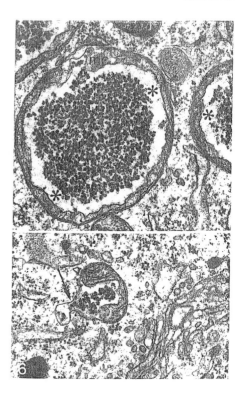

mitochondrial function such as seen in hypoxia and compromised oxidative phosphorylation states. This alteration in mitochondrial function in Gunn rats is viewed
as a result of a prolonged exposure to the toxic effects of intracerebral unconjugated
bilirubin.

In another very interesting study on structural changes in the brains of Gunn
rats (Hanefeld, F., and Natzschka, J. 1971), sulfadimethoxine was administered to
newborn (day 10–14) homozygote Gunn rats. Results from this treatment showed
that all newborns injected with the drug became highly symptomatic within 24 h
following injection. Symptoms included hypotonia, ataxia, opisthotonic posturing,
and clonic tonic seizures. Nine animals died within the first 24 h.

Necropsy results showed that all treated homozygotes had yellow-appearing
brains on the surface, and yellow staining of the basal ganglia, cerebella, and brainstem. Microscopically, bilirubin could be visualized in the cerebellum, cochlear
nucleus, and pontine nucleus. This was confirmed using phase contrast microscopy.
Vacuolization and lysis could be observed in Purkinje cells.

Further studies on the Gunn rats included enzyme histochemical techniques
applied to ganglion cells of homozygous and heterozygous newborns. Results
showed all staining reactions for dehydrogenases (LDH, SDH, IDH) and reductases
(NADH2, NADPH2) were negative in bilirubin-stained cells. Cytochrome oxidase

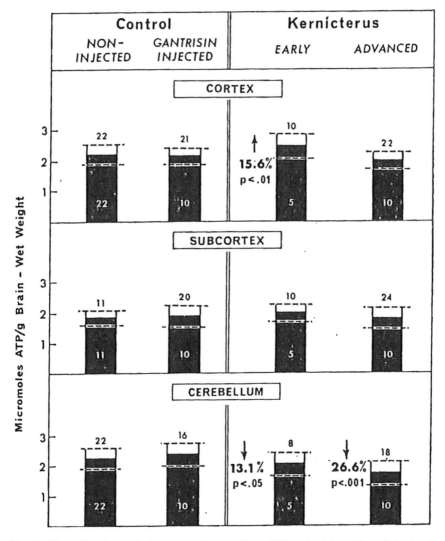

Fig. 5 Effect of kernicterus in Gunn rat brain on regional ATP levels. Adapted from Schenker, S., McCandless, D., and Zollman, P. (1966)

was also reduced in cells from homozygous newborn Gunn rats. All these changes occurred in cerebellar Purkinje cells.

The authors speculate that the histochemical changes are in keeping with similar effects shown in vitro by others. Schutta, H., Johnson, L., and Neville, H. (1970) have demonstrated via electron microscopy that mitochondria from Gunn rats are enlarged and vesicle filled. This suggests impairment of mitochondrial function as reflected in the present study could underly cerebral symptoms. There were also some neurons which had negative enzyme histochemical findings, but no evidence of bilirubin. This could be explained by the fact that not all bilirubin can be visualized. A more meaningful test is histochemistry which judges functionality. Enzyme histo chemistry could have particular significance in the assessment of potential treatment regimes.

An electron microscopy study focused on the effects of bilirubin specifically on its effects on Purkinje cells has been done (Moore, P., and Karp, W. 1980). In this study, homozygous Gunn rats and suitable controls were prepared for electron microscopy. The purpose of this study was to further examine cerebellar Purkinje cell's ultrastructure. Previous studies had shown that mitochondria in Purkinje cells from kernicteric Gunn rats showed mitochondrial hypertrophy and intramitochondrial glycogen deposits. Newborn rats were sacrificed and examined at 28–30 days of age at a period when Purkinje cell development should have been completed.

Results showed that each homozygous Gunn rat had cerebella smaller than controls. Ninety percent of homozygous Gunn rats showed large immature looking Purkinje cells scattered among normally appearing Purkinje cells. There was no correlation between the number of immature Purkinje cells and serum unconjugated bilirubin levels. Frequently the immature cells were in proximity to degenerating Purkinje cells. That the immature looking Purkinje cells were indeed Purkinje cells was indicated by the presence of an indented nucleus, size, organization of cytoplasm organelles, and location. Also present were intramitochondrial glycogen deposition granules, thought to be specific to Purkinje cells in kernicteric brain.

The authors speculate that bilirubin and its effects on energy metabolism may result in a delay in the maturation process of Purkinje cells. The development of Purkinje cells is normally completed by postnatal days 28–30, but the presence of immature Purkinje cells in kernicteric brain suggests developmental delays.

While neuropathological studies were being conducted, biochemical studies on Gunn rats were similarly proceeding. In one study aimed at elucidating the biochemical nature of bilirubin metabolism in Gunn rats, several techniques were applied (Schmid, R., et al. 1958).

In these studies, hemoglobin concentration was within normal limits. Reticulocyte counts were similarly within the normal range. Paper chromatographic studies showed elevated unconjugated bilirubin levels in jaundiced Gunn rats, whereas bilirubin levels in non-jaundiced littermate controls were similar to non-Gunn rat levels. Bilirubin was not present in the urine of any newborn rats studied. Conjugated bilirubin was not demonstrable in bile from jaundiced Gunn rats, and fecal urobilinogen excretion was much reduced as compared to controls. Patency of the biliary system was confirmed using chlorografin and bromsulphalein.

Glucuronide formation in vivo in jaundiced rats was about one half that of non-jaundiced littermate controls. In adult Gunn rats, both crystalline bilirubin as well as bilirubin glucuronide were administered, and bilirubin levels in serum and bile were measured. Results showed that after injection of crystalline bilirubin, only a trivial increase in bilirubin in bile, and a transient increase in bilirubin in serum occurred. When bilirubin glucuronide was injected there was no additional increase in serum bilirubin, but there was a 100-fold increase in bilirubin in bile. Essentially, all the biliary bilirubin was in the form of bilirubin glucuronide.

Histologically, all adult jaundiced Gunn rats exhibited a deposit of an orange material at the tip of the renal papillae. This material was ultimately demonstrated by paper chromatography to be unconjugated bilirubin crystals. In some jaundiced adult rats, these crystals of bilirubin were noted in the brain, adrenals, and intestinal tract.

The authors speculate based on the above findings, that the jaundice in homozygous Gunn rats is non-hemolytic and non-obstructive. Further, no bilirubin glucuronide could be found in serum, bile, or urine. Both in vivo and in vitro, highly significant impairment of bilirubin glucuronide formation occurred due to a defective glucuronyl transferase enzyme. It was found that non-jaundiced Gunn rat littermates had a lowered ability to conjugate bilirubin, but that this level of deficiency was not sufficient to result in overt jaundice.

In another study examining bilirubin glucuranide formation by Gunn rat liver, labeled (^{14}C) aminolevulinic acid added to blood-perfusing Gunn rat livers showed minimal incorporation of label into biliary bilirubin (Garray, E., Owen, C., and Flack, E. 1966). Plasma bilirubin showed a significant increase in label as compared to controls. These data reconfirm that bilirubin is not being conjugated by Gunn rat perfused liver. Indeed, the ^{14}C label recovered on both plasma and bile was low, suggesting that bilirubin has an inhibitory effect on heme synthesis (Billing, B. 1963).

Although most Gunn rat studies focus on effects on the CNS for obvious reasons, other systems are also affected, such as previous observations of bilirubin nephropathy in Gunn rats (Odell, G., Natzschka, J., and Storey, G. 1967). Results from renal function studies showed that homozygous Gunn rats were about threefold less efficient in concentrating their urine as compared to littermate controls. Urinary osmolalities were similiarly reduced. Urinary sodium losses were threefold greater in homozygotes as compared to littermate controls. Regional renal analysis for bilirubin showed a 100-fold greater concentration of bilirubin in the papilla of homozygote Gunn rats as compared to heterozygotes. The authors speculate that bilirubin may interfere with sodium reabsorption in the ascending limb of the loop of Henle.

In terms of possible mechanisms of toxicity of bilirubin, it was early demonstrated that there was a correlation between the level of serum bilirubin and incidence and severity of kernicterus (Hsia, D., et al. 1952 zetterstrom the book). Bilirubin had also been shown to decrease chopped brain respiration in vitro (Day, R. 1954). Soon, the key observation was made that bilirubin uncoupled oxidative phosphorylation in rat liver mitochondria (Zetterstrom, R., and Ernster, L. 1956). Warburg manometer was used to show that increasing concentrations of bilirubin had corresponding decreasing P/O ratios. Respiration was also decreased by bilirubin, but at higher bilirubin concentrations.

Addition of cytochrome C or DPN acted to decrease the depression of respiration by bilirubin, but these cofactors did not change bilirubin's effect on P/O ratios. Neither cytochrome C nor DPN could restore oxidative phosphorylation. The authors went on to show similar effects of bilirubin on brain mitochondria. Later studies (Quastel, J., and Bickis, I. 1959) subsequently showed that bilirubin uncouples oxidative phosphorylation in whole cells as well as mitochondria. The supposition at this point in time was that the uncoupling of oxidative phosphorylation was severe enough to compromise energy production, leading to cellular damage and compromised function in tissue such as brain and kidney.

A study looking at the uptake of ^{14}C bilirubin into brains of Gunn rats was performed (Menken, M., et al. 1966). In this study, 12–16–day-old Gunn rats were utilized. Sulfisoxazole was used to displace bilirubin from albumin-binding sites thereby assuring entry of the labeled pigment into brain tissue. Results showed that animals which were kernicteric as judged by neurologic symptoms such as lethargy, hypotonia, and ataxia had higher levels of labeled bilirubin than did slightly jaundiced but healthy littermates. Both jaundiced groups had higher levels of pigment than did non-jaundiced littermates (JJ). Adult Gunn rats injected for as long as 2 weeks did not take up labeled bilirubin into brain tissue.

The authors state the critical factor for the development of kernicterus is the level of diffusible unconjugated bilirubin in serum, not the absolute serum concentration. Administration of sulfisoxazole to newborn homozygous Gunn rats lowered serum bilirubin levels due to displacement of pigment from albumin-binding sites, and concomitant entry of unbound unconjugated bilirubin into the brain of Gunn rats.

In addition, jaundiced but asymptomatic newborn Gunn rats, while showing a small uptake of labeled bilirubin, had no observable neuropathologic findings. Brain tissue examined for cellular damage in kernicteric symptomatic newborn Gunn rats always showed structural lesions. As a side observation, the hyperbilirubinemia in the kernicteric Gunn rats showed no evidence of liver damage, suggesting a "susceptibility" of certain brain cells for bilirubin damage.

In another study (Menken, M., and Weinbach, E. 1967), newborn (10–14 day old) Gunn rats were injected with sulphisoxazole until evidence of kernicterus appeared: hypotonia, weight loss, ataxia, and seizures. Since some toxins uncouple oxidative phosphorylation in vitro, but may have different in vivo actions, the authors wanted to see if bilirubin resulted in uncoupling of oxidative phosphorylation in vivo in Gunn rat newborns with overt neurological symptoms.

In these studies, jaundiced symptomatic newborn Gunn rats and appropriate littermate controls were decapitated and whole brain mitochondria prepared. Oxygen consumption was measured by polarographic methods using a Clark electrode. P/O values and respiratory control indicators were calculated.

Results showed that there were no significant differences in respiration between kernicteric brain mitochondria and that of littermate controls. P/O values and indicators of respiratory control were also not significantly different when comparing kernicteric symptomatic brain mitochondria to mitochondria of littermate controls.

The authors speculate that the similar indicators of mitochondrial function and oxidative phosphorylation lead to skepticism as regards the concept that bilirubin uncouples oxidative phosphorylation in vivo. Since the lesions in kernicteric Gunn rats and newborn humans with kernicterus are similar, the authors hoped the Gunn rat results might shed light on the nature of the biochemical lesions in humans.

The very nature of the lesion in both Gunn rats and kernicterus in humans is of a highly localized nature. Obviously, any preparation of whole brain mitochondria would, in addition to lesioned areas, include large quantities of normal non-affected tissue. The lesioned material would only be a small fraction of the total. This would effectively dilute out any change in affected tissue. The authors do briefly acknowledge this possibility, but state that uncoupling of oxidative phosphorylation in vivo remains unproved.

In order to circumvent the problems with whole brain analysis in a disorder producing highly localized lesions, regional cerebral neurochemical studies need to be done . In one such study, energy metabolism was studied in newborn Gunn rat brain which was prepared regionally (Schenker, S., McCandless, D., and Zollman, P. 1966).

In this study, 2-week-old Gunn rats were injected with sulfisoxazole in order to produce kernicterus in all jaundiced newborns, and to assure uniformity of onset of symptoms. The sulfisoxazole acts to displace bilirubin from albumin-binding sites, facilitating bilirubin's entry into the brain, as described earlier in this chapter. Newborn Gunn rats at 2 weeks of age were used since bilirubin levels are high at that time, and the brains are of sufficient size to easily permit regional analysis.

Symptomatic newborn Gunn rats were sacrificed in a mixture of dry ice and acetone in order to best preserve labile metabolites such as ATP. In addition to ATP, oxygen consumption and ATPase activities were measured. Both "early" and "advanced" kernicterus in newborn Gunn rats were studied. This was an arbitrary designation based on severity of symptoms. Early animals displayed mild ataxia and righting impairment, whereas advanced kernicteric rats showed drowsiness, stupor, opisthotonus, and seizures. All animals were classified before knowledge of degree of pigment staining, or of ATP level or oxygen consumption.

Results showed that brain cortex and subcortex tissue was not significantly different in early or advanced kernicteric Gunn rat brain. Cerebellar tissue, however, showed a statistically significant changes. Animals displaying signs and symptoms of early kernicterus had ATP levels 13% lower than littermate controls, and cerebella from severely affected (advanced) Gunn rats showed a 26.6% decrease in ATP (see Fig. 6).

In terms of oxygen consumption, values for whole brain were similar in control vs. early kernicterus. Regional studies of the oxygen consumption in the cerebellum showed no change between control and asymptomatic jaundiced homozygote Gunn rats. By contrast, oxygen consumption in symptomatic kernicterin Gunn rats was on average nearly 28% lower than appropriate littermate controls. The decrease in oxygen consumption was present in both alert and stuporous kernicteric brain. ATP levels were similar in the cerebral cortex of kernicteric animals, but was decreased in the cerebellum as compared to jaundiced but asymptomatic littermate controls.

Interestingly, whole brain oxygen consumption was not significantly different from that of controls, yet the oxygen consumption in cerebellum was decreased. This observation emphasizes the inappropriateness of whole brain analysis which can easily miss regional changes by including non-affected tissue.

The authors note that there is a correlation between the type and severity of the symptoms and the site and intensity of cerebral bilirubin deposits. Thus, there was

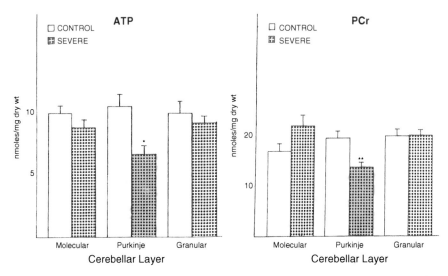

Fig. 6 Effect of kernicterus in Gunn rats on the cerebellar layers of ATP and phosphocreatine. Adapted from McCandless, D. and Abel, M. (1980)

a progressive depletion of cerebellar ATP accompanying the progressive display of cerebellar symptoms and bilirubin deposit. These data, along with the oxygen consumption data, support the concept that bilirubin uncouples oxidative phosphorylation with changes in ATP, and that regional studies are imperative in order to elucidate the biochemical basis of the neuropathology of kernicteric brain.

Since bilirubin has a predilection for Purkinje cells in the cerebellum, localized studies can be carried further by assessing energy metabolism in discreet cerebellar layers using oil well techniques and enzymatic cycling (McCandless, D., and Abel, M. 1980).

In this study, homozygous newborn Gunn rats and controls were injected daily starting on day 14 with sulfisoxazole until symptoms were severe. Animals were sacrificed in liquid N2 to protect labile metabolites, then assayed using enzymatic techniques (Lowry book).

Results showed reproducible neurological symptoms in newborn Gunn rats including rapid running, ataxia, opisthotonus, and stupor. Energy metabolites were measured in three cerebellar layers – molecular, Purkinje cell rich, and granular. Glucose and glycogen were elevated in all three layers to about the same extent. ATP and phosphocreatine showed a more regional effect. While ATP values in the molecular and granular layers were comparable to values in controls, the Purkinje cell layer showed a 40% decrease in ATP as compared to controls. Similarly, phosphocreatine showed a similar level of decrease only in the Purkinje cell rich layer (see Fig. 6).

These data once again show the necessity to look for regional changes in the regions affected in any metabolic encephalopathy. These data are also further evidence supporting the concept, shown many years earlier, that bilirubin's major and initial mode of toxicity is the uncoupling of oxidative phosphorylation, with resultant lowering of available energy in the form of ATP and phosphocreatine. The question of why Purkinje cells are selectively affected is not yet answered.

Other studies have examined bilirubin effects on cerebral energy metabolism. In this study, newborn guinea pigs were infused with high levels of unconjugated bilirubin in a quantity sufficient to produce jaundice and unspecified neurologic injury. Mitochondria from whole brain or from cerebellum were prepared and analyzed for oxygen consumption and P/O ratios.

Results showed that neither whole brain nor cerebellar mitochondria showed evidence of altered P/O ratios, or of uncoupling of oxidative phosphorylation. Electron microscopy failed to show any evidence of mitochondrial structural alterations. There was no loss of pigment during mitochondrial preparation which might account for normal mitochondrial findings. The authors speculate that uncoupling of oxidative phosphorylation may not be a definitive factor to explain bilirubin neurotoxicity. These results are often quoted as evidence that a direct effect on cerebral energy metabolism as a result of hyperbilirubinemia does not exist.

This study, however, suffers from at least two major defects. First, results from whole brain have repeatedly been shown to have limited value because of the highly localized nature of the lesion. Even the case of whole cerebellum, if results are negative, as in this study, the burden of proof lies with the investigators since Purkinje cells (site of the cerebellar lesion) compose only a small fraction of the total cerebellar weight. Finally, the guinea pig model may be flawed in that guinea pigs are born at a much more advanced developmental stage than rats or humans. Newborn guinea pigs have hair, open eyes, are mobile, etc., whereas rats require 2–3 weeks to reach the same level of development. Guinea pig blood–brain barriers, Purkinje cell development, liver function, etc., would all be further developed, and as such, better equipped to respond to bilirubin.

In another study (Katoh, R., Kashiwamata, S., and Niwa, F. 1975), high energy phosphates and some Krebs cycle intermediates were measured in homozygous Gunn rats. Animals were treated with novobiocin to produce neurological symptoms and to render kernicterus consistant. Krebs cycle intermediates and high-energy phosphates were measured in whole brain following decapitation and rapid brain removal.

Results showed bilirubin levels were more than tenfold higher in the brains of homozygotes than controls. Krebs cycle intermediates such as lactate, glutamate, aspartate, malate, succinate, isocitrate, and citrate were unchanged in homozygous kernicteric Gunn rat whole brain. Only pyruvate was decreased. NADH, NAD, and ATP were measured and found to be similar between kernicteric rats and littermate controls in whole brain.

The authors speculate that a direct effect of bilirubin on mitochondrial structure and function might not be the definitive biochemical lesion. The authors cite other negative whole brain studies to support their conclusions. They ascribe the changes in pyruvate levels to be related to the use of novobiocin in the generation of kernicterus in homozygous Gunn rats.

This study, although technically well done, fails to be convincing due to the brain sampling techniques. The studies were performed on whole brain, which includes tissue not stained by the pigment. This dilutes the damaged areas with healthy tissue, rendering interpretation of the derived data impossible.

In 1976, a paper was published (Nakata, D., Zakim, D., and Vessey, D. 1976) which examined the defective nature of UDP-glucuronyltransferase in Gunn rats and controls. Evidence had accumulated that glucuronidation of various compounds by liver microsomes was not always equal. For example, UDP-glucuronyltransferase is near zero in Gunn rats for bilirubin, but for other aglycones such as aminopherol, the rate is 20% of normal. The purpose of the study was to determine the characteristics of liver microsomes ability to conjugate various aglycones.

Results showed that the slow rate of glucuronidation of p-nitrophenol was caused by a low affinity of UDP-glucuronyltransferase for UDP-glucuronic acid by Gunn rat liver microsomes. This rate was not as slow as other aglycones such as aminophenol and bilirubin. These various rates can be explained by the existence of different sites of glucuronidation. There may also be different time points in development of activities of various UDP-glucuronyltransferase activities.

It would appear from the data that normal quantities of defective UDP-glucuronyl- transferases are produced by Gunn rat liver microsomes. It is likely that there are different UDP-glucuronyltransferases all having a common type of UDP-glurcuronic acid-binding site. A defect in this common binding site might result in defective function in several glucuronyltransferases. Differences in interaction of these binding sites with binding units on the various aglycones would explain variability in the glucuronidation of these aglycones.

The above-cited studies on Gunn rats represent what might be called "primary" defining mechanisms of toxicity, and important characteristics of Gunn rats regarding liver and brain biochemical and structural changes. Evidence supporting these primary lesions is overwhelming. Many other alterations in the brains of Gunn rats have been elucidated, but most of these might be viewed as secondary to the initial alterations described in this chapter. Other neurochemical changes will be described in other chapters when appropriate.

Crigler–Najjar Syndrome

In 1952 (Crigler, J., and Najjar, V. 1952), the initial report of a congenital abnormality of the liver-conjugating system was first described by the two investigators whose name the disorder carries. This paper reports seven patients, all descendents of two people, six generations back. Five of the patients were studied directly, one was in another hospital, and one was diagnosed by a careful history, but not examined by the authors. Complete genetic studies were not completed at the time of publication. It was suggested that 15 probable cases existed out of 105 family members over the six generations.

There were no less than 16 stillbirths. Only one of the six patients examined died of kernicterus, and one was still alive at 18 months of age. The one who died had an autopsy, and the results are described below.

Case 1 was a female admitted at 5 weeks of age with jaundice, but asymptomatic until 5 days before admission. She was reported to have had several seizures, showed no hepatosplenamegaly, and showed few neurological symptoms. Soon after admission, she demonstrated neurological symptoms including spasticity, and rigidity. She developed uncontrolled athetoid movements, was discharged at 16 weeks, and died 1 week later.

Case 2 was a male admitted at 16 days of age because of intense jaundice. A family history showed that three siblings had died at home because of jaundice. On admission, the patient appeared ill, dehydrated, and very jaundiced. Twenty-four hours after admission, the patient went into shock, vomited blood, and died in about 12 h. This patient had an autopsy.

Case 3 was a male, brother of case 2, admitted to the hospital at 6 days of age because of jaundice and a positive family history. This patient showed only minimal signs of extrapyramidal involvement. No definitive other neurological signs were seen, and this child was alive, 18 months old, when the paper was written.

Case 4, a male, was admitted at age 9 weeks with persistent jaundice since newborn day 2. This patient was hospitalized for 43 weeks, and during that time signs of kernicterus became more evident. These included hypertonia, opisthotonic posturing, jerking movements of arms and legs, and an expressionless face.

Case 5 was a sibling of cases 4 and 6. He was markedly jaundiced. Neurological signs included expressionless face, opisthotonic posturing, athetosis, writhing movements, flexed wrists, and extended fingers. This patient died at 29 weeks.

D.W. McCandless, *Kernicterus*, Contemporary Clinical Neuroscience,
DOI 10.1007/978-1-4419-6555-4_7, © Springer Science+Business Media, LLC 2011

Case 6 was a sibling of cases 4 and 5, and was hospitalized elsewhere. He was jaundiced at birth, with neurological signs appearing shortly after birth. The patient exhibited opisthotonic posturing. He lived for 8 weeks after admission. An autopsy showed diffuse atelectatic pneumonia, but the brain was not examined (see Table 1).

Table 1 Summary of common Crigler–Najjar symptoms

	Bilirubin indirect/total	Blood smear	Jaundice onset	Opisthotonus Posturing	Extra pyramidal disease	Died w/signs of Kernicterus
Case 1	33/37	Normal	2nd Day	+	+	+
Case 2	23.8/25.8	Normal	2nd Day	−	−	+
Case 3	27.4/27.4	Normal	3rd Day	−	+	alive @ 18 mos.
Case 4	32/34	Normal	2nd Day	+	−	+
Case 5	44.8/44.8	Normal	2nd Day	+	+	+
Case 6	total 50	Normal	at birth	+	+	+

Adapted from Crigler, J., and Najjar, V. (1952)

Laboratory data showed no incompatibilities in blood groups except in case 1 in which there was an anti-A titer of 1:4000, which was not believed to be of significance in the patient's jaundice. Serum bilirubin levels ranged up to 44 mg/100 ml, with the major component being indirect reacting In these patients, after the physiologic jaundice period was over (weeks 4–12), the serum bilirubin levels remain more less fixed until later in the first year, or until just before death. In two patients, a bilirubin excretion test was done in which 5 mg/kg of bilirubin was dissolved in 0.1 M sterile sodium carbonate and injected into the patients. Results showed an abnormal retention of the pigment.

The pigment was isolated in crystalline form some of the patients. This isolated pigment had the same chemical and physical characteristics as that isolated from other patients with other types of indirect hyperbilirubinemia. In addition, bilirubin in serum was removed from one of the six cases and injected into a normal 2-month-old infant, and the injected bilirubin was rapidly cleared by the normal liver.

Of the five patients who died when the paper was written, only one had a proper autopsy. As regards the liver, there were no remarkable histological findings. There were small thrombi in the hepatic canaliculi, and also in hepatic ducts. This abnormality was not judged to be of consequence as regards the overall nature of this disorder.

The brain was only examined in one case (case 2). Areas of jaundice were noted in the cerebral cortex, thalamus, corpus striatum, mammillary bodies, dentate nucleus, and inferior olives. Cells of the basal ganglia showed fat droplets in neurons, but the changes were deemed unremarkable. Hemolytic disease was not noted in either of the two cases autopsied (see Fig. 1).

The authors comment that in the absence of any other possible cause such as hemolytic disease or primary biliary obstruction, these cases seem to be a result of

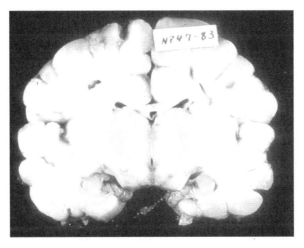

Fig. 1 Brain slice from a case of Crigler-Najjar syndrome. Obvious staining in the hippocampus, thalamus, etc. Courtesy of Dr, Michael Norenberg. See acknowledgements

a functional defect in the newborn liver. Evidence supports a genetic basis. This genetic defect materializes as a persistently elevated indirect-reacting bilirubin level That this pigment is bilirubin is supported by crystalline isolation, and by clearance of the material by normal liver.

The authors state that all but one case had severe brain damage suggestive of a metabolic defect. The authors could not predict whether the brain damage was directly related to hyperbilirubinemia or an associated co-existing anomaly. The authors note that the prognosis is poor. Four of the cases died during a febrile episode. The overall mortality rate in this untreated series of patients was 83%. The authors do speculate that the extent of CNS damage is a key factor in predicting the patient's life span.

The authors point out that this heretofore undescribed disorder closely resembles other hyperbilirubinemia disorders of late childhood or early adult age including Gilbert disease, familial non-hemolytic jaundice, etc. These patients have fatigue, and chronic non-hemolytic acholuric jaundice. It is suspected that these disorders may be the same, thereby having a genetic basis as regards acquisition, although the precise nature of the defect is unclear.

Over several years, it had been shown that a variety of drugs, hormones, etc., were capable of stimulating the activities of microsomal and soluble enzymes in vivo (Fouts, J., and Hart, L. 1965 in Crigler/gold paper 1965; Conney 1965). This phenomenon has been documented with more than 100 compounds. The enzyme glucuronyl transferase is one enzyme among many which can be stimulated to increase activity (Arias, I., et al. 1963a). Arias also showed that treating pregnant women with chloroquine prior to delivery did not change aspects of hyperbilirubinemia following birth. The paper by Crigler and Gold (Crigler, J., and Gold, N. 1969) undertook to examine the effects of phenobarbital on unconjugated bilirubin in a newborn infant with a congenital absence of activity of the conjugating enzyme glucuronyl transferase, and accompanying hyperbilirubinemia.

This report centers around an infant hospitalized from 2–23 months of age. Birth was early, weight was 4.5 lbs, and when seen at 2 months of age was markedly jaundiced, and showed neurological signs consistent with a diagnosis of kernicterus. Serum bilirubin at this time ranged between 17 and 25 mg/100 ml, and was nearly all indirect reacting. The highest observed bilirubin was 30 mg/100 ml.

Results in this case showed that a phenobarbital dose of 5 mg/kg/day decreased hyperbilirubinemia from 20 to 6 mg/100 ml. This drop in bilirubin took about 30 days. Later reduction of Phenobarbital to 2.5 mg/kg/day did not alter bilirubin, but stopping Phenobarbital treatment resulted in a slow increase in pigment concentration in serum. It was shown that the rate of decline in bilirubin concentration with treatment, and the rate of increase in bilirubin when phenobarbital was stopped, were almost identical. These rates of change were statistically significant. Using injected ^{14}C and ^{3}H bilirubin, the change in bilirubin over time proceeded showing first-order kinetics. Most of the labeled bilirubin was found in the patient's stool.

Examination of liver biopsies both before and after phenobarbital treatment using light microscopy, were comparable to controls. With electron microscopy, there was a significant increase in smooth endoplasmic reticulum following phenobarbital treatment. This drug induced in ultrastructure had been previously described in animals, but not man (Orrenius, S., Ericsson, J., and Ernster, L. 1965).

The authors comment that their data show that phenobarbital acted to reduce the half-life of bilirubin this hyperbilirubinemic infant by two- to threefold. The removal of bilirubin from the body pool decreased bilirubin concentration in both the vascular and extravascular pools. The authors note that the 30 days required to reduce bilirubin from a high of as much as 30 mg/100 ml was several times what might have been expected. This implies that the phenobarbital acted on a slow rate-limiting step rather than on a rapid process. The similar increase when phenobarbital is withdrawn suggests the same slow reverse steps might be essential in the bilirubin rise.

The data suggest that phenobarbital acted to induce glucuronyl transferase. Interestingly, some studies showed no change in bilirubin levels in Gunn rats treated with phenobarbital (DeLeon, H., Gartner, L., and Arias, I. 1967). Further studies were necessary to determine the mechanism of phenobarbital-induced lowering of high serum unconjugated bilirubin.

In a review paper (Karp, W. 1979), the possible mechanisms of action of bilirubin on cerebral tissue were discussed. Karp references a paper by Katoh, et al. (Katoh, R., Kashiwamata, S., and Niwa, F. 1975) in which ATP, NAD, and NADP were all measured in whole brain of Gunn rats. Results led Katoh and coworkers to conclude that uncoupling of oxidative phosphorylation was probably not of importance in Gunn rat brain since the levels of high-energy metabolites were not changed.

Karp points out correctly that studies on whole brain may not show changes since bilirubin encephalopathy is a highly localized metabolic encephalopathy (as are most metabolic encephalopathies). Inclusion in any assay sample of contiguous normal tissue dilutes the lesioned region. This makes impossible any meaningful interpretation of data. Karp correctly points out that in newborn Gunn rats as young as 2–3 days old, show conclusive damage and alteration of brain mitochondria, the earliest observable change (Jew, J., and Williams, T. 1977; Schutta, H., Johnson, L., and Neville, H. 1970).

Karp in his review also comments on the selective nature of bilirubin for brain. He states that the blood–brain barrier of newborns is relatively immature allowing easy passage of bilirubin into cerebral tissue. In support of this concept is the fact that liver mitochondria do not show as much uncoupling of oxidative phosphorylation at a concentration of 8 μmolar as do brain mitochondria. At 8 μmolar, brain mitochondria are almost completely uncoupled, whereas liver mitochondria are uncoupled less than 50%. Other studies show liver mitochondria subjected to elevated bilirubin levels have normal cell respiration and phosphorylation as compared to controls (Schenker, S., McCandless, D., and Wittgenstein, E. 1966).

Another area discussed by Karp is the mechanism by which bilirubin acts in a highly selective manner. The author states that ligandin is a cytoplasmic protein in high concentration in liver, kidney, intestine, etc. Ligandin is not present in brain. In other organs, the ligandin affectively binds bilirubin thereby inhibiting or preventing its toxic effects.

A paper by Rubboli, et al. (Rubboli, G., et al. 1997) reported results from a neurophysiological study in patients with type 1 Crigler–Najjar syndrome. While phototherapy, plasmapheresis, and blood transfusion have all acted to reduce the occurrence of kernicterus in Crigler–Najjar syndrome, later onset of neurological signs and symptoms may develop. In order to study this phenomenon, patients diagnosed with type 1 Crigler–Najjar were followed up in terms of neurophysiological criteria. A total of five patients were evaluated with reference to course of hyperbilirubinemia and treatment.

Case 1 was a 15-year-old boy who received phototherapy since birth. At various times during his childhood, phototherapy was not consistently provided. At age 4 years, minor motor changes were noted, and later he showed dysarthria and diffuse hypotonia. By 14, he had impaired levels of consciousness, myoclonic jerks, and generalized seizures. At 15 years of age, mental retardation, dysarthria, motor impairment of upper limbs, and a mild ataxia. At this time, total serum bilirubin was 26 mg/100 ml. The EEG showed irregular 5–6 Hz theta background with bursts of generalized spike and wave discharges.

The second case was a 23-year-old woman with a family history of Crigler–Najjar syndrome (younger brother case 4). In this case, phototherapy was maintained since birth, such that bilirubin levels were kept below 20 mg/100 ml. Since age 14, phenobarbital was given during the winter seasons. By 19 years of age, she became apathetic and sleepy. At this time, bilirubin levels ranged between 30 and 35 mg/100 ml. Other liver function tests were normal. EEG showed sporadic generalized spike–wave discharges. With time, paroxysmal abnormalities increased. At age 20, the patient had generalized convulsive seizures and myoclonic jerks. She appeared mentally retarded, but no focal neurological defects were noted. The EEG at this age showed 9–10 Hz background, with 6–7 Hz paroxysmal abnormalities which were present during sleep. Photostimulation succeeded in producing a paroxysmal response accompanied by myoclonic jerks. Bilirubin levels ranged from 30 to 35 mg/100 ml.

This patient had a liver transplant at 21 years of age with resultant improvement. She became more active and alert. The ataxia completely resolved, and the dysarthria also improved. The EEG also improved, and 5 months after the liver transplant, the EEG was normal.

The third case was a 15-year-old girl who had phototherapy for hyperbilirubinemia since birth. This patient was placed on phenobarbital therapy at age 10. Serum bilirubin levels were always kept below 29 mg/100 ml. At age 13, the patient started showing spike–wave discharges in her EEG during sleep. At this time, bilirubin was between 15 and 20 mg/100 ml. This patient's EEG showed no further discernible worsening.

Two other patients were studied as above. These two, after 2 and 4 years of follow up, and at ages 7 and 8, showed no alteration in EEG. Furthermore, the neurological exam was not remarkable. In both of these patients, phototherapy was administered since birth, and bilirubin levels were always below 20 mg/100 ml.

The authors note that phototherapy seemed effective in keeping bilirubin levels low, except for case 1 in whom the treatment may have been intermittent. There seemed to be a correlation between elevated bilirubin levels and neurological signs and symptoms, and in EEG findings. Supporting the hypothesis that bilirubin was responsible for cerebral deterioration, the patient (#2) who had a liver transplant enjoyed a nearly complete reversal of both neurological and EEG abnormalities. The authors state that the EEG abnormalities seen in these cases differ from those described in hepatic failure. The key feature seems to be a slowing and disorganized background activity accompanying triphasic waves. This is not unexpected since in hepatic failure other toxins such as ammonia, phenols, short chain fatty acids are all elevated in addition to bilirubin. It is therefore a completely different example of metabolic encephalopathy.

What is remarkable is that after a liver transplant in a patient with long-standing hyperbilirubinemia, and definite signs and symptoms consistent with kernicterus could undergo a reversal of symptoms, Not only that, reversal of EEG abnormalities occurred. So, what happened to the focal staining of discreet cerebral areas? What happened to mitochondrial and cellular structural damage? These findings seem to stress the concept of metabolic encephalopathy – these disorders can be reversed, perhaps after even longer insult than was believed.

The authors state that the data from their study support the benefit of neurophysiological methodology for monitoring the cerebral effects of hyperbilirubinemia in the Crigler–Najjar syndrome, type 1. The data also support the concept that these techniques and EEG could be a positive and sensitive method for evaluating cerebral function in older patients with hyperbilirubinemia.

In the above paper, liver transplantation seemed the most effective treatment modality for hyperbilirubinemia. Surprisingly, transplantation seemed to reverse long-standing neurological and EEG changes. In some measure, this flies against considerable evidence showing that unconjugated bilirubin uncouples oxidative phosphorylation in the brain, lowering ATP, and producing catastrophic morphologic changes including the disappearance of Purkinje cells, and death. All this, in a matter of hours to days in some patients.

In another case report involving liver cell transplantation (Fox, I., et al. 1998), a 10-year girl who had severe unconjugated hyperbilirubinemia due to Crigler–Najjar type 1 was given a hepatic transplantation. This patient's high bilirubin levels were lowered somewhat by phototherapy, but not by phenobarbital. The patient had a streptococcal infection which seemed to precipitate kernicterus with neurological features such as slurred speech, ataxia, and stupor and coma. Treatment with plasmapheresis, and intense phototherapy produced recovery with no sequelae. Subsequently, 10–12 h per day of phototherapy was required to maintain bilirubin levels at 24–27 mg/ 100 ml.

The patient received FDA approval for hepatocyte transplantation. Increased phototherapy served to reduce serum bilirubin to 18 mg/100 ml. After finding a donor, 7.5×10 to the ninth power of hepatocytes were infused through a catheter in the portal vein. Infusion took 25 h. The patient was discharged a day later.

In terms of the infusate, 95% of the cells were hepatocytes following a processing procedure. The donor hepatocyted were normal in terms of the glucuronyl transferase enzyme. The efficiency of the donor hepatocytes was tested and proved effective before infusion into the patient. The investigators goal was to provide about 2.5% of the existing liver as transplant. The assumption was that about 50% of infused cells would adhere.

Results showed that serum total bilirubin fell to 13 mg/100 ml. There was a spike in bilirubin levels by day 7 after hepatocyte transplantation to 26 mg/100 ml, after which the levels fell to around 10–14 mg/100 ml.

The authors state that their patient was successfully treated with the hepatocyte infusion from a donor, and bilirubin levels were reduced significantly. A plateauing of serum bilirubin to around 14 mg/100 ml was achieved. Phototherapy was still continued at a level of about 6–7 h per day. Glucuronyl transferase activity in the "new" liver and conjugated bilirubin in the bile provided conclusive evidence of engraftment of infused hepatocytes. Bilirubin glucuronide was the predominant conjugate excreted, matching that of normal controls. It was, however, different from other partial enzymatic deficient patients such as those with Crigler–Najjar type 2, or of those with Gilbert disease (Bloomer, J., and Sharp, H. 1984).

In this study, the function of engrafted hepatocytes was evaluated. The hepatic glucuronyl transferase levels of activity after transplantation of hepatocytes of about 5% of normal indicates that most infused hepatocytes were engrafted. Animal studies indicate the possibility of lifelong attachment and function.

Multiple infusions of hepatocytes might provide enough cells to lower bilirubin levels such that phototherapy might not be needed. Alternatively multiple infusions might result in rejection of the donor hepatocytes. The levels in this patient of unconjugated bilirubin would appear to be low enough as to carry no risk for the development of kernicterus.

The authors note that previous studies on transplantation have not been as well controlled as was theirs. At the time of publication (1995) the world registry showed 19 patients having liver transplants with Crigler–Najjar syndrome type 1. Of these, three required a second transplant, and two died. Risks associated with whole liver transplant include pulmonary thrombosis, portal hypertension, and portal vein thrombosis with concomitant liver damage. These risks do not occur with hepatocyte transplantation.

Another study of partial liver transplantation in Crigler–Najjar type 1 disorder was published shortly after the hepatocyte study above (Rela, M., et al. 1999). In this study, a partial orthotopic liver transplantation technique was developed. Orthotopic liver transplantation has been shown to be successful in type 1 Crigler–Najjar syndrome, but experimental studies in Gunn rats had shown that only 1–2% of normal hepatic mass was required to metabolize bilirubin (Jansen, P., et al. 1989). An advantage of a partial transplant would be that some original liver would be left which could serve as a base in case of rejection. There would also be original liver ready in the event of gene therapy or enzyme replacement development at a later date.

In this study, seven partial orthotopic liver transplantations in six patients were performed in Crigler–Najjar type 1 cases. The mean age was 10.5 years, range 8–18. All had received phototherapy from birth in order to keep serum bilirubin below 20 mg/100 ml. Patients were receiving 12–16 h of phototherapy per day. One child had neurological signs, the rest were neurologically intact.

The procedures consisted of six left-lobe partial transplants and one right side transplant. All were size and ABO blood group matched donors. Briefly, the procedure involved dissecting the porta hepatitis and isolating the portal vein. The left lateral segment was resected to the left of the falciform ligament. A portion of the caudate lobe was resected for access reasons. The donor portal vein was sewn to the recipient portal vein by end-to-end anastomosis. The right side transplant was modified slightly to adjust to the anatomic features of the right lobe.

The results of this procedure were satisfactory in all cases except one, which required a retransplantation due to rejection problems. The remaining five patients survived without problems, and with serum bilirubin levels below 3 mg/100 ml. None of the patients have required phototherapy since transplantation.

The authors comment that orthotopic partial liver transplantation has corrected the hyperbilirubinemia caused by the absence of hepatic glucuronyl transferase. The authors state that it is best to perform this procedure before CNS damage occurs. The authors also state that severe brain damage and mental retardation is irreversible. The question as to when severe brain damage occurs remains unanswered. Also, the "plasticity" of young brain is as yet undefined.

The authors reiterate that Gunn rat studies showing only 1–2% of liver function is needed for conjugation of bilirubin. This indicates that partial liver replacement should be more than enough to correct for the defect in Crigler–Najjar type 1 patients.

Another advantage of a partial transplant is that if rejection becomes a problem, there remains some original liver to revert to along with phototherapy. In the event of future gene therapy, or enzyme replacement therapy, there would be liver left that would be receptive. The authors point out that with hepatocyte transplantation, bilirubin levels remain high enough to require 6–7 h per day of phototherapy. That was not the case with partial hepatic transplantation.

The authors note that the long-term success of partial transplantation remains to be proven. The diagnosis of graft rejection was not made early in the first child due to normal liver function tests. The diagnosis was made by liver biopsy. In future cases, suspicion should increase on the onset of rising unconjugation levels in serum. The authors suggest that partial orthotopic liver transplantation might be considered in other liver-based inborn errors of metabolism such as urea cycle defects, fatty acid metabolism defects, and familial hypercholesterolemia.

In a recent paper from Johns Hopkins and published online (Kniffin, C. and Wright, M. 2002), Crigler–Najjar type 2 is discussed. The authors state that unconjugated hyperbilirubinemia can result from Gilbert disease, and Crigler–Najjar types 1 and 2. Conjugated hyperbilirubinemia may result from Rotor syndrome, Dubin-Johnson syndrome, and intra- and extrahepatic cholestasis of several types. Variations of the above may also result from bacterial or viral infection.

The authors note that Crigler–Najjar types 1 and 2 can be distinguished based on several features. Type 1 is thought to be a complete inactivity of hepatic glucuronyl transferase, whereas type 2 is a partial inactivity of the enzyme. This translates to serum unconjugated bilirubin rising at birth to levels of 20–45 mg/100 ml, whereas in type 2 cases, serum unconjugated bilirubin levels range from 6 to 20 mg/100 ml. Phenobarbital is effective in type 2, but results in type 1 are equivocal. Also, in type 2 Crigler–Najjar syndrome, bilirubin glucuronides are found in bile, indicating some conjugation, whereas very little is found in bile from type 1 patients. Patients with the Gilbert syndrome are characterized by a lack of morbidity and a low total serum bilirubin from 1 to 6 mg/100 ml.

The inheritance patterns of Crigler–Najjar syndrome type 2 is not totally resolved, but seems to be autosomal recessive. Consanguinity and the occurrence in brothers argues in favor of autosomal recessive (Gollan, J., et al. 1975). In another example of Crigler–Najjar syndrome type 2, a 34-year-old woman born to first cousin parents was one of four out of six siblings to have the disorder. The four jaundiced siblings had a total of 11 children, all unaffected (Guldutuna, S., et al. 1995).

In terms of the molecular genetics of Crigler–Najjar syndrome, type 2, a point mutation in the uridine diphosphate glucuronosyltransferase has been described in a patient who was the offspring of a consanguineous marriage (Gollan, J., et al. 1975).

In another study (Yamamoto, K., et al. 1998), seven patients from five separate families diagnosed with Crigler–Najjar type 2 were studied. It was found that double homozygotes missense mutations in exons 1 and 5 were present in five of the patients. One patient had only a single mutation in exon 1, while the final patient had an insertion mutation. Finally, in a recent study (Petit, F., et al. 2006), it has been shown that at least 15 point mutations can be identified in patients with Crigler–Najjar type 2 syndrome.

Further expounding as regards the genetics associated with defects in the conjugation of bilirubin has been published (Kaplan, M., Hammerman, C., and Maisels, M. 2003). The authors state that unconjugated hyperbilirubinemia can result from increased bilirubin production, defective and decreased bilirubin conjugation, or a combination of the two. The conjugation process requires the active participation of an enzyme, glucuronyl transferase. The production of this enzyme is under control of a uridine diphosphogluconate glucuronosyl transferase gene locus. This locus is composed of various exons, many of which may be vulnerable to mutation. This gene determines the actual structure of glucuronyl transferase, which is produced in liver cells, the site of bilirubin conjugation.

The exons are portions of DNA which codes for the enzyme. The exact amino acids and their locations are encoded in DNA. This sequence is transferred from the nucleus to the cytoplasm by messenger RNA. Mutations in the gene coding areas will obviously result in errors in enzyme formation, and subsequent function. In the case of the Crigler–Najjar syndrome this alteration in enzyme results in hyperbilirubinemia due to an incapacited bilirubin-conjugating system. The transferases are a class of enzymes; alteration, depending on site, may produce various phenotypic outcomes. The gene encoding this transferase is located on the second chromosome at 2q37.

In terms of the Crigler–Najjar syndrome,unconjugated hyperbilirubinemia can reach very high levels if left untreated. This level of hyperbilirubinemia may well cross into the newborn brain, producing kernicterus. In the case of type 1 Crigler–Najjar syndrome, the activity of glucuronyl transferase is at zero, whereas in type 2 disorder, the enzyme activity is reduced, but can be "activated" by phenobarbital administration. Phenobarbital may not act directly on glucuronyl transferase, but acts on an enhancer module which stimulates the appropriate gene to induce the enzyme. The Crigler–Najjar syndromes 1 and 2 are caused by 1 or more mutations in any of the five exons which are in the gene which encodes for glucuronyl transferase (Labrune, P., et al. 1994).

Gilbert syndrome will be covered in a later chapter, but briefly, this syndrome is an adolescent onset inherited syndrome. It is characterized by unconjugated hyperbilirubinemia. Other liver function tests are normal. There is no evidence of hemolysis, and the elevated serum bilirubin levels are usually less than 10 mg/100 ml. The genetic alteration appears to an additional TA insertion in the TATAA box of the gene promoter. The additional TA decreases the affinity of the TATAA-binding protein. There is an increase relation between the number of TA repeats and the promoter activity. This in turn acts to decrease glucuronyl transferase activity.

Gilbert syndrome is not thought of as involving hemolysis, but it has been shown that a decrease rbc half-life may be present in some Gilbert patients (Okolicsany, L., et al. 1978). A recent study (Kaplan, M., et al. 2002) showed definitively that the uridine glucuronide transferase promoter polymorphism does increase rbc catabolism as well as decreasing bilirubin conjugation.

The possibility that some unconjugated hyperbilirubinemic patients actually had Gilbert syndrome instead of some other variation such as Crigler–Najjar type 2 could not be completely discerned until the genetics of these disorders could be determined. The nature of this has been described (Kaplan, M., et al. 1997). In this study, three different uridine glucuronyl transferase dehydrogenase-deficient newborn patients were examined. Results showed that no significant difference in the hyperbilirubinemia incidence was shown in neonates with different promoter genotypes. The authors concluded that hyperbilirubinemia was dependent on the interaction between the promoter genotypes and the variant UGT promoter. Neither of these two factors alone increased serum bilirubin – only the two acting simultaneously.

Glucuronyl transferase alterations and breast milk jaundice has also been studied. A correlation with Gilbert syndrome has been noted (Monaghan, G., et al. 1999) as regards the similarities in the 7/7 promoter genotype. It was shown that about one third of newborns who were breast feeding with serum bilirubin levels greater than 5.8 mg/100 ml for more than 14 days had the 7/7 promoter genotype. Other studies have shown that newborns with prolonged jaundice associated with breast milk may show a mutation in the UGT gene. These findings suggest that the UGT gene coding region is a key factor in breast milk jaundice.

The authors point out the importance of awareness of identifying newborns with hemolysis, and of the complexities of newborn jaundice. Blood tests and careful history taking can explain most cases of hyperbilirubinermia. Ultimately in unexplained hyperbilirubinemia, genetic counseling may be appropriate.

An interesting paper was published in PNAS examining the possible correlation of the glucuronyl transferase defect in homozygous Gunn rats using chimeric RNA/DNA oligonucleotide for a site-directed insertion of a single guanosine in genomic DNA from hepatocytes and Gunn rat intact liver.

The chimeric ONs were designed with hybrid RNA/DNA targeting the non-transcribed DNA sequence of the uridine glucuronyl transferase gene. The sequence was identical to the mutant gene except that an extra G was placed as the center N+ within the DNA residues. The ON sequence is complimentary to 28 residues of genomic DNA spanning the mutation site with the exception of a G base at position 1206. The cell's endogenous DNA repair process controls insertion of the new G at the target site, affecting a correlation of the mutation (Kren, B., et al. 1999).

Results showed significant cell uptake and nuclear localization of the labeled chimeric molecules into primary Gunn rat hepatocytes. Chimeric ONs complexed with polyethylenimine or encapsulated in liposomes were injected by the tail vein. Seven days, and 4 and 6 months later liver was obtained and DNA isolated. PCR studies showed that G insertion at the target site was about 20% even as long as 6 months. Bilirubin-conjugating enzyme after treatment was shown by Southern and Western blot analysis to be restored by about 25%, and following a second treatment dropped to less than one half of the pretreatment levels, indicating an effective restorative outcome on glucuronyl transferase activity.

The authors comment that the genomic insertion of G at the target site was not an artifact. Although there was a complete disappearance of ONs from the liver at 48 h, the frequency of G at 4 and 6 months after injection was unchanged from that seen at 1 week. Correction of the glucuronyl transferase defect was confirmed by Southern blot analysis and by a reduction of serum hyperbilirubinemia. The decrease in serum bilirubin was gradual, simulating that seen in hepatocyte transplantation, rather than that seen in whole liver transplantation. There was a greater proportion of bilirubin monoglucuronide than diglucuronide in bile simulates partial glucuronyl transferase states such as Crigler–Najjar type 2 syndrome, Gilbert syndrome, etc.

The authors estimate that about 100,000 ONs were delivered to hepatocyte nuclei with each administration. The authors state that using RNA/DNA ONs as a treatment form to correct genetic/metabolism liver defects has advantages as compared to viral-mediated transgene expressions. It obviates random genomic integration, and immunogenicity and lack of persistent gene expression. The percent decrease in serum unconjugated bilirubin is enough to render levels in type 1 Crigler–Najjar syndrome to levels seen in type 2 Crigler–Najjar syndrome. This in effect lowers these levels to levels which ordinarily do not enter the brain and cause kernicterus.

The use of chimeroplasty for the correction of metabolic defects has been called into question (Taubes, G. 2002). A company called Kimeragen was formed to develop chimerplasty as a means to treat various disorders including sickle-cell anemia and Crigler–Najjar syndrome.

Within a couple years, members of the Kimerogen company were resigning, and most of the chimeroplasty position of the company was shut down. The license for chimeraplasty was sold to a small company in California to apply the technology to plants. The end results were equivocal. At least nine laboratories, some famous, published papers in which the technology seemed to work, or was at least partially positive. Two of these laboratories stopped this line of research. Meanwhile, dozens of other laboratories experienced in gene repair failed to replicate the dramatic positive results. Only a handful of the studies which were negative were published, but word of the unreliability of the technique was quick to spread. It is beyond the scope of this chapter to detail any more about this controversy, which is spelled out by Tauber. It does, however, show how papers published in excellent journals, can launch a concept into the spotlight. Subsequent conflicting reports and failure to replicate data produces skepticism. The result is several years and untold hours of work down a dead end pursuit.

In a paper (Labrune, P. 2001), the various "theme and variations" of newborn and later jaundice were defined. The incidence of Crigler–Najjar syndrome was placed at 1:1,000,000 live births. This represents those cases with a deficiency in the conjugating enzyme glucuronyl transferase. Other causes of hyperbilirubinemia, such as Dubin-Johnson syndrome, Rotor syndrome, breast milk jaundice, etc., are not variations of Crigler–Najjkar syndrome. Gilbert disease is only a partial deficiency in glucuronyl transferase. Two types of Crigler–Najjar syndrome are distinguished by their response to phenobarbitol treatment. Crigler–Najjar type 1 is unresponsive to phenobarbital treatment, whereas the jaundice of Crigler–Najjar type 2 undergoes a rapid diminution when phenobarbital therapy is initiated. Phenobarbital is found to be effective if it induces a fall in unconjugated hyperbilirubinemia equal to a 65% drop within 2–3 weeks of treatment. The rise in serum bilirubin after birth can be dramatic, and since phenobarbital treatment is not as quick to work, transfusion may be needed. Following transfusion, a minimum of 12 h per day exposure to blue light may be required. Treatment of Crigler–Najjar syndrome type 1 may not be effective, necessitating liver transplantation. The phenobarbital treatment in type 2 syndrome is able to maintain hyperbilirubinemia levels below that thought of as being a risk for kernicterus. It must be remembered that the risk for kernicterus is not dependent upon an arbitrary level of serum bilirubin, but upon several other factors such as prematurity, hypoxia, albumin-binding capacity, etc.

The risk for kernicterus can extend beyond the childhood period, into adolescence or adulthood. This can occur if phototherapy is stopped, or if stress is imposed such as fever, infection, fasting, etc. Proper treatment (phenobarbital) in type 2 Crigler–Najjar syndrome effectively prevents kernicterus by keeping unconjugated bilirubin levels low.

Before a diagnosis of Crigler–Najjar syndrome is made, other treatable causes such as hemolytic disease must be ruled out. Diagnosis of Crigler–Najjar syndrome in utero is difficult due to the fact that UDP-glucuronyl transferase activity cannot be detected in amniocytes or fetal blood. Therefore, possible prenatal diagnosis is most likely made by history, and a study of the family of the possible case. Prenatal diagnosis serves to alert the neonatalogist of the impending complication

The author of this paper suggests that Crigler–Najjar type 1 patients would be good candidates for gene therapy (or enzyme replacement therapy) because of the relentless progression of the disorder without transplantation. Transplantation or hepatocyte infusion are not without significant risks, and may not always be particularly effective. It is estimated that a "resumption" of glucuronyl transferase activity of only 2–3% can provide enough conjugating capacity to allow a normal existence. This is added impetuous to look for gene, or other therapy which might partially ameliorate the enzyme deficiency. This has, and is being done – see chapter on gene therapy below.

In a recent paper by Toietta, G., et al. (2005), elimination of unconjugated hyperbilirubinemia in Crigler–Najjar syndrome Gunn rat model was achieved by administering helper-dependent adenoviral vector (HD-Ad). Earlier studies had shown the ability of early-generation adenoviral vectors to correct Gunn rat hyperbilirubinemia for periods up to 4 weeks (Li, Q., et al. 1998). Attempts to prolong the duration of lowered bilirubin levels were limited because of the immune response. The present paper describes studies in which HD-Ad vectors were used, devoid of viral coding sequences.

In this study, Gunn rats were utilized. The HD-Ad vector preparation was injected into the Gunn rat's tail vein. Phenobarbital was also administered to a group of Gunn rats. Blood was analyzed for bilirubin levels, and complete blood counts were done. Biliary bilirubin glucuronides were measured in bile by cannulating the bile duct. The liver was analyzed for total RNA and protein.

Results demonstrated a complete return to control levels of total serum bilirubin within about 1 week following the injection of 3×10 to the 12th UP/kg HD-Ad. Non-treated homozygote Gunn rats in this study had total serum bilirubin levels from 5.8 to 10.5 mg/kg, while HD-Ad-treated Gunn rats had values ranging from

0.4 to 1.4 mg/kg. Remarkably, this correction of hyperbilirubinemia in homozygous Gunn rats remained stable for 2 years. A lower dose (6×10 to the 11th vp/kg) dropped bilirubin levels to about 3.5 mg/100 ml or less, which also remained stable for a comparable period. The effect of phenobarbital on this process was assessed by giving phenobarbital for 8 days to the low-dose HD-Ad group. Results from this showed no increase in bilirubin reduction with bilirubin treatment. There was also no effect of phenobarbital on Gunn rat controls.

Analysis of bile by chromatography examined bilirubin glucuronides. Large amounts of bilirubin monoglucuronides and diglucuronides were present in HD-Ad-treated homozygous Gunn rats, while these glucuronides were not present in homozygous non-HD-Ad-injected controls. To examine the long-term capacity for bilirubin conjugation, 2-year-old Gunn rats treated with HD-Ad as newborns were injected with a dose of bilirubin to see if clearance was compromised. These results showed that the conjugating capacity remained high over time as the pulse of bilirubin was rapidly cleared.

Hepatic expression of glucuronyl transferase mRNA at both 20 and 52 weeks after an intermediate dose of HD-Ad in homozygous Gunn rats was confirmed by Rt-PCR. No Rt-PCR was detected in any other organ. Immunoblot analysis of protein from both homogenates and microsomal preparations from liver biopsies at 20 weeks showed the HD-Ad-treated animals had a protein expressed of 52 kDa. This represents the molecular mass of glucuronyl transferase. Controls did not show this protein. Immunohistochemical analysis reconfirmed this finding in liver sections from the same treated Gunn rats.

HD-Ad vectors proved not to be toxic to the liver as compared to first generation Ad vectors. To indicate this, the low-dose group showed no significant drop in platelet count. The intermediate dose group showed a transient thrombocytopenia and platelet reduction, which returned to normal 3 weeks after HD-Ad injection.

The authors comment that patients with type 1 Crigler–Najjar syndrome can now survive the newborn period with phototherapy, but are at risk for sudden kernicterus due to various unpredictable factors. Liver transplantation, not without serious risks, is available. Crigler–Najjar patients may not be at the top of transplantion lists unless they are in neurological crisis.

Crigler–Najjar syndrome patients would seem to be ideal for gene therapies for a variety of reasons including ease of monitoring efficacy, well-defined biochemical defects ideal animal in the Gunn rat, etc. Several gene-transfer methodologies have been tried in the Gunn rat. These methods have all been encouraging, but not totally effective due to only a partial correction of the coenzyme deficiency, toxicity, invasive administration techniques, etc.

The authors further comment (2-year Gunn rat study) that the present findings using the much less toxic HD-Ad vector which is devoid of all viral protein coding sequences, reduced hyperbilirubinemia to control levels, and further this was maintained for 2 years – near the life span for rats. A single injection of the HD-Ad vector achieved this result. Unlike first generation HD-Ad injections, severe long-lasting thrombocytopenia was not observed. Toxicity in the present study was mild and transient, and there was no chronic toxicity. Even a low dose of HD-Ad vector was effective in reducing hyperbilirubinemia to a level of 3.5 mg/100 ml or less without any sign of toxicity. The ability of only about 2.5% of normal liver hepatocytes to regulate bilirubin conjugation is an advantage for treatment as opposed to other liver metabolic disorders such as urea cycle diseases.

In a recent paper (Hansen, T., et al. 2009), the potential reversibility of bilirubin encephalopathy/kernicterus, six case studies were examined in which patients had elevated total serum bilirubin (477–792 μmol/l equivalent to about 24–40 mg/100 ml). This is an important paper as it reaffirms earlier suggestions that there is a potential capacity for reversal of kernicterus symptoms.

In all six cases, aggressive therapy was started when the extent of hyperbilirubinemia was realized. All six infants were given phototherapy, and four of the cases also received exchange transfusion. Before the initiation of aggressive therapy, all of these patients showed signs/symptoms of kernicterus including seizures, opisthotonus, lethargy, hypotonia, high-pitched cry, etc.

Follow-up results (ages 17 months to 6 years old) showed mostly normal development, except two patients had delayed speech development. The authors state that they believe that quick initiation of exchange transfusion and phototherapy was responsible for the reversal/prevention of severe neurological sequelae. Similar results have been described earlier, which speaks to the metabolic encephalopathy concept of a biochemical lesion early, followed later by structural and possibly irreversible structural changes. There are no studies aimed at defining the length of time one could sustain kernicteric symptoms before reversibility would not be possible.

The authors stress the importance of no hesitation in initiating fast aggressive treatments such as exchange transfusion in order to drop the hyperbilirubinemia as quickly as is possible.

The decreased speech development in two of the six patients may be reflective of the early effects of unconjugated bilirubin on the auditory system (see ABR chapter 17). Such an effect might produce some long term on hearing, shown to have an adverse effect on speech development. So, coupled with phototherapy, fast resolution of "missed" hyperbilirubinemia might just about eliminate kernicterus in the USA.

Neuropathology of Kernicterus

In one early paper (Zuelzer, W., and Mudgett, R. 1950) two groups of a total of 55 kernicteric newborn infants were compared as regards etiology, laboratory data, and neuropathology. One group of patients consisted of 23 cases, all kernicteric and associated with erythroblastosis fetalis, and a second group of 32 cases, not associated with erythroblastosis, but associated with a variety of other suspected causes. Neurological symptoms were not present in all cases, probably because death ensued quickly in many instances. Kernicterus was predominant in all 55 cases, as demonstrated by postmortem exam, and was the basis for inclusion in this study. The aim was to compare the distribution of bilirubin-stained areas in the two groups to see if cause of kernicterus had any effect on which areas were stained.

Results showed that overall, the gross distribution of pigmented brain areas was nearly similar between the two groups. The frequency of stained areas in decreasing order of incidence are depicted in Fig. 1. The areas most frequently affected were the corpus striatum, and the hippocampus. Least often stained were the cerebral cortex and spinal cord (see Figs. 2, 3, and 4).

Following fixation in formaldehyde, the pigment was bright yellow, or orange-yellow, and after exposure to light (!! 1950), or alcohol, the color fades. There was no difference in these characteristics, or in location of stained brain areas between the groups.

Microscopic examination of stained brain areas did show rather trivial differences between the two groups. The main feature seen microscopically was that of nerve cell necrosis in stained brain regions. The damaged neurons had shrunken irregular nuclei, and the cytoplasm appeared to be homogeneous and eosiniphilic. The nerve cell damage was not 100% in any area stained by bilirubin. About 22/23 brains showed these features in group 1 (erythroblastosis), whereas 22/32 showed these alterations in group 2.

The authors comment that prematurity was present in 75% of cases in group 2, and since prematurity per se is not considered by many to be a cause of kernicterus, it most surely is a contributing factor. The authors suggest that kernicteric type brain damage in these infants may contribute to mental retardation in those who survive.

The authors also comment that hyperbilirubinemia and nuclear staining seem to be a common denominator in an otherwise heterogeneous group of newborn patients. The authors state that kernicterus would appear to be completely limited

D.W. McCandless, *Kernicterus*, Contemporary Clinical Neuroscience,
DOI 10.1007/978-1-4419-6555-4_8, © Springer Science+Business Media, LLC 2011

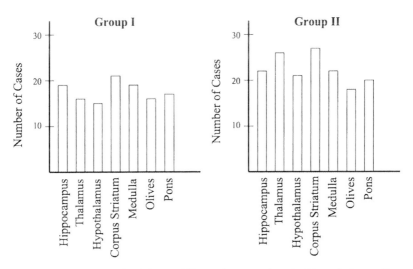

Fig. 1 Representitive sites of staining by bilirubin in cases of erythroblastosis (group 1), a non erythroblastosis (group 2). There was little difference between groups. Adapted from Zuelzer, W., and Mudgett, R. (1950)

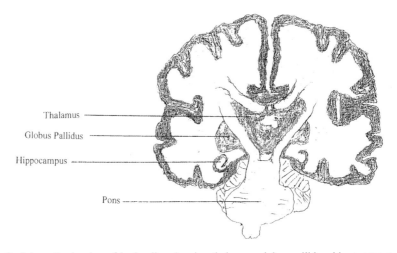

Fig. 2 Schematic drawing of brain slice showing thalamus, globus pallidus, hippocampus, and pons

to the newborn patient, probably due to the early characteristics of the developing blood–brain barrier. Cases of kernicterus in adults, however, are not unknown. The elevated incidence of kernicterus in premature infants is no doubt linked to hypoxia and respiratory failure, cerebral hemorrhage, and to immature liver function. One observation indicating the above is that death from kernicterus is rarely seen under 24–48 h, because that length of time is required in order for unconjugated bilirubin levels to reach over 20 mg/100 ml, even in cases of Crigler–Najjar syndrome.

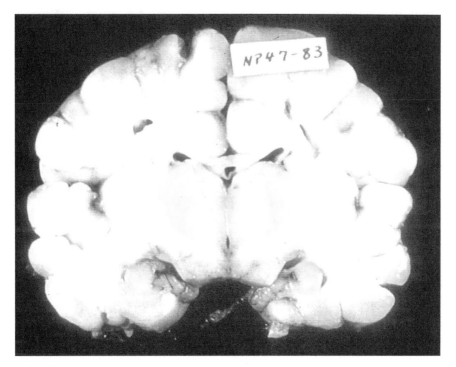

Fig. 3 Brain slices showing highly regional bilirubin staining. These slices are from Crigler-Najjar patients. Courtesy of Dr. Michael Norenberg. See acknowledgements

The neuropathology of hyperbilirubinemia has been described by investigators for many years. Early descriptions of the yellow cerebral staining are described in the chapter on the early history of bilirubin. Causes of newborn infant hyperbilirubinemia are varied, and can be mild or severe jaundice due to hemolytic disease, or the absence of the bilirubin-conjugating enzyme glucuronyl transferase. Regardless of the immediate cause of hyperbilirubinemia, if the albumin-binding capacity is surpassed, bilirubin can readily enter the brain.

Once bilirubin has crossed the blood–brain barrier, it can be highly toxic to selective brain areas. Like most metabolic encephalopathies, the effects are usually bilateral, symmetrical, and highly selective and consistent for certain brain areas. Each metabolic encephalopathy has a rather specific pattern of neuropathologic alterations, although some areas are more frequently affected than others.

The appearance of fresh brain from newborn infants who succumbed to kernicterus may be different depending on the course of the disease (Claireaux, A. 1961). If the patient died in an acute phase, one might find focal areas of staining. If death occurs in a more chronic or late phase, then the brain might appear more normal as regards its coloration. Nuclear areas characteristically stained in the acute phase may have reverted to a more normal color. There may be noticeable shrinkage of some cerebral areas such as subthalamic nuclei. Table 1 is an adaptation from

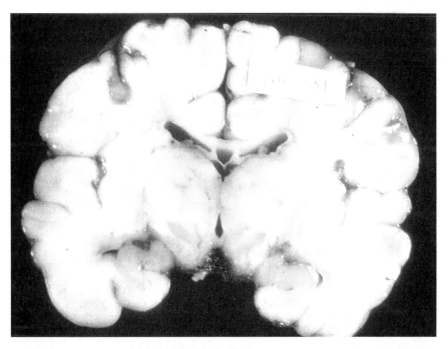

Fig. 4 Brain slices showing highly regional bilirubin staining. These slices are from Crigler-Najjar patients. Courtesy of Dr. Michael Norenberg. See acknowledgements

Table 1 Distribution of bilirubin from 35 cases of Kernicterus

Basal ganglia	32
Cerebellum	25
Hippocampus	24
Medulla	24
Subthalamus	19
Thalamus	18
Corpus striatum	18

Adapted from Claireaux, A. (1961)

Claireaux showing the incidence of staining of various areas in a series of 35 cases of kernicterus. Since the thickness of the brain slice technique is such that some small nuclear areas might not always be visible, some focal areas of jaundice could be missed. In addition, the process of dying might act to permit a diffuse staining of areas not actually stained during life. In addition, not all areas listed in the table are stained in all cases, but many are characteristic. The author comments that with more effective treatment, the incidence of kernicterus is falling. Also, there seems a reluctance to give permission for autopsies.

As regards histological findings, the staining of bilirubin is seen in nerve cells, nerve cell processes, and may be noted in surrounding glia. Intracellularly, pigment can be seen in cellular ground substance and in vessels. Pigment is also noted in the choroid plexus and leptomeninges.

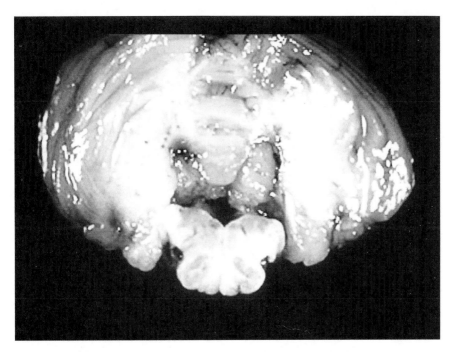

Fig. 5 Brain slices showing highly regional bilirubin staining. These slices are from Crigler-Najjar patients. Courtesy of Dr. Michael Norenberg. See acknowledgements

Nerve cell damage may be seen as described above, but nerve cells may also appear normal. This might be expected in cases where the progress was acute, and the kernicterus is still in the metabolic phase where irreversible structural damage has not occurred. Microglial proliferation is not so obvious in the acute phase, but later, microglial proliferation is evident and pronounced.

In the late chronic stage, areas of obvious necrosis are noted. Demyelination is usually seen, and may be severe. Some areas of staining may show only a few neurons, while the majority is absent, and have been replaced by glia. An example cited was the dentate nucleus in a 20-month kernicteric infant. Normally packed with neurons, this dentate had only a scant few neurons. The spaces were filled with astrocytes. The symptoms of spasticity and athetosis can be attributed to the lesions in the basal ganglia, and hearing loss to the lesioned eighth cranial nerve. The reason for the highly specific and bilaterally symmetric nature of the lesioned areas is as unclear as it is in the rest of the metabolic encephalopathies.

The availability of the superb animal model of human kernicterus, the Gunn rat, has proven beneficial many times. The neuropathology of the hyperbilirubinemic Gunn rat has been well studied (Blanc, W. 1961). In these investigation, the author was trying to determine whether the Gunn rat model of kernicterus and the human disease were similar, permitting meaningful comparison of various criteria.

Five criteria were selected to make such comparisons. First, there should be clear evidence of CNS involvement during life. Second and third, there should be staining of nuclear masses, and persistence of the stain even after formalin fixation. Fourth and fifth should be the presence of pigment within the cells, and evidence of cell degeneration. The author states all five criteria are met in kernicteric Gunn rats. The author states that in Gunn rats, areas stained include cerebellum, hippocampus, thalamus, substantia nigra, subthalamus, dentate nucleus, globus pallidus, colliculi, brainstem, and sometimes, the cerebral cortex. All Gunn rats displaying kernicterus upon neuropathological examination were neurologically symptomatic. Microscopically, sections contained neurons with yellow pigment which survived formalin fixation. Pigmented cells showed microscopic features of cell damage. Neuronyl cell loss was evident in affected areas. Areas of cell necrosis were associated with glial reactions. Spectrophotometric studies showed the cellular pigment had absorption values similar to that seen in human kernicteric brain cells. These data, taken together, suggest that the Gunn rat is indeed an excellent animal of human kernicterus, and data derived from Gunn rat studies should have an important and significant relation to the human counterpart.

In another study, newborn animals – kittens, puppies, and rabbits were injected with bilirubin sufficient to raise serum levels of bilirubin to about 50 mg/100 ml, and the results were analyzed (Rozdilsky, B. 1961). If necessary, repeat injections were performed so as to keep serum levels at 20 mg/100 ml or higher. Most animals died within 24 h of the first injection. The remainder died within another 12 h. Many had neurological symptoms consisting of opisthotonus, rapid running, muscle twitches, and stupor and coma.

In kittens, neuropathology was found in the thalamus, subthalamus, colliculi, floor of the fourth ventricle, inferior olives, and cuneate nuclei. In kittens who lived from 18 to 24 h, yellow pigment was seen within nerve cells in the cuneate nucleus and the cochlear nuclei. There was a positive correlation between length of time until death and the amount of visible neural pigment. Gross nuclear jaundice was seen in 75% of jaundiced kittens. The most severely affected neurons showed a loss of chromatin, contained pyknotic nuclei, and were pale.

In puppies, areas which were jaundiced included periventricular gray around the fourth ventricle, thalamus, cuneate nuclei, colliculi, and olives. The staining was not as sharply delineated as that in kittens, and no neural cells with pigment were seen in frozen sections. In another group of puppies, insulin was administered, producing hypoglycemia. These animals had intense jaundice in selected nuclei, including the inferior colliculi. Neurons in these nuclei were pigment stained. In rabbits, only 2 out of 28 developed a selective nuclear staining by bilirubin.

The author notes that in puppies and rabbits, severe jaundice seemed to only occur in animals who had other brain trauma (hypoglycemia). In newborn kittens no preexisting injury was necessary for neurological symptoms, and neuropathologic damage similar to that in human cases of kernicterus. In addition, the areas of bilirubin staining are similar. There are differences, for example the hippocampus is usually stained in newborn human cases of kernicterus, but was rarely stained in kittens. It is speculated that the neuronal and nuclear damage in both human cases of kernicterus and that in newborn kittens is caused by unconjugated bilirubin.

A close examination of the yellow (bilirubin) crystals seen in autopsy-obtained tissues has been performed (Vietti, T. 1961). The material was obtained from several human newborn infants, most of whom died in the second to fourth post-natal day with signs of kernicterus. The mean serum bilirubin was 22 mg/100 ml. Bilirubin crystals were seen in previous cases of kernicterus, and found in brain, kidney, and adrenal glands. The crystals have also been found in homozygous Gunn rats, and by chemical analysis identified as unconjugated bilirubin (Schmid, R., et al. 1958). These pigmented crystals may dissolve from the tissue during the process of embedding in paraffin for microscopic examination. The crystals fade during long exposure to formaldehyde, and also are noted to fade when exposed to sunlight.

Gunn rat brain has been examined using electron microscopy (Jew, J., and Williams, T. 1977). Results from these studies have been described in the Gunn rat chapter, so will only be briefly mentioned again. In this study, 21 homozygous Gunn rats aged from 2 days to 7 months were examined using electron microscopy. The areas selected for study were the ventral cochlear nuclei and the dorsal cochlear nucleus. Areas were prepared using conventional electron microscopy techniques, stained with uranyl acetate, and examined.

Results showed ultrastructural changes in all age groups. The cytoplasm of neu-rons from jaundiced Gunn rats contained many membrane-bound vesicles. The most prominent vesicles were located within the mitochondria. While the inner and outer mitochondrial membranes were usually intact, the cristae and matrix were reduced in size. The intramitochondrial vacuoles were probably composed by membranes of the cristae. The vesicles appeared to be filled with electron dense granules which were most probably glycogen. In the cochlear nuclei neurons, glycogen particles were found in the rough endoplasmic reticulum. Some cells seemed to be "taken over" by glycogen-filled endoplasmic reticulum. Distorted mitochondria were seen in most neurons (see Figs. 6, and 7).

The authors state that previous studies have described changes in oxidative phosphorylation and even depleted energy metabolites such as ATP in brains of kernicteric Gunn rats. Results from the present study provide structural evidence of damage to mitochondria.

The authors also suggest that these electron microscopic studies provide evidence that the neuronal damage was initially that of mitochondrial involvement consistent with the concept that the initial primary biochemical lesion is one involving energy metabolism and oxidative phosphorylation.

In a retrospective study (Ahdab-Barmada, M., and Moossy, J. 1984), the records from 102 cases of kernicterus confirmed at autopsy were examined. The purpose of the study was to determine if premature asphyxiated newborn infants who developed kernicterus had any neuropathologic or clinical differences compared to full-term kernicteric newborn infants.

Results from this 7-year retrospective study showed that 97 of 630 autopsied infants had neuropathological evidence of kernicterus (15.4%). The greatest number of cases were in newborns aged 25–32 weeks. Nearly all age groups had simi-lar instances. Serum bilirubin levels were low due to initiation of therapy. Most were treated with phototherapy, and many with exchange transfusion. Localization of lesions was similar as seen in other studies. Lesions were found in the globus pallidus, hypothalamus, pons, cuneate nucleus, cranial nerve nuclei, and the

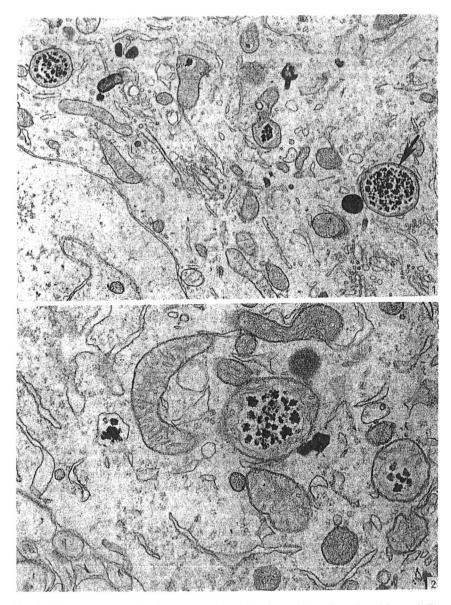

Fig. 6 Electron microscopic micrographs of cytoplasmic membrane bound vesicles and disrupted mitochondria These mitochondrial changes were judged to be the initial structural alteration indicating damaged mitochondrial function. Permission, Jew, J., and Williams, T. (1977)

cerebellum. The results as regards localization of lesions in premature infants was not significantly different from that seen in full-term infants. There were minor differences between the two groups such as more pronounced staining of the brainstem cranial nerve nuclei in the premature group of kernicteric infants.

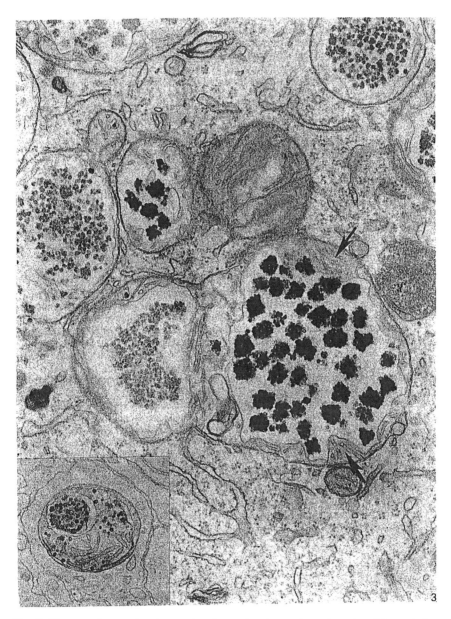

Fig. 7 Electron microscopic micrographs of cytoplasmic membrane bound vesicles and disrupted mitochondria These mitochondrial changes were judged to be the initial structural alteration indicating damaged mitochondrial function. Permission, Jew, J., and Williams, T. (1977)

Microscopic examination of lesioned areas showed vacuolated spongy neuropil. Alteration in cytoplasmic and nuclear membranes, and loss of Nissl substance was also observed in lesioned nuclei. Yellow pigment granules were seen in the neuronal vacuoles. The cerebellar Purkinje cells were affected, showing swelling, and cytoplasmic granules which were periodic acid Schiff positive.

Microscopic lesions interpreted as irreversible were those in which damage to neurons had progressed to cellular dissolution, diffused hyperchromasia, and fragmentation of nuclear and cytoplasmic membranes. Another neuronal cell change was the mineralization of the cytoplasm. This change, called incrustation of the neuron' progresses to a fully mineralized cell. These cells may be seen many weeks after the start of post kernicteric encephalopathy in infants who survive the initial phase of the disease.

The authors comment that the retrospective study shows that kernicterus in premature neonates with asphyxia is essentially the same as the kernicterus associated with full-term non-asphyxiated newborn infants. The authors note the association in previous studies of kernicterus and high serum bilirubin levels. The present retrospective study shows that asphyxia plus moderate hyperbilirubinemia also can produce kernicterus in premature newborn infants. Other concomitant insults leading to kernicterus in premature infants include hypoglycemia and acidosis. In some cases, the yellow pigment may be washed out during fixation. When this occurs, the diagnosis of kernicterus may be questioned, as kernicterus is partially defined as the intraneuronal retention of pigment even after formaldehyde fixation of the brain tissue.

The hyperbilirubinemia was only moderate in patients in this retrospective study, being highest (13.9 mg/100 ml) in the 37–40 week gestational age group, and next highest in (12.4 mg/100 ml) in the 33–36 week age group. The presence of acidosis may act to enhance bilirubin tissue levels while acting to lower serum concentrations. This leads to nuclear staining and resultant cellular damage and necrosis. Secondary bilirubin staining of brain regions may follow in previously damaged areas. Insults such as cortical infarcts may predispose brain tissue to bilirubin staining. This could be seen as bilirubin–lipid complexes which are phagocytized by macrophages. This would appear as pigment staining.

Finally, the authors note that the lesions of anoxia/ischemia and hypoxia seen in newborn infants differ from those seen in kernicterus. The cytopathology seen in kernicteric brain areas is that of damage to cytoplasmic and nuclear membranes. Mitochondria show a somewhat later degeneration, and the presence of glycogen granules. Bilirubin-associated neuronal damage in the premature newborn infant was in this study, found in the same topographical locations as was bilirubin in full-term kernicteric brain.

It is important to note that in the premature newborn infant with a degree of asphyxia, bilirubin entry into the brain and kernicterus occurred at a lower level of serum bilirubin (9–14 mg/100 ml) than the accepted danger level of 20 mg/100 ml used to evaluate kernicterus risk.

The reason for the highly selective localization of unconjugated bilirubin in various brain regions in kernicterus remains unclear. One hypothesis is that there was a slower clearance of pigment from certain areas, which would result in increased deposition. To examine this hypothesis, labeled bilirubin was infused into young Sprague-Dawley rats (150 g wt), and the clearance of pigment from the basal ganglia was measured (Hansen, T., and Cashore, W. 1995). Animals were sacrificed at several time points from 15 to 360 min following infusion of the bilirubin isotope.

Serum and brain regions (medulla, hippocampus, hypothalamus, cerebellum, mid-brain, striatum, cerebral cortex, etc.) were counted and the half-life of the isotope was determined.

Results showed that bilirubin cleared from brain areas associated with kernicterus at a rate similar to that in brain regions not associated with kernicterus. This study was performed in 150 g rats, at a time at which development of kernicterus, for example, in Gunn rats, has passed. The blood–brain barrier in 150 g rats is well developed.

In another paper relating the blood–brain barrier to kernicterus (Wennberg, R. 2000), various theories regarding entry of bilirubin into brain are explored. The author states that when bilirubin reaches a toxic level in brain, neurological symptoms develop. These consist of high-pitch crying, alterations in brainstem auditory-evoked responses, opisthotonus, hypertonia, seizures, stupor, coma, and death. Residual sequelae may consist of choreoathetosis and hearing loss. Patients dying of kernicterus show all the localized pigment staining as described above. With minor variations, probably due to methodological differences, the pattern of neurolopathology is similar between studies.

One accounting of the pathogenesis of kernicterus relates to the amount of uncon-jugated bilirubin bound to the carrier albumin. The free serum bilirubin, that is the amount of serum bilirubin not bound to albumin is thought to pass quickly into brain. Bilirubin bound to albumin cannot enter the brain. Under in vivo conditions the binding capacity of albumin is never completely saturated, but the levels of unbound free bilirubin increase as saturation is approached. The unbound free serum levels of unconjugated bilirubin are very low compared to that bound to albumin (10–60 nmoles vs. 100–300 nmoles bound).

Bilirubin is lipophilic, and "single pass" estimates show bilirubin entering the brain at a rate of from 12 to 28%. Metabolic acidosis did not alter entry into brain, whereas hypercarbic acidemia did increase brain entry of bilirubin. The idea of an immature blood–brain barrier, found throughout the literature, may not be signif-icant for bilirubin. The blood–brain barrier is functional by the intrauterine fourth to fifth month. And the risk of kernicterus in premature infants is similar to the risk of full-term infants. This not the case in animals; for example, in monkeys the susceptibility of damage by bilirubin to the cerebral auditory system is higher in prematurely born animals than in full-term monkeys. Other animal studies have shown the blood–brain barrier is weaker in 2-day-old piglets than in 2-week-old piglets (Lee, C., et al. 1995). Other factors such as asphyxia have been mentioned as factors in increasing bilirubin deposition in brain areas, but the data is not clear.

The nature of bilirubin entry into brain and its relation to serum albumin are criti-cal. The author points out that in essentially all studies looking at the risk factors for kernicterus, albumin concentrations are not measured. Not only are the binding sites and capacities important, but the presence of other substances such as sulfonamides in the serum are critical. The presence of such substances which competitively and successfully compete for albumin-binding sites, increases the concentration of free bilirubin. Methods of measuring unbound free unconjugated bilirubin may be a bet-ter predictor of kernicterus risk than more conventional techniques which do not assess free bilirubin (Funato, M., et al. 1994). It is very important to be aware of the subtleties involved in the assessment of risk factors for free vs. bound serum unconjugated bilirubin levels. In addition are the problems associated with the early release from hospitals of newborn infants even before hyperbilirubinemia can be assessed. In addition is the potential problems generated by a general relaxation of criteria for treatment of hyperbilirubinemia. All of this in light of increased reports of bilirubin-induced auditory problems (Shapiro, S. 2003).

A recent paper (Gupta, A., and Mann, S. 1998) has examined the auditory brainstem response to increasing levels of serum bilirubin. This was a retrospective study in which 60 newborn infants in which the auditory brainstem response in hyperbilirubinemic neonates were compared to normal neonates. Results showed that auditory brainstem conduction times were prolonged in hyperbilirubinemic infants as compared to controls. The results were most pronounced in the group with the highest levels of hyperbilirubinemia (22 mg/100 ml). Upon retest 1 month later, 33% of neonates in the high bilirubin group had totally recovered, while 80% in the low group had totally recovered. In contrast, after 6 months, four patients failed to show any recovery. These results indicate that bilirubin deposition is taking place in the brainstem, and may be doing permanent damage.

In another very recent paper (Chang, F-Y., et al. 2009), the authors looked at the induction of long-term synaptic plasticity in hippocampal slice cultures exposed to conjugated hyperbilirubinemia. The hippocampal slice method maintains the cytoarchitecture of the in vivo brain. This technology is excellent for prolonged pharmacological treatments allowing investigation of possible effects of bilirubin on hippocampal CA1 long-term potentation (LTP), and long-term depression (LTD).

Results showed that unconjugated bilirubin at a dose in the culture of 10 μmol, produced a decrease in the stimulus/response of Schaffer collateral CA1 synapses. The effect was due to a decrease in basal synaptic transmission, and the excitability of afferent fibers was not altered.

Other results showed that exposure of hippocampal slices to unconjugated bilirubin impared LTP at a dose of 1 μmol for 48 h, or 10 μmol for 24 and 48 h. The effect of bilirubin on LTD delivered to Schaffer collateral afferent fibers only had an effect at a dose of 10 μmol frequency stimulation at this dose/time, as it was induced at all other times and doses. The authors state that this rules out a possible residual role of bilirubin in the slices after long-term exposure. These physiological data serve to show that bilirubin is producing an effect in a brain region which is invariably pigment stained in newborn infants suffering kernicterus.

The effect of prolonged exposure of unconjugated bilirubin on the NMDA/AMPA ratio was examined. The ratio was lowered in slices treated with ten micromolar bilirubin for 24 and 48 h. The nature of these results leads the authors to state that prolonged exposure to unconjugated bilirubin causes a decrease in presynaptic transmitter release and also impairs the postsynaptic NMDA receptor. Results also showed that 24 and 48 h treatment of hippocampal slices with unconjugated bilirubin at a dose of 10 μmol caused proteolytic cleavage of NMDA receptor subunits.

The authors state that overall, their results demonstrate that prolonged exposure to bilirubin at clinically relevant concentrations acts to partially block the induction of the hippocampal CA1 LTP and LTD. These results are in part mediated by overstimulation of NMDA receptors, which in turn leads to cleavage and degradation of the NMDA subunits.

In addition, and in keeping with the above findings, the authors showed that exposure to unconjugated bilirubin results in decreased protein levels of NR1, NR2A, and NR2B subunits.

Stimulus/response relationships of postsynaptic potentials suggest that prolonged exposure to unconjugated bilirubin may lead to an inhibition of glutamatergic synaptic transmission. The authors state that this inhibition is most likely caused by a reduction in presynaptic glutamate release. As regards the bilirubin-related decrease in NMDA receptor subunit proteins, there was a correlation with an increase in calpain activity. The block of calpain activity abolished most of the effect of unconjugated bilirubin.

Finally, the authors point out that many previous studies of the effects of unconjugated bilirubin on regional brain neurophysiological and biochemical alterations use bilirubin at concentrations so high that the levels seen in kernicteric infants are exceeded. To avoid that, the present investigations were performed at lower bilirubin levels, comparable to what is seen in kernicteric patients. This indicates that the bilirubin levels in this study are clinically relevant, and supports the conclusion that bilirubin has an inhibitory action on the induction of CA1 hippocampal long-term synaptic plasticity in rat hippocampal slices.

As regards the actual incidence of kernicterus in the USA., results are somewhat confusing, in part due to the lack of an official registry of cases. In one very recent report looking at a possible increase in frequency, the numbers were assessed based on hospital discharge data (Burke, B., et al. 2009). Neonatal discharges with primary or secondary diagnoses codes for jaundice or kernicterus occurring in the initial 30 days of life were used. The data established trends for 1988–2005.

Results showed an overall incidence in the USA of 2.7 kernicterus cases per 100,000 newborn infants. The incidence of jaundice was placed at 15.6% of newborn infants. The trend for kernicterus was from 5.1 cases per 100,000 in 1994–1996, to 1.5 cases per 100,000 now. This study then clearly demonstrates a decline in the incidence rate of kernicterus among newborn infants in the USA.

Bilirubin and Energy Metabolism

This chapter on bilirubin and its effects on energy metabolism will examine many studies, both in liver and brain relating to significant mitochondrial alterations induced by bilirubin. These studies were conducted by early investigators, and confirmed by later workers. Changes in a multitude of other neurochemical metabolites and compounds have been described, and are the subject of the next chapter. Many of these changes may be secondary to primary alterations in energy metabolism. The following represent some of the key investigations regarding liver and brain energy metabolism changes in hyperbilirubinemia.

By the early 1950s, it had been determined that bilirubin was the cytotoxic agent responsible for brain damage and death from kernicterus. It was generally recognized that several conditions of newborn infants were associated with jaundice, but whether jaundice was directly responsible for pathology was not clear. Furthermore, it was shown that there was a correlation between the incidence of kernicterus and levels of hyperbilirubinemia (Hsia, D., et al. 1952). Before the description of the genetic rat model of hyperbilirubinemia first described by Dr. Gunn (Gunn, 1938), a variety of animal models were utilized. In 1954, Day (Day, R., and Zetterstrom, R. 1956) was able to demonstrate that bilirubin lowered respiration in chopped rat brain. A single cell organism, *Tetrahymena pyriformis*, also showed toxicity when exposed to bilirubin, and unlike rat brain, the toxicity could be reversed by administration of cytochrome *c.T. pyriformis* is a unicellular fresh water protozoan. At about the same time, Waters and Bowen (Waters, W., and Bowen, W. 1955) made a similar finding, and in addition, noted that the deleterious effects of bilirubin could be reversed by adding DPN to the incubation media.

By 1956 (Zetterstrom, R., and Ernster, L. 1956), the nature of the toxicity of bilirubin had been shown to be an uncoupling of oxidative phosphorylation in isolated mitochondria. The studies were performed in rat liver because many previous studies on oxidative phosphorylation had been done in liver, and therefore there was already a base of knowledge. The liver findings were later extended by the authors to brain (Ernster, L., Herlin, L., and Zetterstrom, R. 1957).

At this point in time, it was already known that the mitochondria represent the location of the chemical components where oxidative phosphorylation occurs, and that it is also the major site where ATP is formed. The passage of electrons "down" the respiratory chain results in the formation of three molecules of ATP by coupling

D.W. McCandless, *Kernicterus*, Contemporary Clinical Neuroscience, DOI 10.1007/978-1-4419-6555-4_9, © Springer Science+Business Media, LLC 2011

ADP and inorganic phosphate. The relation between the consumption of oxygen and esterified phosphate is called the P/O ratio (see Fig. 6 in the chapter "Biochemistry and Physiology of Bilirubin," this volume).

Extensive studies comparing the effect of bilirubin on oxidative phosphorylation to the effect of 2,4-dinitrophenol were performed. Results showed that the loss of phosphorylation in response to bilirubin was different than that of 2,4-dinitrophenol. In comparison, 2,4-dinitrophenol, while depressing phosphorylation to about the same extent as bilirubin, had no effect on respiration. Bilirubin, however, did decrease respiration. The addition of DPN and cytochrome c was able to reverse the effect of bilirubin. Bilirubin was able to induce swelling of mitochondria, but only at a concentration of bilirubin necessary for uncoupling of oxidative phosphorylation.

It was also shown that bovine serum albumin provided a protective effect on the ability of bilirubin to uncouple oxidative phosphorylation. When bilirubin levels were constant, increasing concentrations of albumin resulted in increasing levels of protection. A slight protective effect was noted with the addition of manganese or EDTA. ATP was not able to prevent or reverse the inhibitory effect of bilirubin.

In summary, these studies proved evidence that bilirubin acts on mitochondria by uncoupling phosphorylation and inhibits the ADP/inorganic phosphate reaction. Mitochondria are induced to swell by bilirubin, but at a higher concentration than that required for uncoupling of oxidative phosphorylation. The effects were prevented by albumin, and also manganese and EDTA. It was also shown that biliverdin had no deleterious effects on mitochondrial function.

In a later study (Odell, G. 1959), observations were made on the effects of protein (albumin) binding with bilirubin of certain organic anions. The results of these studies showed that many organic anion as well as bilirubin are bound to serum albumin. Further studies showed that many anions are able to displace bilirubin from albumin-binding sites. Thus uncoupling of protein-bound bilirubin by organic anions may result in a large change in free unbound bilirubin, which would then be readily diffusible. A shift in the diffusible gradient would likely result in movement of bilirubin out of extracellular fluids and into cellular cytoplasm. This study shows that salicylate and sulfisoxazole were effective in displacing bilirubin from albumin.

Because of the similarity of newborn Rhesus monkeys and newborn humans with respect to bilirubin metabolism, a study of oxygen metabolism in the monkeys was undertaken (Behrman, R., and Hibbard, E. 1963). Polarography was used to measure oxygen concentrations in the peritoneal cavity. Newborn Rhesus monkeys were subjected to hyperbilirubinemia by infusing bilirubin, and over a 3.5–10-h period, oxygen levels were measured, and blood was withdrawn to monitor serum bilirubin levels. Controls consisted of monkeys infused with Na_2CO_3.

Bilirubin was infused sufficient to raise and maintain serum bilirubin levels between 30 and 45 mg/100 ml. Oxygen, as measured polarographically, decreased within an hour following bilirubin infusion. Oxygen levels decreased from 20 to 54%. During this time frame, heart rate, body temperature, and respiration rates all were comparable to controls.

The animals were sacrificed at the end of the experiment by the rapid infusion of a solution of formaldehyde in order to allow postmortem examination of tissues. These results showed many tissues, including the peritoneum, to be diffusely stained yellow. The brains were described as having a diffuse yellow color, but nuclear areas were not more heavily stained. Microscopically, there was intravascular deposition of bilirubin, but intracellular bilirubin was not noted, nor were any microscopic structural changes observed.

The authors state that in their study kernicterus was produced in a newborn Rhesus monkey model which closely resembles human kernicterus. High serum bilirubin levels (30–45 mg/100 ml) assured reproducible and consistent induction of kernicterus. The hyperbilirubinemia resulted in a decrease in oxygen concentrations, and a diffuse yellow staining of the brain. The decrease in oxygen concentration in tissues would render cells more vulnerable to the entry and damage by bilirubin. This in turn may proceed the intracellular action of bilirubin on oxygen consumption and phosphorylation by mitochondria. These studies in primates are in keeping with the concept that bilirubin acts by uncoupling oxidative phosphorylation.

Tissue culture was also used in elucidating the mechanism of bilirubin toxicity (Cowger, M., Igo, R., and Labbe, R.F. 1965). Based on many earlier studies, these authors chose to examine NADH2 oxidase activity, as well as several other cellular criteria such as glucose utilization and oxidative metabolism.

Results showed that a 50% inhibition of NADH2 oxidase activity resulted from a bilirubin concentration of $1–1.5 \times 10^{-5}$ M. Succinate oxidase was inhibited by 50% at a bilirubin concentration of 2.5×10^{-4} M. The addition of cytochrome c, unlike other studies, did not alter these results. However, the addition of albumin totally prevented the changes produced by bilirubin in the absence of albumin. Localization of the inhibition showed that NADH2–cytochrome c reductase was inhibited by 40% at a bilirubin concentration of 1×10^{-5} M., and 80% at a concentration of 5×10^{-5} M. NADH2 dehydrogenase was similarly progressively inhibited by increasing concentrations of bilirubin. By contrast, biliverdin and stercobilin required a 16-fold increase in concentration over bilirubin, and a 47-fold increase over bilirubin, respectively, in order to produce a 50% inhibition of enzymes.

Metabolites were also measured in these experiments. The amount of "excess Lactate," or the lactate produced by anaerobic pathways was found to be 57% of total lactate at a bilirubin concentration of 1.0×10^{-5} M, and 63% at a concentration of 2.5×10^{-5} M. Biliverdin and stercobilin at a concentration of 2.5×10^{-5} M did not increase lactate production.

The effects of both bilirubin and biliverdin on cellular ATP were measured. After 2 hours exposure to bilirubin at a concentration of 2.5×10^{-5} M, cellular ATP concentrations were decreased by 76%. Bilirubin at a similar concentration had no effect on cellular ATP levels. At the time of ATP measurement, cells still had maintained excellent viability.

There was an inverse relationship between oxygen consumption and lactate production. Thus, when oxygen consumption dropped due to increasing concentrations of bilirubin, lactate production increased.

The authors note that their experiments show that bilirubin is a powerful metabolic toxin. Albumin was effective in preventing the inhibitory effect of bilirubin. There was no significant effect of bilirubin on cytochrome c oxidase, pointing to a bilirubin effect on the initial portion of the respiratory chain. As with Amytal, bilirubin had almost no effect on the succinate oxidase system. This and other results indicate that Amytal and bilirubin have similarities in action via inhibition of terminal oxidation. Conversely, bilirubin seems more effective uncoupler of oxidative phosphorylation than Amytal.

Biliverdin and stercobilin were not as toxic as bilirubin, and levels of these two compounds had to be many times that of bilirubin to have any effect. This lack of toxicity of biliverdin and stercobilin is attributed by the authors as being due to the basic pH as compared to the acidic nature of bilirubin. Biliverdin is a weaker base than stercobilin, and therefore is toxic at a lower concentration as compared to stercobilin.

Finally, the authors state that tissue culture is an effective means for studying toxic effects. They found a significant effect of bilirubin on viability, and a depression on cell growth. The cell death which resulted from bilirubin exposure was attributed to uncoupling of oxidative phosphorylation. The significance of this uncoupling was shown by the dramatic decrease in measurable ATP concentrations. Comparisons between bilirubin and other oxidative phosphorylation uncouplers as regards cell growth, viability, and death should prove beneficial in examining the primary effects on cells of bilirubin.

In another study (Odell, G. 1966), the relationship between albumin and bilirubin was further examined in vitro. This study was performed in rat liver mitochondria. The mitochondria were prepared by conventional techniques such that 1 ml of mitochondrial suspension contained the mitochondria from 1 g of rat liver. These basic mitochondrial suspensions were utilized to study the distribution of bilirubin between mitochondria and albumin. The morphology of the mitochondria was examined using electron microscopy.

Results showed that as the concentrations of albumin were increased, bilirubin partitioning between the aqueous phase and mitochondria was decidedly away from mitochondria. Thus the quantity of bilirubin associated with mitochondria decreased as the albumin in solution increased. The effect of salicylates on bilirubin distribution showed that as levels of the salicylates increased, so increased, so increased the distribution of bilirubin toward mitochondria. The presence of salicylates in a suspension of mitochondria and albumin had a similar effect. Thus, the salicylates increased the amount of bilirubin distributed with mitochondria. Substitution of human albumin in place of bovine serum albumin showed similar results. It was apparent that albumin bound to bilirubin until the molar ratio of 1:1 was passed. After that, association of bilirubin and mitochondria decreased. The dissociation of bilirubin from albumin by salicylates was greater with bovine serum albumin than from human albumin.

Electron microscopic studies showed that mitochondria which were exposed to bilirubin in the absence of albumin became swollen. Intramitochondrial inspection showed a loss of intraluminal electron electron dense material, and a loss of the cristae terminalis. The mitochondrial structural changes could be prevented provided albumin in sufficient quantities are present.

The author notes that the study shows that up to a molar ratio of albumin to bilirubin of 1:1 is reached, bilirubin does not associate with the mitochondria. When this molar ratio is exceeded, bilirubin may gain entry, and damage the mitochondrian. The albumin concentrations in the CNS are lower than other organs. This may act to allow bilirubin relatively easy entry into brain, and also to prevent the deleterious effects of bilirubin. Thus, when the blood–brain barrier may not be mature, more damage from hyperbilirubinemia in brain may occur.

Another paper (Diamond, I., and Schmid, R. 1966a) examined how binding of plasma bilirubin or lack thereof, may influence the entry of the pigment into brain.

In this study, two animal models were utilized. First were adult Gunn rats and second were premature fetal guinea pigs. Both have an inability to conjugate bilirubin. The fetal guinea pigs were delivered by Caesarian section about 2 weeks prematurely. Animals were cannulated in order to infuse ^{14}C-bilirubin into a leg vein. At the end of the experiment, animals were decapitated and the brain prepared for analysis.

Results showed that when ^{14}C-bilirubin was infused with human albumin sufficient to bind the bilirubin, serum levels were 36 mg/100 ml and the brain concentration was 3.4 μg/g. When infused with guinea pig serum, the albumin-binding capacity was exceeded, and the serum concentration was only 18.2 mg/100 ml. In this case, the brain levels were 10.3 μg/g – 3 times that of ^{14}C-bilirubin infused with albumin. When sodium salicylate was added, there was a substantial shift of bilirubin out of plasma. In this experiment, brain ^{14}C-bilirubin levels reached 16.5 μg/g tissue. When saline was substituted for salicylate, brain levels of bilirubin were only 4.4 μg/g, but blood levels were 34.9 mg/100 ml. Making the serum more acid by exposing the animals to 25% CO_2 in oxygen resulted in a blood pH of 7.1, and a decrease in serum bilirubin from 36 to 26 mg/ml. At that time brain bilirubin concentrations rose from 3.4 to 7.9 μg/g. This represented a threefold increase in the brain to serum ratio.

In Gunn rats, similar results were obtained as in premature guinea pigs as regards results with albumin and bilirubin. In some Gunn rats, regional analysis of cerebral deposition of ^{14}C-bilirubin was performed, and the highest concentrations were found in the cerebellum and brainstem.

In an in vivo portion of the study, newborn guinea pigs were infused with bilirubin at a rate of 100 μg/hour, and 11 of 30 animals developed severe neurological symptoms and appeared moribund or died. Fifteen animals infused with bilirubin plus human albumin survived without any evidence of neurological involvement. When the brains of guinea pigs who died with kernicterus were examined, focal deposits of yellow pigment were found in the basal ganglia and nuclei around the fourth ventricle.

The authors note that use of radiolabeled bilirubin permitted quantitative analysis as opposed to qualitative visual inspection. These experiments show that human albumin has a higher affinity for bilirubin than either rat or guinea pig albumin. Administration of unbound bilirubin results in lower serum concentration of the pigment, and higher brain levels since lack of albumin facilitates entry into bodily tissues. In newborn guinea pigs, increased brain deposition of bilirubin was associated with neurological symptoms.Subsequent treatment with human albumin increased the survival rate. These findings suggest that only unbound bilirubin can cross the blood–brain barrier. The study provides direct unequivocal support for the concept that the use of albumin for the treatment of unconjugated hyperbilirubinemia is justified. The authors were able to move bilirubin in and out of the brain by simply varying serum albumin.

This study also supports the concept of the importance of using caution in administrating agents which might successfully compete with bilirubin/albumin-binding sites. Salicylates competitively displace bilirubin from albumin-binding sites, which in turn elevates brain levels of the pigment. In this study, a regional cerebral localization of bilirubin to areas which when damaged, may explain neurological systems.

One problem with tissue culture and in vitro techniques to assess oxidative phosphorylation is that levels of unconjugated bilirubin may be greater than in vivo levels, and protective mechanisms may not be in place in in vitro studies. In order to circumvent these possibilities, the effects of unconjugated hyperbilirubinemia in intact guinea pig's capacity for liver oxidative phosphorylation was assessed (Schenker, S., McCandless, D., and Wittgenstein, E. 1966).

In these studies, guinea pigs were infused via the inferior vena cava with unconjugated bilirubin in saline. One hour later liver was sampled for measurement of ATP, oxygen consumption, and blood was obtained for bilirubin measurement.

The average serum bilirubin level was 27.9 mg/100 ml, a level frequently seen in newborn infants with evidence of kernicterus. There was almost no conjugated bilirubin in the serum. In spite of the high level of unconjugated bilirubin, hepatic ATP levels were not decreased as compared to control guinea pigs. Similarly, oxygen consumption, measured by Warburg monometry, was not changed as compared to controls. These data were confirmed in a small number of Gunn rats – hepatic ATP was similar between jaundiced/symptomatic Gunn rats, and heterozygous littermate controls.

The authors note that although in vitro studies have shown conclusively that bilirubin uncouples oxidative phosphorylation, these studies show that in vivo hepatic ATP levels are unchanged. The authors speculate that levels of bilirubin accessible to mitochondria in vitro may be greater than in in vitro studies, or that some protective mechanism is in action in vivo to prevent damage to liver cells.

Another in vivo study (Schenker, S., McCandless, D., and Zollman, P. 1966) describes experiments examining energy metabolism in the kernicteric brains of Gunn rats. In this study, neurological signs, ATP levels, oxygen consumption, and ATPase activity were all measured in kernicteric brain. Regional cerebral assays were performed.

Results showed that ATP unchanged or slightly elevated in the cerebral cortex and subcortex of symptomatic 2-week-old Gunn rats. By contrast, results from the cerebellum showed a 13% decrease in ATP in the early symptomatic stage, and a 27% decrease in ATP levels in advanced kernicteric rats. Neurological symptoms included in coodination, ataxia, impaired righting response, opisthotonus. stupor, coma, and death. Many of these symptoms can be attributed to cerebellar dysfunction. Examination of brain for yellow pigment showed some staining in the basal ganglia, and pigmentation in the cerebellum.

Results from oxygen consumption and ATPase studies in whole brain (because of the amount of tissue needed) showed no effect in newborn kernicteric Gunn rats as compared to littermate controls. When ATPase was measured in cerebral cortex and cerebellum, a slight but statistically significant decrease was noted in the cerebellum.

This study, in newborn kernicteric Gunn rats is significant in that it demonstrates a decrease in ATP in vivo in a brain region associated with many of the symptoms. Thus, the cerebellum plays an important role in coordination and motor control. The decrease in cerebellar ATP was greater in the more neurologically advanced animals, in keeping with the concept of continuing decreasing functional capacity of the cerebellum.

While the cerebellum showed yellow staining, small areas in the basal ganglia were also stained. One might imagine that even more regional analysis of the basal ganglia, or of the cerebellum might show more dramatic changes in energy metabolites from the uncoupling of cerebral oxidative phosphorylation. Measuring net levels of ATP does not answer the question of whether the decrease was due to impaired formation or from increased utilization. Turnover studies are necessary to examine that question.

In a later study (McCandless, D., and Abel, M. 1980), the further "regionalization" of bilirubin effects in the cerebellum was performed. In this study, homozygous newborn Gunn rats and non-jaundiced littermate controls were utilized. Consistent hyperbilirubinemia and kernicterus were assured by injecting Gantrisin (sodium sulfisoxazole), which acts to displace unconjugated bilirubin from serum albumin-binding sites. This facilitates movement of bilirubin from serum into the brain.

Glucose, glycogen, ATP, and phosphocreatine were all measured using methods of enzymatic analysis described by Passonneau and Lowry (Passonneau, J., and Lowry, O. 1993). These methods involve weighing samples on quartz fiber fishpole balances, oil well techniques, and enzymic cycling. Use of these techniques permits analysis in samples as small as single cells. In this study, the cerebellar layers molecular, granular, and Purkinje cell rich were analyzed. Samples weighed from 200 to 800 ng.

The newborn Gunn rats in this study showed reproducible and constant neurological symptoms consisting of rapid running, ataxia, opisthotonic posturing, stupor and death. Results from energy metabolite studies showed that both glucose and glycogen increased by 25–60% in all three layers of the cerebellum in severely kernicteric newborn kernicteric Gunn rats. By contrast, ATP showed a more selective effect. ATP in the molecular and granular layers of symptomatic Gunn rats was similar to that in heterozygous control littermates. However, ATP values were about 40% depleted in the Purkinje cell-rich layer in severely kernicteric Gunn rats. Phosphocreatine was similarly depleted in the Purkinje cell-rich layer (see Fig. 1).

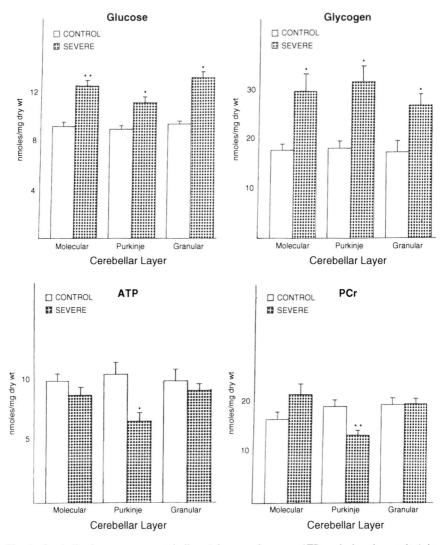

Fig. 1 Cerebellar layer energy metabolism (glucose, glycogen, ATP, and phosphocreatine) in kernicteric Gunn rats. Adapted from McCandless, D., and Abel, M. (1980)

These studies support earlier studies showing a selective effect of hyperbiliru-binemia on the cerebellum (Schenker, S., McCandless, D., and Zollman, P. 1966). In keeping with the idea that the most damaged areas are most likely to have alter-ations in energy metabolism, results from this study show energy deficits limited to the Purkinje cell-rich layer, the area of the functional cells. The finding of increased glucose and glycogen in all layers probably reflects some compensatory mechanism acting to ensure enough substrate during stress. This mechanism has previously demonstrated (McCandless, D., et al. 1979). In any case, these studies emphasize the importance of considering local neurochemical changes in any metabolic encephalopathy. These data lend further credence to the concept that bilirubin acts to uncouple oxidative phosphorylation with concomitant effects on ATP and phosphocreatine. The drop in energy metabolites is of sufficient extent as to likely affect other cellular energy-requiring processes.

Another method employed to examine regional cerebral effects of bilirubin has to use cerebellar tissue culture (Silberberg, D., and Schutta, H. 1967). In these studies, newborn rat cerebellum was placed on collagen-coated coverslips and placed in nutrient media. To this, various compounds including unconjugated bilirubin were added. Subsequently, both light and electron microscopy were used to evaluate the cerebellar structure.

Results using bright field microscopy to examine living cultures showed that there was a consistent pattern of change in Purkinje cells exposed to 15 mg/100 ml of unconjugated bilirubin. Macrophages with pigment in them were seen by 1 h after onset of exposure. In 3–4 h, the nuclei of Purkinje cells were eccentric, and the cytoplasm was more granular than controls. Effects on myelin sheaths appeared by 6 hours. By 12 h, most intact myelin was gone. Exposure to 5 mg/100 ml produced similar results but at a slower pace. Exposure to 1 mg/100 ml left the cultures unchanged until 36 h when minimal neuronal changes began to occur. These graded changes coupled to the dose argue in favor of a primary direct effect.

With electron microscopy, nerve cells showed complex intracytoplasmic inclusions. Mitochondria in cerebellar Purkinje cells exposed to 5 mg/100 ml of bilirubin for 12 h were enlarged. Controls showed no such changes. Macrophages were plentiful in bilirubin-exposed cultures. Cultures exposed to hemin showed changes similar to those seen in bilirubin-exposed samples.

The authors note that one significant finding in these studies was that there was dose- related effect. Thus, cultures exposed to high bilirubin levels had greater and faster alterations to cytoplasmic and nuclear structures. Biliverdin added to cerebellar cultures did not produce any changes.

The structural changes seen in mitochondria in tissue culture cells exposed to bilirubin are most likely the structural counterpart to biochemical changes noted be induced by bilirubin – changes such as uncoupling of oxidative phosphorylation.

As a possible cause of the altered myelin,seen in the above experiments, studies of the effects of hyperbilirubinemia on the activity of the hexose monophosphate shunt (pentose phosphate pathway) have been carried out (Thong, Y., and Rencis, V. 1977). In these studies, polymorphonuclear leukocytes from control subjects were prepared by centrifugation. Warburg flasks were prepared with the appropriate media, and unconjugated bilirubin was introduced. Flasks were incubated and $^{14}CO_2$ was captured and counted in a scintillation counter. Hexose monophosphate shunt activity was estimated in control and experimental (bilirubin) flasks by the evolution of $^{14}CO_2$.

Results showed that bilirubin induced a significant inhibition of hexose monophosphate shunt activity (down 51%) on stimulated neutrophils. The concentration of bilirubin which achieved this level of inhibition was 10^{-5} M. Another set of experiments showed that albumin was capable of reversing the inhibitory action of bilirubin.

The authors note that the mechanism by which bilirubin exerts its deleterious effect on the hexose monophosphate shunt is not clear. One possible mechanism could be an effect by bilirubin on NADH oxidase (Cowger, M., Igo, R., and Labbe, R.F. 1965). The actual level of bilirubin in culture (10^{-5} M), corresponds to a serum level of only 0.6 mg/100 ml. This is an almost insignificant amount in humans in vivo. However, when the binding capacity of bilirubin to albumin is exceeded, small increases may prove toxic to brain cells, and result in kernicterus. The increase in bilirubin over albumin-binding sites result in the altered function of neutrophils, and may result in increased susceptibility of a patient to intercurrent infection.

In another study examining the effects of bilirubin on brain energy metabolism, in mitochondria, results showed an uncoupling of oxidative phosphorylation (Vogt, M., and Basford, R. 1968). In the study, mature rats and 21-day-old weanling rats were decapitated and brain mitochondria prepared by centrifugation. Oxidative phosphorylation was estimated using polarographic determination.

Results showed that whether from adults or newborn rats, bilirubin at a concentration in the incubation media containing mitochondria, completely uncoupled oxidative phosphorylation. It acted to block state 3 respiration completely. When albumin was added along with bilirubin, the toxic effect on oxidative phosphorylation was abolished. Further, mitochondrial ATPases from mature and 21-day-old rats were reduced by bilirubin. Magnesium-stimulated ATPase was not affected by bilirubin, but DNP-activated ATPase was inhibited from 80 to 95%.

The authors state that their experiments show bilirubin to be a powerful inhibitor of cerebral mitochondrial respiration, and a potent uncoupler of oxidative phosphorylation. The uncoupling was prevented by the addition of albumin to the media. These results are somewhat in disagreement with those of others (Menken, M., Waggoner, J., and Berlin, N. 1966) who showed that in their system, albumin had no effect on cerebral oxidative phosphorylation. The present authors attribute this discrepancy to possible impurities in the albumin used by Menken, et. al. Weanling rat brain was not completely protected by albumin from the effects of bilirubin. Bilirubin caused an almost complete inhibition of ATPase activity, but with albumin present, ATPase were increased toward normal levels of activity.

In yet another study of the effects of bilirubin on brain mitochondria, results showed a significant effect (Mustafa, M., Cowger, M., and King, T. 1969). In this study, brainfrom rats, chicks, monkeys, and rabbits were prepared by standard differential centrifugation techniques.

Results from this study confirmed many earlier studies showing that bilirubin has a depressing effect on brain mitochondria. Brain mitochondria showed a dose-dependent bilirubin effect which was continuous. These studies indicated that respiratory control was more sensitive than phosphorylation to the action of bilirubin. The sensitivity of brain was significantly more noticeable than in any other organ tested (heart/liver). Serum albumin was capable of preventing the uncoupling of oxidative phosphorylation completely when the proper ratio was maintained. Results also showed that in the presence of an oxidizable substrate such as succinate, bilirubin produced mitochondrial swelling.

The authors note that high concentrations of bilirubin interfere with the electron transport system, decreasing respiratory control. Further, the authors comment that bilirubin acts differently from conventional uncouplers. Bilirubin seems to act as a transport-inducing agent. Bilirubin also acts on mitochondria by inducing swelling of the cytoplasmic organelles. The bilirubin-induced swelling occurs on intact mitochondria, but not on mitochondrial particles. The authors speculate that swelling of mitochondrial is initiated when bilirubin binds with lipid moieties in the mitochondrial membrane. This in turn alters the membrane permeability allowing an energy-dependent entry of ions and water into the mitochondrion. The amount of bilirubin required for this process is very small. The movement of ions and water is unidirectional, that is into the mitochondrion. The authors speculate that binding of bilirubin to lipid components of the mitochondrial membrane is the initial event precipitating further events.

Taking into account the Mitchell chemiosmotic theory of oxidative phosphorylation (Stumpf, D., Eguren, L., and Parks, J. 1985), the effect of bilirubin on rat liver mitochondria was revisited. The Mitchell hypothesis is that the electron transport chainmoves hydrogen ions out of the mitochondrial matrix, in turn producing a proton motive force which is dependent on pH and transmembrane potential. The study was undertaken to examine effects of bilirubin on these parameters.

Results show that bilirubin stimulated oxygen consumptionin brain in state 4. There was an increase in the proton motive force at high bilirubin concentrations. Liver mitochondrial membrane showed an increased conductiveness when bilirubin was added. Essentially all bilirubin was associated with the mitochondrial pellet.

The authors note that their polarographic studies show abnormal coupling of oxidative phosphorylation in liver mitochondria using proton motive force data. These data indicate a loose coupling which may be associated with normal production of ATP. No change in matrix volume was seen in this study suggesting that the inner membrane of the liver mitochondrion was intact. Liver was utilized for these studies because methodology for brain studies were not available. These investigations actually support previous studies on the effects of bilirubin on oxidative phosphorylation even though not performed in brain.

In another study, newborn piglets were subjected to hyperbilirubinemia, and cerebral oxygen, glucose, and lactate were measured. Results showed no major effects of 11.0 nmoles/g bilirubin on the parameters measured in the cerebral cortex of the piglets. The authors state that these negative results do not support that bilirubin has an effect on mitochondrial function. It must be noted that: (1) piglets are born much more mature than mice, rats, and humans and (2) neurobiochemical changes occur, or do not occur in the cerebral cortex probably do not reflect alterations areas which should be studied such as the cerebellum. In another study (Ives, N., et al. 1988), the effects of hyperbilirubinemia was assessed in adult guinea pigs using 31-P nuclear magnetic resonance spectroscopy (31P-NMR) to evaluate energy metabolism. In these studies, bilirubin was infused into a brain slice preparation with albumin in a molar ratio of 5:1, assuring significant bilirubin would be free of albumin. Cerebral cortex was chosen the brain region for study; subcortical tissue was discarded.

Results indicated that exposure of the cortical tissue to bilirubin to a level of 120 μmoles/l had no effect on the 31P-NMR spectrum under normoxia. When tissue was exposed to hypoxia, a decrease in the phosphocreatine/inorganic phosphate (PCr/Pi ratio) occurred. These changes were reversible by reperfusing the tissue with control media. The reduction in PCr/Pi ratios in the presence of hypoxia and high bilirubin concentrations were highly statistically significant.

The authors comment that advantages of these methods include the ability to make continuous measurements of energy metabolism, but do not evaluate turnover. The authors correctly point out the measurements were made on cerebral hemispheres from adult animals, and this might not reflect changes in areas highly stained by bilirubin in vivo in newborn animals. In spite of these limitations, bilirubin alone did not change energy metabolism in the cerebral hemispheres. Only in the presence of reduced oxygen tension (hypoxia) did addition of bilirubin result in a further significant decrease in the PCr/Pi ratio indicating a further effect by bilirubin in addition to that caused by hypoxia alone. The authors speculate that decreasing brain pH may also play an important role in these observed changes.

A report of experiments designed to examine the effects of bilirubin on cerebral energy metabolism after opening the blood–brain barrier has been published (Wennberg, R., et al. 1991). This study was designed to see if bilirubin caused a rapid change in energy metabolism in rats whose blood–brain barrier had been opened using arabinose. The study also examined whether energy metabolism changes would continue after bilirubin was withdrawn. The idea of opening the blood–brain barrier and allowing a rapid equilibrium between cerebral tissue and blood would circumvent the sampling problems experienced in other studies (see above).

Results showed that in animals whose blood–brain barrier was opened with arabinose, and who were infused with bilirubin had a significant and immediate effect. Within 15 minutes, phosphocreatine fell by 60%, and ATP dropped by about 40%. ADP and AMP levels rose. This lowered the calculated adenylate charge. Accompanying these changes, glucose and glycogen dropped, and lactate increased. The metabolic changes were not related to peak bilirubin levels suggesting there may have been regional variations in the effectiveness of the arabinose-induced opening of the blood–brain barrier. Four hours after infusion of bilirubin/arabinose, when these two compounds should have been cleared, animals still were lethargic, suggesting possible long-term damage could have occurred.

The authors state that since the effect of bilirubin was so rapid (15 min) in rats with open blood–brain barriers, this effect on energy metabolism may be primary. After hyperosmolar opening, no change in pH was observed, ruling out any indication of ischemia. This further implies that the increased lactate and normal pyruvate shows intracellular accumulation of NADH.

The authors comment correctly that extrapolation of these results to human newborns must be cautious. The cells studied in this experiment might not have responded as would immature newborn brain cells. Examining energy metabolism in cells from brains of rats whose blood–brain barriers were not opened was not done because it seemed unlikely that such bilirubin treatment would result in significant changes. The importance of the blood–brain barrier in newborn infants is not clear. The present data do suggest that the rapid and dramatic changes in energy metabolism in rats with open blood–brain barriers suggest an initial primary effect of bilirubin.

In a recent paper (Park, W., Chang, Y., and Lee, M. 2002), the effect of 7-nitroindazole (a neuronal nitric oxide synthetase inhibitor) was examined as to bilirubin-induced alteration of brain cell function and energy metabolism in newborn piglets. Results showed that bilirubin induced cerebral Na, K-ATPase in neurons and lipid peroxidation products were decreased by 7-nitroindazole administration. Energy metabolites, ATP and phosphocreatine, which were decreased in animals treated only with bilirubin, showed a lesser effect when treated with both bilirubin and 7-nitroindazole. The authors note that an increase in nitric oxide produced by introducing nitroindazole was responsible for attenuating the effect of bilirubin. The mechanism of this attenuating/protective effect remains unclear.

In another paper (Grojean, S., Vert, P., and Daval, J. 2009), the effects of a combined insult of hyperbilirubinemia and hypoxia were examined in developing rat neurons. In this study, cell cultures of newborn rat neurons were exposed to bilirubin (0.25–5 μmoles/l), and/or to a gas mixture of 95% N_2/5% CO_2 (hypoxia) for 3–6 h. Over a period of 96 h, energy metabolism, protein synthesis, and cell death were analyzed.

Results showed that bilirubin alone acted to increase apoptotic cell death. When the cells were exposed to both hypoxia as defined above, and bilirubin, cell death was increased. Bilirubin alone decreased energy metabolism. Hypoxia alone served to increase energy metabolism, and this increase was not altered by adding bilirubin. Bilirubin alone or with hypoxia acted to increase protein synthesis. These studies serve as further evidence, in terms of apoptotic cell death, that the combination of bilirubin and hypoxia may be more deleterious than either alone.

As a brief summary (see epilog for more comment), the paper by Bo Siesjo's group (Wennberg, R., et al. 1991) is a key study. Opening the blood–brain barrier using arabinose eliminates the blood–brain barrier protection, which is no doubt in various brain regions, and is age related. This alone makes a comparison of 14–day-old Gunn rats vs. 2-day-old guinea pigs vs. adult Wistar rats, etc., highly problematic. Then, having "leveled the playing field," these investigators chose to look at early changes in energy metabolism. Infusion of bilirubin into rats with opened blood–brain barriers produced dramatic decreases in cerebral glucose, glycogen, ATP, and phosphocreatine. Lactate and the lactate/pyruvate ratio was markedly increased.

These changes in energy metabolism had occurred by 15 min after initiating the bilirubin infusion. The authors state that the rapidity of metabolite change implicates these energy changes as being primary and direct.

In a very recent study (Vaz, A., et al. 2010), the effects of unconjugated bilirubin were examined in neuron tissue culture. Samples were obtained from 16 to 17 day fetuses,and cerebral cortex was used to obtain tissue samples. Cultured neurons were incubated in the presence or absence of 50 μmolar (25 mg/100 ml) of unconjugated bilirubin. Mitochondrial respiratory chain complex activities, oxygen uptake, and various related metabolites were all measured using standardized techniques.

Results showed selective changes. Thus, 50 μmolar unconjugated bilirubin plus 100 μmolar human serum albumin had an effect on cytochrome c oxidase activity. The inhibition equaled about 50%. Further, unconjugated bilirubin resulted in an increase in reactive O_2 and a decrease in NADPH in the cultured cortical neurons.

In spite of the reduction in the respiratory chain components, cortical neuron levels of ATP remained unchanged. There was an increase in lactate and the fructose-1-6 – phosphate/fructose-6-phosphate ratio indicating an increase in glycolysis. Observation by fluorescence microscopy demonstrated neuronal condensed and fragmented nuclei, and cell death via apoptosis. This was an indicator that the respiratory chain alterations were toxic to the viability of the cultured cortical neurons.

The authors state that these data show for the first time that a 1-h exposure of prenatal cortical neurons to bilirubin alters respiratory chain components. Further, structural changes to the neurons occurs, and cell death results. The authors speculate that the effect on the respiratory chain component levels may induce an increase in glycolysis which acts to maintain ATP levels. This may therefore represent an early response to unconjugated hyperbilirubinemia which might proceed other changes later. In any case, this treatment did lead to structural changes and cell death in cultured neurons from the cerebral cortex. It should be noted that, in utero, fetuses are not exposed to elevated bilirubin levels due to the transfer of bilirubin across the placenta. Also, the neuronal demands for ATP would be expected to be lower than in the adult.

Jaundice and Other Biochemical Changes

Many other metabolic disturbances have been described besides those involving energy metabolism as described in the preceding chapter. For example (Youngs, J., and Cornatzer, W. 1963), the effects of bilirubin on mitochondrial phospholipids in rat liver have been comparatively studied. Bilirubin, arsenate, dinitrophenol, and oligomycin were all evaluated and compared as regards their ability to alter phospholopid synthesis in rat liver mitochondria. Results showed that bilirubin, arsenate, and oligomycin inhibited synthesis of lecithin, phosphatidyl serine, and phosphatidyl ethanoamine. Dinitrophenol, in addition to the phospholipids above, also inhibited sphingomyelin and phosphatidyl inositol.

These results suggest different mechanisms of action for dinitrophenol as compared to bilirubin, arsenate, and oligomycin. Other studies have suggested a different mode of action between dinitrophenal and bilirubin. This study does not evaluate the possibility that energy-dependent synthetic pathways might be blocked due to a decrease in energy metabolism.

Effects of bilirubin on DNA synthesis have shown that in the cerebellum, there is an inhibitory effect of bilirubin on synthesis of the nucleotide (Sawasaki, Y., Yamada, N., and Nakajima, H. 1973). The hypothesis was suggested that perhaps bilirubin accumulation in the cerebellum was greater than in other brain areas, which might explain why the cerebellum showed these effects. In order to examine this concept, bilirubin levels were measured in several brain regions in developing Gunn rats (Sawasaki, Y., Yamada, N., and Nakajima, H. 1976).

In this study, heterozygous and jaundiced homozygous Gunn rats were sacrificed, and bilirubin levels measured in the cerebellum, cerebral cortex, caudate, thalamus, colliculi, hippocampus, and brainstem, among other areas. Measurements were made at ages 8, 16, and 30 days. Results showed mean values of the three time points of 27.5 at day 8, 23.5 at day 16, and 9.9 at day 30. Values are expressed as micrograms/gram tissue, wet weight. It is apparent that the mean levels drop from day 8 to about one third those levels by day 30. None of the bilirubin values were significantly different, one area from the other. That is to say that the concentration in the cerebellum was not higher than the concentration, for example, in the cerebral cortex.

Some comments are pertinent to these data. First, whole cerebellum bilirubin levels may not accurately reflect bilirubin concentration in key cells such as the

D.W. McCandless, *Kernicterus*, Contemporary Clinical Neuroscience,
DOI 10.1007/978-1-4419-6555-4_10, © Springer Science+Business Media, LLC 2011

Purkinje cells. Second, it has been suggested by several workers that at the time of death in Gunn rats, there is a generalized movement of bilirubin into all brain areas. This explains the observation that kernicteric Gunn rat brain may show a diffuse cerebral pigment staining.

The authors of this paper conclude that the impaired DNA synthesis seen in the cerebellum of kernicteric Gunn rats is not attributable to bilirubin concentrations higher than other brain regions. The cerebellum seems to have unique neurochemical attributes as regards several phenomena. It would seem again that the cerebellum is uniquely sensitive in some yet to be defined way. The progressive drop in bilirubin in all regions as the animals age may reflect blood–brain maturation.

In another paper (Hansen, T., Mathiesen, S., and Walaas, S. 1996), the effects of bilirubin on neural protein phosphorylation systems were investigated. These studies were performed in a controlled media. Bilirubin was added to the reaction media in concentrations of from 0 to 320 μmol/l. Bovine serum albumin was also added to selective preparations.

Results showed that bilirubin inhibited cyclic nucleotide-dependent protein kinases, Ca-calmodulin-dependent protein kinases, and Ca-dependent protein kinases. Fifty percent inhibition occurred between concentrations of 20 and 125 μmol of bilirubin.

Results also showed that bilirubin decreased the Vmax inhibiting the catalytic domains of protein kinases by non-competitive mechanisms. Results also showed that bovine serum albumin was able, in high concentrations, to reverse the inhibitory effect on protein phosphorylation of phospholemman peptide, with one half maximum inhibition at a bilirubin concentration of 25–50 μmolar.

The authors comment that these results were achieved in the absence of respiratory enzymes or organelles. This implies that the previously observed effects of biliurubin on protein phosphorylation may not be solely due to the effect of bilirubin on oxidative phosphorylation, and that there may be a direct effect on protein phosphorylation by bilirubin. The authors state that the concentrations of bilirubin required for a 50% inhibition of protein phosphorylation are about that thought to be in pigmented foci in the brains of kernicteric newborn infants. This is a highly in vitro system, and so the possibility that inhibition of protein phosphorylation occurs in vivo without a primary effect on oxidative phosphorylation remains to be demonstrated.

The rates of protein synthesis and degradation have been investigated in the cerebellum of symptomatic Gunn rats (Katoh-Semba, R., and Kashiwamata, S. 1984). The animals were sacrificed on days 13 and 30 at a time when the cerebella showed hypoplasia. Heterozygous littermates served as controls.

Results showed that protein degradation rates were about tenfold greater in homozygous Gunn rats at day 13 than in heterozygous littermate controls. By day 30, protein synthesis was significantly lower in the jaundiced Gunn rats as compared to non-jaundiced littermates. The authors comment that lower protein levels observed in the cerebellum of symptomatic Gunn rats may result from a dual mechanism: increased protein catabolism around day 13 and decreased synthesis of protein moieties by day 30.

Yet another study (O'Callaghan, J., and Miller, D. 1985) examines the relation between cerebellar hypoplasia and neurotypic and gliotypic proteins in Gunn rats. In this study, several cell-specific proteins were evaluated. These included Purkinje cells-"G" substance (serves as endogenous substrate of cyclic GMP-dependent protein kinase); PCPP-260, also a Purkinje cell-specific phosphoprotein associated with cyclic AMP; synapsin 1, a synapse-specific phosphoprotein present in all neurons; glial fibrillary acidic protein, specific to astrocytes; and myelin basic protein, a substance unique to myelin.

Results showed that homozygous kernicteric Gunn rat hypoplastic cerebellum had alterations in the levels of neurotopic and gliotopic proteins. These changes were consistent with the neuropathologic effects seen in hyperbilirubinemic Gunn rats. The study also showed a decrease in cyclic GMP associated with changes in responsiveness to a nociceptive stimulus. Cyclic AMP levels were unchanged in the jaundiced Gunn rats.

The authors note that generally speaking, the protein changes were in good correlation with other biochemical and structural alterations. This indicates that neurotypic and gliotypic proteins may be useful indicators of the degree of severity in kernicterus.

The uptake of neurotransmitters into synaptic vesicles for storage is a process which is ATPase dependent. A study (Roseth, S., et al. 1998) was undertaken to see if bilirubin interfered with this process. In this study, synaptic vesicles were prepared essentially by routine centrifugation techniques. Radiolabled glutamate and/or dopamine uptake was estimated in vitro. Bilirubin was added to these incubation medias.

Results showed that bilirubin did inhibit the uptake of glutamate and dopamine into synaptic vesicles, and the uptake occurred in a dose-dependent fashion. Hundred micromolar bilirubin achieved an inhibition of over 70% of both glutamate and dopamine uptake. The uptake inhibition was dependent on the amount of vesicular material present at high concentrations, the inhibitory action of bilirubin on uptake was eliminated. Studies further indicated that the inhibitory effect was not caused by any effect of bilirubin on ATPase.

The authors note that previous studies have shown that bilirubin has an effect on neuronal membranes, possibly with polar lipid head groups. Further, bilirubin may build up when the membrane capacity for bilirubin is exceeded (Vazquez, J., et al. 1988). The data presented in this report suggests that bilirubin may act at the interface between membrane lipids and transporter proteins.

The authors state that although they cannot rule it out, they believe that their results indicate that bilirubin inhibition on neurotransmitter uptake is not associated with a concomitant effect on energy metabolism.

Another paper examining other possible effects of bilirubin, examines the concept that bilirubin stimulates enzymes in the inner mitochondrial membrane which in turn oxidizes the bilirubin (Hansen, T., Allen, J., and Tommarello, S. 1999). The speculation is that brain cells may be able to "protect themselves" from the toxic effects of bilirubin through oxidation.

In this study, mitochondria were isolated by centrifugation techniques, and the ability of the mitochondria to oxidize bilirubin was measured. Results showed that, using change in optical density, bilirubin had an interaction with cytochrome P450 oxidase. The activity was inhibited by loss of cytochrome c from the incubation media, and reconstituted by adding additional cytochrome c to the incubation media.

The authors conclude that oxidation of bilirubin, a protective effect, may be mediated through cytochrome c oxidase, and is cytochrome c dependent. Whether this effect is directly attributable to cytochrome P450 oxidase is unclear.

The effects of both hyperbilirubinemia and/or hypoxia on glutamate-mediated apoptosis has been examined (Grojean, S., et al. 2001). Since both elevated bilirubin and hypoxia are significant risk factors in newborn infants, these authors looked at possible synergistic effects. Six-day-old cultured neurons were exposed to hypoxia or to unconjugated bilirubin (0.5 μmolar), or to a combination of both hypoxia and increased bilirubin concentrations.

Results showed that at 96 h, viability of cells was reduced 23% by hypoxia, 25% by bilirubin, and cell viability was reduced 34% by the combination of both hypoxia and elevated bilirubin levels. Examination of nuclear structural changes showed apoptotic cells present (13%) following hypoxia, and after bilirubin alone, 16% of cells were apoptotic. With both hypoxia and bilirubin treatments, 23% were apoptotic. Bilirubin action was blocked by the glutamate receptor antagonist MK-801. This antagonist was without effect on hypoxic damage.

The authors comment that the data may suggest free radical scavenging. Measurements of intracellular free radical production did not confirm an antioxidant role for bilirubin. It was clear that taken together, hypoxia and bilirubin can have a synergistic relation in terms of deleterious effects on newborn rat brain.

Further studies on mechanisms of bilirubin toxicity other than direct effects on energy metabolism have centered on brain protein phosphorylation (Morphis, L., et al. 1982). Additional studies (Hansen, T., Bratlid, D., and Walaas, S. 1988) were able to demonstrate that bilirubin inhibited the phosphorylation of synapsin 1, a presynaptic vesicle protein. Since synapsin 1 in its dephosphorylated form is inhibitory to neurotransmitter release, this could explain the mechanism in which bilirubin inhibits synaptic activation in rat hippocampal slices (Hansen, T., et al. 1988).

Yet another possible effect of bilirubin on the function of synapses in brains of hyperbilirubinemic infants is inhibition by bilirubin and transport proteins. Such inhibition can be thought of as of altering neurotransmitter uptake into these vesicles (Roseth, S., et al. 1998). Vesicular storage of brain catecholamines is also blocked, and this effect seems to be independent of ATP.

These changes in brain metabolism may explain in part, the symptoms of lethargy in hyperbilirubinemic newborn infants. It may also have a bearing on the often noted symptom of auditory-evoked responses in these hyperbilirubinemic newborn infants.

Since the brainstem auditory pathways are well known to be affected by hyper-bilirubinemia, auditory brainstem-evoked responses were measured in newborn infants with bilirubin levels between 15 and 25 mg/100 ml (Perlman, M., et al. 1983). In this study, 24 newborn infants with high serum bilirubin levels were compared to 19 infants whose bilirubin levels were not elevated. Results showed wave complex IV–V was absent in 10% of the 24 jaundiced infants as compared to controls. There was also decreased brainstem transmission time in jaundiced patients as compared to control infants. The authors state that neonatal hyperbiliurubinemia was associated in an alteration in auditory nerve/brainstem-evoked responses. These results were reversible, but indicate that alterations in brain function may occur at somewhat lower bilirubin levels than previously believed.

In another interesting study (McDonald, J., et al. 1998), intra striatel injection of a glutamate analog, N-methyl-D-aspartate caused an increased incidence of atrophy in the striatum and hippocampus in jaundiced Gunn rats. One previously unknown feature of the neuropathology of bilirubin was the greater susceptibility of astrocytes over neurons in these anatomic regions. N-methyl-D-aspartate produced an increased atrophy of the striatum and hippocampus, and results indicated that mitochondrial function was decreased. Neurons were more susceptible than glia as regards cell viability and apoptotic cell death.

In a more recent study (Genc, S., et al. 2003), results showed that unconjugated bilirubin decreased oligodendrocytes' viability. This decrease was both time and concentration dependent. This tissue culture study further showed that unconjugated bilirubin induced nitric-oxide synthase mRNA expression, increased nitrate production, and increased apoptotic cell death. The authors note that the in vitro data may indicate that oligodendrocytes could also be a target of unconjugated bilirubin, in addition to astrocytes and neurons. This is supported by previous studies showing an effect on myelin by hyperbilirubinemia in Gunn rats (O'Callaghan, J., and Miller, D. 1985).

An interesting feature of bilirubin was the observation (Stocker, R., Glazer, A., and Ames, B. 1987) that in low and physiologic levels, unconjugated bilirubin might be acting as a potent antioxidant. This action was seen as a scavenging of peroxyl radicals. Alpha-tocopherol is another scavanger, and unconjugated bilirubin may be at least as efficient. When the bilirubin is bound to human albumin at physiological, it may prevent the in vitro oxidation of albumin bound fatty acids, and also prevent the oxidation of albumin. One role of unconjugated bilirubin is that it is capable of scavenging 2 moles of peroxyl radicals. These cytoprotective effects of low physiologic levels of unconjugated bilirubin are novel. This protective effect has been described as occurring in the heart (Neuzil, J., and Stocker, R. 1994), and also in the liver (Chiu, H., Brittingham, J., and Laskin, D. 2002).

The extent of the ability of unconjugated bilirubin to protect other systems such as the immune system, neural cells, and smooth muscle is being actively explored. These protective effects occur in low normal unconjugated bilirubin levels.

In a recent study (Lanone, S., et al. 2005), the effect of unconjugated bilirubin on NADPH oxidase was examined in Gunn rats. Results from this study showed that the inhibition of NADPH was probably secondary to an effect by bilirubin on the expression of NOS2 protein. This may work through an attenuation of NOS2 in cardiovascular and other tissues. This newly described mechanism of cytoprotection by unconjugated bilirubin extends knowledge of its antioxidant properties. It is noteworthy that these studies were done in homozygous Gunn rats which have relatively high serum unconjugated bilirubin levels. Nevertheless, the Gunn rats in this study were more resistant to endotoxin-induced hypotension and death than were heterozygous littermate controls.

In another paper (Fernandes, A., et al. 2006), the possible complexity of the effects of unconjugated bilirubin are described. In this study, the possibility that unconjugated bilirubin stimulates the production of cerebral cytokines was examined. Astrocyte cultures were used and the profile of cytokene production was measured. Tumor necrosis factor (TNF), tumor necrosis factor-alpha receptor 1 (TNFR1), mitogen-activated protein kinase (MAPK), and nuclear factor kB (NF-kB) were all examined for possible activation by unconjugated bilirubin.

Results showed that TNFR1 and MAPK were activated in astrocyte culture by bilirubin. These two signal effectors rose prior to an early upregulation of TNFR1 and interleukin mRNA, and later secretion of TNFR1. Loss of cell functionality and cytokine secretion were lowered when the NF-kB signal transduction pathway was decreased.

While the above results suggest a complex cascade of events may result in astrocyte tissue culture, the possibility that these events occur in vivo remains to be elucidated. And the relation of these events, either directly or indirectly, to the effects of unconjugated hyperbilirubinemia are far from being identified. There are many unanswered questions relating to the deleterious effects of unconjugated hyperbilirubinemia and its effects on newborn infant brain. One must always ask whether changes might be primary or not. One must ask questions independent of methodologies at hand. The application of methodology to any experimental question must be a secondary consideration. Far better to ask cogent questions, then design experiments which will provide answers, than to use the labs "pet" techniques searching for non-critical answers.

Breast Milk Jaundice

The occurrence of breast milk jaundice, or an exaggerated hyperbilirubinemia associated with breast milk, was first described by Arias and colleagues in 1963 (Arias, I., et al. 1963a, b, 1964a, b). The suggestion was made that the hyperbilirubinemia was a result of the presence of the steroid pregnane-3-20-diol in breast milk which acted to inhibit glucuronyl transferase, the bilirubin-conjugating enzyme.

In a letter to the editor of the *British Medical Journal* (Arias, I., and Gartner, L. 1970), the authors reiterate that they had isolated pregnane-3-20-diol from inhibitory breast milk, as well as from the urine of women whose children were jaundiced for no other apparent reason. The authors point out that this unusual isomer has only been detected in the milk from mothers whose infants are jaundiced. Obviously, not all jaundiced newborns receive this isomer from their mothers milk.

Some investigators made a link between breast milk jaundice/pregnane-3-20-diol and oral contraceptives (Wong, Y., and Wood, B. 1971). This was a prospective study in which women who had been on contraceptive pills before becoming pregnant were compared to those who never had been on the pill, as regards jaundice development.

Results from 116 mothers showed that there were 69 non-jaundiced infants and 47 jaundiced subjects. Of the 69 non-jaundiced infants, 35% of the mothers had been on the pill, and of the 47 jaundiced infants, 70% of mothers were on the contraceptive pill. This was a statistically significant difference. In 18 severely jaundiced (over 15 mg/100 ml) newborn patients, 14 were born to mothers who were taking contraceptive pills, while only 4 were born to mothers not taking the pill. These data suggest the pill may have a component with a causative effect on bilirubin metabolism and jaundice.

The effect of breast milk on the conjugation and excretion of bilirubin by rat liver slices was reported (Hargreaves, T., and Piper, R. 1971). Results showed that the addition of breast milk from the mother of a breast milk jaundiced infant acted to decrease the rate of conjugation of bilirubin by the slices. The effect was on the conjugating enzyme, glucuronyl transferase. Addition of pregnane-3-20-diol to liver slices only reduced conjugation by 55% as compared to 100% inhibition by breast milk. These results suggest there may be something else acting with the pregnane-3-20-diol in breast milk which results in a 100% inhibition of bilirubin conjugation.

D.W. McCandless, *Kernicterus*, Contemporary Clinical Neuroscience,
DOI 10.1007/978-1-4419-6555-4_11, © Springer Science+Business Media, LLC 2011

The idea that pregnane-3-20-diol was the inhibitor of bilirubin conjugation in maternal breast milk and a cause of neonatal jaundice was not without controversy (Murphy, J., et al. 1981). These authors used gas chromatography–high resolution mass spectrometry to examine breast milk for the presence of pregnanediols. Results showed no presence in any milk samples collected from mothers of jaundiced infants of pregnanediols. This finding was not unique, as other reports of a lack of pregnane-3-20-diol in breast milk had been reported (Ramos, A., Silverberg, M., and Stern, L. 1966). The reasons for not detecting pregnanediols were not clear, although the actual incidence of breast milk-induced jaundice is low. The authors speculate some other factors must be present that contributes to the generation of hyperbilirubine-mia. The authors do state that the jaundice in the ten infants in this study was higher than usually seen in physiological jaundice, and lasted longer (see Fig. 1).

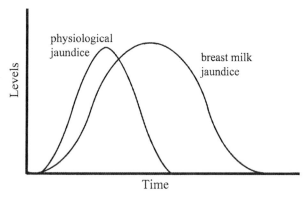

Fig. 1 Schematic representation of the difference between physiological jaundice and breast milk jaundice. Breast milk jaundice starts several days later than physiological jaundice, and may last a few weeks. Levels also tend to be higher

In keeping with the concept that bilirubin levels were higher and lasted longer in breast milk jaundice than in newborn infants fed formula, a retrospective study was performed (Sirota, L., et al. 1988). In this retrospective study, 40 preterm infants who received a combination of breast milk plus formula were compared to 60 preterm infants on formula only. Results showed that the group fed a combi-nation diet had, on average, a higher serum bilirubin level than did the group fed formula only. Further, the combination group had higher bilirubin levels on the day of discharge from the hospital. Seventy-six percent of the combination diet group received phototherapy, whereas only 45% of the preterm formula newborn infants were candidates for phototherapy. The authors speculate that breast milk may cause an early and late increase in bilirubin levels in preterm infants.

In a follow-up study to the above findings, 60 infants with breast milk jaundice were further evaluated (Grunebaum, E., et al. 1991). Results showed that there were two peaks in serum bilirubin – one on the fourth to fifth day, and another on the

14th–15th day. In infants with continuous breast feeding, the elevated bilirubin levels slowly abated, but were still high 12 weeks after birth. The familial incidence was 14%, and no hearing deficit was detectable at a later time.

In a more recent paper looking at breast milk jaundice (Maruo, Y., et al. 2000), 17 breast-fed newborns were examined as regards their glucuronyl transferase gene. In this study, the 17 newborns had prolonged jaundice, and total bilirubin levels were above 10 mg/100 ml. When breast feeding was stopped, bilirubin levels began to drop. Upon resumption of breast feeding, bilirubin levels began to rise in some but not all infants. Serum bilirubin levels had reached normal concentrations by 4 months of age. Using direct sequencing, the polymerase chain reaction amplified exon, promotor, and enhancer regions of the bilirubin uridine diphosphate-glucuronyl- osyltransferase gene (UGT 1A1) were evaluated.

Results showed that all but one newborn had one or more mutations of the UGT 1A1 gene. Fifteen of the newborn infants had missense mutations. A multitude of other mutations occurred, none of which was in a majority. Missense mutations in the current study were identical to those seen in previous studies of Crigler–Najjar syndrome and Gilbert syndrome (10, 11, 19 in this paper). In in vitro studies, the authors found 32 and 60% of normal enzyme activity in homozygous and heterozygous cases, respectively.

The authors state that despite extensive investigation, no specific component or combinations of components have been demonstrated to defiantly cause breast milk jaundice. In contrast, the authors note, bilirubin UGT 1A1 activity is 1% or less of adult values, which are then reached by 3 months of age (Onishi, S., et al. 1979). Thus the UGT 1A1 mutations noted in this study may be an underlying factor in the development of breast milk jaundice. As yet unidentified (conclusively) factors or components, or combinations of factors in breast milk in turn may serve to induce breast milk jaundice in genetically susceptible newborn infants.

The possible connection between breast milk jaundice and Gilbert disease is also noted. The authors, based on gene mutation results, speculate that breast milk jaundice may be an "infantile" form of Gilbert syndrome. Most breast milk jaundice is mild, and the hyperbilirubinemia is not a clinically significant problem. There are, however, cases in which unconjugated hyperbilirubinemia can exceed 20 mg/100 ml, risking kernicterus, and the clinician must be mindful of climbing pigment levels. One estimate indicates that 12.9% of breast milk-jaundiced newborns had bilirubin levels of 15 mg/100 ml or greater. Many factors regulate and influence cerebral staining by bilirubin, and in some cases, 15 mg/100 ml may be a cause for concern and treatment (Gourley, G. 2000).

In a recent review of neonatal jaundice, the unique characteristics of breast milk jaundice are discussed (Gartner, L., and Herschel, M. 2001). It is noted that in breast milk-fed infants, the curve of onset and diminution of jaundice is more drawn out. Thus, the rise may take 10 or more days for the peak to occur, and several weeks to 3 months for the hyperbilirubinemia to return to adult levels. Of the elevated serum bilirubin levels, less than 10% is conjugated. The authors state that in rare cases, breast milk-jaundiced infants may have bilirubin concentrations as high as 24 mg/100 ml. This is a dangerous level, especially given that many factors govern entry into brain regions. Indeed, kernicterus has been reported in breast-fed infants who had no other specific pathologic condition which could have contributed (Newman, T. 1995). Borderline premature infants may also be at risk for kernicterus in cases of breast milk jaundice.

The authors note that early studies of breast milk jaundice showed a metabolite of progesterone (pregnane-3-20-diol) was present in breast milk, and much evidence was gathered to suggest that this metabolite was key in the development of jaundice. The mode of action was to interfere with the conjugation of bilirubin. Further studies have not always confirmed either the role of human milk, or of progesterone metabolites as the definitive cause of this form of neonatal jaundice.

It is certain that breast-fed infants have a different bilirubin pattern as compared to Formula-fed infants. For example, 13% of breast-fed infants have a bilirubin peak of about 12 mg/100 ml during the first 5 days of life, whereas formula-fed infants show a similar peak in only 4% of cases. Breast feeding jaundice and/or inadequate human milk intake is similar to an adult disorder termed "starvation jaundice." In this condition, 24 h of fasting leads to an approximate doubling of serum bilirubin levels in normal adults. This may be related to an increased activity of intestinal absorption of bilirubin. This is similar to breast milk jaundice in which there may be a lower intake as compared to formula-fed infants, and subsequent recycling of bilirubin through enterohepatic circulation.

In terms of diagnosis and management, concerns are raised due to the sometimes mandated early discharge of mother and newborn in less than 72 h after birth. At this time, the efficiency of breast feeding, and bilirubin levels may not have been properly evaluated. It has been suggested that all early released infants be seen and examined in 2–3 days post hospital discharge as to feeding efficiency and level of jaundice. The development of a device for the transcutaneous estimation of hyperbilirubinemia facilitates this process in a setting outside the hospital.

In a recent paper (Gourley, G. 2009), it was noted that there is a strong association between breast feeding and hyperbilirubinemia in newborn infants. Several mechanisms may be operational including inhibition of glucuronyl transferase activity, decreased coloric intake, and increased enterohepatic circulation. It is noted that almost all reports of kernicterus in the past 15 years have been in breast-fed infants. While kernicterus is rare, health care workers must always be alert to this possibility, since rapidly increasing hyperbilirubinemia must be met with aggressive therapeutic intervention. The early release of mother and newborn after birth does not serve this goal.

In another very recent study (Huang, A., et al. 2009), maternal and neonatal risk factors were assessed in a multiethnic Asian cohort of healthy full-term newborn infants.

This study was undertaken to examine factors such as ethnicity, weight loss after birth, breast feeding, etc., on hyperbilirubinemia. The authors state that early onset hyperbilirubinemia can be a high-risk event due to the rapid rise in bilirubin levels and the various conditions which can result in kernicterus In fact, kernicterus has reemerged as a possibility in jaundiced newborn infants, and 98% of recent cases, as stated above, have been noted in infants exclusively breast fed.

Some studies have shown that in spite of breast feeding, more critical factors are the inability to establish optimal breast feeding regimes with resultant decreased caloric intake, and weight loss.

The cause of neonatal jaundice is unclear in as many as 60% of all cases. There has been an hypothesis that newborn infants of Asian descent may have a higher incidence of neonatal jaundice. The study reported involved 1034 neonates of which 56% were Chinese, 24% Indian, and 9% were Malay, the rest being of miscellaneous descent. Basically full-term newborns were studied; low birth weight, those requiring intensive care, etc., were excluded. Breast feeding was carefully supervised. Mode of delivery was recorded as vaginal or Caesarean. At birth, cord blood was collected for determination of ABO grouping and for measurement of glucose-6-phosphate dehydrogenase activity.

Results showed a mean gestational age of 39.1 weeks. During hospitalization, 37% of infants were visibly jaundiced, and 27% had serum bilirubin levels over 9 mg/100 ml, and 6% were over 13 mg/100 ml. Twenty-three percent of the 1037 infants were placed on phototherapy. None of the factors evaluated correlated with the hyperbilirubinemia. Bivariate analysis did show that risk factors for jaundice included Chinese ethnicity, breast feeding, blood incompatibilities, and weight loss. Sixty-six percent of jaundiced newborns were of Chinese background, whereas other ethnic groups were not at an increased risk for jaundice. Infants with jaundice had a larger weight loss than those without jaundice. Breast feeding was another factor associated with an increased risk of jaundice.

The authors comment that their study shows that the concept that some East Asians had a higher susceptibility for developing jaundice than other groups. Further, the authors' study narrows this to newborn infants of Chinese origin. Breast feeding was also a factor related to the risk for hyperbilirubinemia, and breast feeding was supervised by one of the authors to be sure it was performed properly. The study also showed that caloric intake and associated weight loss also increased the risk of jaundice. Losing 7% or more of birth weight was associated with a 1.4-fold greater risk for jaundice, and jaundiced infants had a 12.6% greater weight loss. There was a strong correlation between weight loss and breast feeding. Breast-fed infants on average have the maximum weight loss by postnatal day 3, and have a 6.6% weight loss as compared to 3.5% for formula-fed infants. This suggests that it is critical to closely monitor the efficiency of breast feeding.

Infants with glucose-6-phosphate dehydrogenase deficiency were at a threefold higher risk for neonatal jaundice. ABO incompatibility was, of course, a high-risk factor for developing jaundice. Newborns delivered by Caesarean section were afforded some protection from jaundice. This could be due to stress acting to stimulate enzymes, or of an older gestational age. Infants delivered by Caesarian section were usually formula fed before breast feeding was established.

The overall conclusion was that independent risk factors consisted of weight loss, breast feeding in Chinese ethnic newborns, glucose-6-phosphate dehydrogenase deficiency, and ABO incompatibilities. These results raise health care cost issues. The fact that breast feeding improves the health of the newborn should be balanced against the cost of neonatal jaundice. This cost may be based on both

diagnosis and subsequent therapy. The sequelae of hyperbilirubinemia and associated kernicterus may be staggering. Early identification of hyperbilirubinemia and
aggressive therapy – both phototherapy and transfusion if necessary – are essential
to lower mortality and morbidity. The identification of risk factors in Asian populations (and the identification of non-factors) alert health care workers as to groups
at risk. Knowledge of these factors in turn should lead to early diagnosis and treatment. Awareness of these risk factors is especially important in light of the current
"fashion" which encourages breast feeding, and encourages early dismissal from
the hospital. Education of parents as to signs and symptoms of hyperbilirubinemia
is essential.

Jaundice in Malaria

Malaria is a leading cause of death and morbidity in the world. Malaria is caused by the infection of humans by the Plasmodium species of protozoan parasites. *Plasmodium falciparum, Plasmodium vivax, Plasmodium ovale,* and *Plasmodium malariae* represent the parasites which produce malaria. *P. falciparum* is the protozoan which produces the most severe cases of malaria, whereas the other three produce a more mild form of malaria. Historically, malaria was described by early writers in China and India, and Hippocrates described clinical features of malaria in the fifth century BC.

The incidence of malaria is staggering. As many as 200 million cases of malaria occur each year, and over 1,000 cases enter the USA. annually. Worldwide, 1–3 million die each year from malaria. Endemic areas include Tropical Asia, and sub-Saharan Africa. Malaria is transmitted by the female Anapheles mosquito. There have been recent advances in the prevention of malaria, centered on dispersal of mosquito netting, and increased spraying procedures. Nevertheless, the incidence remains high. Some countries have made progress in this area, whereas other countries, because of political and other reasons, have not been cooperative.

There are significant differences in malaria expression in various world endemic areas. For example, in Southeast Asia (Narian, J. 2008), there are about 120,000 deaths from malaria each year. The social, cultural, and economic impact of this level of malaria can hardly be estimated. In this region, Plasmodium falciparum accounts for as many as 50% of all cases.

There are two stages in the life cycle of Plasmodium. The first stage (sexual stage) develops in the female Anopheles mosquito. This stage gives rise to the sporozoites. The second stage occurs in infected humans, initially in the liver, where massive multiplication occurs, then in erythrocytes. Erythrocyte hemolysis increases dramatically, which in turn allows further infection of red blood cells.

The most severe form of malaria is caused by *P. falciparum*. In this form, hemolysis is high, organ systems such as renal, pulmonary, and digestive may be severely damaged. And, the so-called cerebral malaria consequences of *P. falciparum* may be a key factor. Cerebral malaria may present with signs of increased intracranial pressure, seizures, and encephalopathy.

D.W. McCandless, *Kernicterus*, Contemporary Clinical Neuroscience, DOI 10.1007/978-1-4419-6555-4_12, © Springer Science+Business Media, LLC 2011

It is in severe cases of *P. falciparum* that elevated bilirubin levels may be noted. Usually in any of the other three forms of malaria, hyperbilirubinemia may be mild (less than 5 mg/100 ml). In severe cases of *P. falciparum*, due to increased hemolysis and liver failure, bilirubin may reach very high levels (greater than 50 mg/100 ml).

The incubation period for the development of malaria following a bite by the Anopheles mosquito is about 2 weeks. At this time, newborn infants are especially susceptible to elevated bilirubin. The hyperbilirubinemia of *P. falciparum* is mixed – that is, both indirect and direct bilirubin levels are elevated in serum. The incidence of malaria in newborn infants (under 6 months of age) is lower than might be expected. But keeping in mind the overall numbers, that still could translate to thousands of infant cases per year.

The jaundice associated with malaria is due to hemolysis, decreased numbers of functional hepatocytes, and alteration in small bowel function conducive to reentry of conjugated bilirubin into the circulation. Altering the normal function of organ systems, including brain and liver, is the occlusion of capillaries by infected red blood cells. Cerebral malaria is the most serious consequence of infection by *P. falciparum*. That hyperbilirubinemia may play a deleterious role in malaria, especially in children, is undeniable.

The purpose of this chapter is not to examine and review details of malaria, but to try to put into perspective the role of hyperbilirubinemia in the morbidity and mortality of malaria. This task involves some speculation absenting well-controlled clinical studies, modern hospital facilities (including laboratory tests), and compliance/follow up of patients.

P. falciparum malaria is frequently divided into two groups: non-complicated and complicated malaria. Complicated malaria is defined as malaria which presents with one or more of the following clinical features: cerebral malaria, hyperbilirubinemia, renal failure, seizures, hypoglycemia, pulmonary edema, circulatory collapse, or spontaneous bleeding. In a clinical study aimed at examining these symptoms of complicated severe *P. falciparum* malaria, a prospective study of 200 patients was undertaken (Bag, S., et al. 1994).

Results from this study identified 50 (25%) patients who fulfilled the criteria for complicated malaria. This was achieved by a thorough clinical examination, including laboratory workup. Blood smears were carefully examined for *P. falciparum*, a hematological profile was examined, bilirubin was measured, urine examined, bleeding times measured, etc. (see Fig. 1).

Results showed a total of 50 patients qualified for the description of complicated malaria. The group consisted of 36 males and 14 females, age 8 months to 12 years. Of 50 patients, 35 (70%) had evidence of cerebral malaria. Twenty-four patients with cerebral malaria due to *P. falciparum* had seizures. Eight patients with cerebral malaria showed meningeal signs. Eight patients died on the second and third day of quinine therapy. In these moribund patients, unconsciousness lasted 24–48 h preceding death. Of the 15 non-cerebral malaria cases, fourdied, two of circulatory failure, and two died from renal failure.

Jaundice was significant in four patients (8%), a frequency similar to that seen in other studies. One patient had elevated conjugated bilirubin, another had elevated

Fig. 1 A CDC provided blood smear showing falciparum malaria. Note evidence in red blood cells

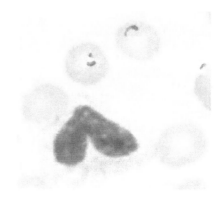

unconjugated bilirubin, and two had elevations of both. The authors note that elevated bilirubin concentrations in plasma might be due to hemolysis of affected red blood cells. Other factors could be decreased transport of bilirubin into hepatocytes. Hepatocyte necrosis is a histological feature in *P. falciparum* malaria.

Cerebral malaria was the most common occurrence in complicated *P. falciparum* malaria. Several factors contributed to the high mortality rate. These included seizures, unconsciousness, and a delay in instituting quinine therapy. The delay in instituting therapy was attributed to the slow response of some patients to recognize the severity of the disease, and to seek medical help (see Table 1).

In a study of liver profile changes and jaundice in *P. falciparum* cases, 390 patients were studied in Thailand (Wilairatana, P., Looareesuwan, S., and

Table 1 Common symptoms of cerebral malaria

Criterion	Patients with admission data	Prevelance	Mortality	Relative risk (95%)
Defining	No.	No. (%)		
Coma	1844	185 (10.0)	31 (16.8)	12.6 (7.2–22.0)
Severe malarial anemia	1816	320 (17.6)	15 (4.7)	1.4 (0.8–2.5)
Respiratory distress	1833	251 (13.7)	35 (13.9)	9.4 (5.5–16.2)
Hypoglycemia	698	92 (13.2)	20 (21.7)	5.4 (2.9–10.2)
Circulatory collapse	1844	7 (0.4)	5 (71.4)	47.7 (2.9–10.2)
Renal failure	1844	2 (0.1)	0	–
Spontaneous bleeding	1843	2 (0.1)	0	–
Repeated convulsions	1842	338 (18.3)	23 (6.8)	2.9 (1.7–4.9)
Hemoglobinuria	1844	2 (0.1)	1 (50.0)	30.7 (1.8–504.6)
Supporting				
Impaired consciousness	1844	151 (8.2)	9 (6.0)	2.0 (0.9–4.1)
Jaundice	1806	84 (4.7)	10 (11.9)	4.6 (2.2–9.4)
Prostration	1571	192 (12.2)	10 (5.2)	1.8 (0.9–3.6)

Adapted from White, N. (1996)

Charoenlarp, P. 1994). Results showed that aspartate aminotransferase and alanine aminotransferase were elevated significantly in jaundiced patients as compared to non-jaundiced patients. About 124 of the 390 patients in this study had hyperbiliurubinemia (32%). The majority of the 124 patients had elevated unconjugated bilirubin levels. The range of total serum bilirubin levels was 3–64 mg/100 ml. These results show that elevated serum bilirubin levels are a significant feature of *P. falciparum*, and signal a complicated form of malaria. Levels of hyperbilirubinemia as high as 64 mg/100 ml clearly will lead to kernicterus when occurring in newborn infants. Remembering an incubation rate of about 2 weeks for *P. falciparum*, an infant infected in the first day or 2 would become jaundiced at a time of significant vulnerability for kernicterus.

Treatment of malaria may be quite straight forward, or be complicated, depending on the type of malaria, the patient's age, relative immunity, and availability of antimalarial drugs (White, N. 1996). The author states that the three "benign" malarias, *P. vivax, P. malariae,* and *P. ovale,* should all be treated with chloroquinine. A 3-day course of treatment is usually effective. A 2-week treatment course may be required in cases of *P. vivex* and *P. ovalve.* This serves to eliminate the parasite from the liver which might survive chloroquinine treatment.

Treatment for *P. falciparum* depends in part on the parasite's sensitivity to antimalarial drugs, which in turn centers on the area of the world where infection occurred. Areas in which the parasite is sensitive to antimalarial drugs, such as North Africa, Central America, the Middle East, should be treated with chloroquinine. Areas resistant to antimalarial drugs such as sub-Saharan Africa are treated with one single dose of a combination of sulfadoxine and pyrimethamine. For particularly resistant *P. falciparum*, the treatment choices are mefloquine, halofantrine, or quinine/tetracycline.

In pregnant women, sulfadoxine given in the third trimester may present a theoretical risk of kernicterus. In newborn infants, sulfadoxine could certainly promote kernicterus. since the drug competes successfully with albumin for bilirubin-binding sites. Unconjugated bilirubin displaced from albumin in the serum readily enters the brain. This method (sulfadoxine administration) can be used in experimental studies in Gunn rats to assure consistent and reliable production of symptoms of kernicterus (see Gunn rat chapter).

The complicated malaria treatment problem is that of *P. falciparum* malaria. Hyperglycemia, jaundice, seizures, anemia, renal failure, and pulmonary edema all may occur in both children and adults, and threaten life. Cerebral malaria is a very ominous symptom. These multiorgan system failures can occur in a matter of hours or a couple days. Seizures may occur in 80% of children, and hypoglycemia occurs in 25%. Hemolysis can be extensive and transfusion should be initiated if the hematocrit drops below 20%. This level of hemolysis produces significant hyperbilirubinemia, with levels exceeding 50 mg/100 ml. Vital signs, development of pneumonia in comatose patients, urinary output, blood glucose levels, and arterial pH should all be monitored often. Following initiation of therapy, the parasite count should be performed, and if not decreasing, alternative antimalarial therapy should be initiated. Finally, exchange transfusion should be considered if the parasite level exceeds 15% This serves to reduce parasitemia, and also to reduce serum hyperbilirubinemia.

The issue of transfusion in cases of *P. falciparum* has been challenged (Mordmuller, B., and Kremsner, P. 1998), who present a study of 113 West African children with a median level of parasitemia of 12%. This group of children were deliberately not transfused. The mortality rate of this group was only 1.8%.

In 1996, a review of transfusions in severe *P. falciparum* was published. The authors state that transfusions should only be performed if the facilities for performing transfusions are good, the patient is quite ill and would benefit from a transfusion, and if the parasitemia is over 15%. These criteria emphasize the importance of considering several factors in the decision to transfuse. The chances for the introduction into the patient of viral pathogens such as HIV must be minimal. The level of parasitemia is important – some studies show a dramatic mortality associated with an increase in *P. falciparum*. The point made in this commentary (Panosian, C. 1998 edit response) is that many factors determine the benefit of transfusion, and all must be considered, especially in children with *P. falciparum*.

A review and prospective study in the *New England Journal of Medicine* (Marsh, K., et al. 1995) examines the effects and outcomes of *P. falciparum* in children. The authors studied 1,844 children diagnosed with malaria. The diagnosis was determined based on parasitemia, and the absence of any other factors which could have produced similar symptoms. A total of 15 criteria based on a WHO report were the basis of clinical assessment.

Children with neurological involvement (cerebral malaria) or with respiratory distress were treated with parenteral quinine. They also received chloroquinine or pyrimethamine/sulfadoxine if able to take oral medications. Children with neurological involvement had a lumbar puncture, and CSF was examined to exclude a meningitis diagnosis. Blood transfusions were performed in patients with hemoglobin levels below 5 gm/100 ml, or if the parasitemia was high enough to suspect a rapid rate of hemolysis (Burchard, G. et al. 1997).

Results showed that of a total of 7,538 children admitted, 40% had *P. falciparum* present in their blood, and 1,866 had a primary diagnosis of malaria. The mean age was 26 months. Eighty-six children died (fatality rate 4.6%). Most deaths occurred within 24 h of admission to the hospital. The incidence of malaria declined in patients over the age of 4 years; males and females were equally affected.

Using the criteria established by WHO, four variables were identified as key prognostic indicators of outcome in childhood malaria. These are cerebral malaria, (impaired consciousness), hypoglycemia, respiratory distress, and jaundice. The patients with cerebral malaria were all deeply comatose. The mortality rate in this group of patients was 12%. More mildly affected patients – conscious, but drowsy – had a lower mortality rate of 5.2%. In this study, severe anemia and jaundice were not statistically linked to mortality. Patients with severe anemia and hyperbilirubinemia received transfusions. Those with transfusions amounted to 35% of anemic patients.

The authors note that while the importance of recognizing cerebral malaria is well known, this study also emphasizes the importance of the presence of respiratory distress. In fact, the authors state that good results can be obtained using only two criteria of those listed by WHO – neurological involvement, and respiratory distress.

In a study examining the pathophysiology of *P. falciparum* in African children, three major clinical findings are associated with most deaths from malaria (Newton, C., Taylor, T., and Whitten, R. 1998). The three factors are cerebral malaria, severe anemia, and hyperpnea. All three of these malaria-associated syndromes can occur alone, or in combination, and can result in rapid death.

The hyperpneic syndrome is best described as consisting of respiratory distress associated with chest recession and/or abnormally deep breathing. In this study, 19% of children in respiratory distress died. In a group of patients judged to be in severe respiratory distress, the mortality rate was 25%. The respiratory distress was assumed to reflect systemic acidosis.

The signs and symptoms of cerebral malaria may be as minor as diffuse cortical involvement, to brainstem abnormalities (altered consciousness and/or coma). Seizures, decerebrate rigidity, opisthotonic posturing, nystagmus, etc. The mortality rate of children with cerebral malaria ranges from 15 to 30% depending on the study (Field, J. 1949; Molyneux, M., et al. 1989; Ramchandran, S., and Pereeria, M. 1976; Snow, R., et al. 2005; Walker, O., et al. 1992). About 10% of survivors leave the hospital with sequelae.

Anemia in malaria is defined in this paper as a hemoglobin concentration below 5 gm/100 ml. This value is coupled to *P. falciparum* parasitemia in excess of 10,000 trophozoites/cubic mm of blood. Mortality rates in malaria infected children range from about 5 to 15%. This incidence is increasing in light of increasing resistance of *P. falciparum* to chloroquinine. Destruction of red blood cells occurs in malaria, possibly by erythrophagocytosis in the spleen. There is evidence that unparasitized red blood cells have shorter life span during malaria infection. Autoimmune hemolysis may also be a component of malarial anemia. It is certain that severe anemia serves to increase unconjugated hyperbilirubinemia to levels where jaundice is evident. Further, some studies, cited earlier, show total bilirubin levels can climb to over 50 mg/100 ml.

In a study of severe malaria in Indian children (Tripathy, R., et al. 2007), 1,682 cases of pediatric malaria were identified in a medical school hospital, and of these, 374 were classified as severe.

The authors note that as many as 1 million childhood deaths occur annually from malaria. Early estimates indicated that as many as 90% of these deaths occurred in sub-Saharan Africa, but more recent estimates show as many as one third occur outside Africa, and most of these are in South and Southeast Asia. In either case, the majority of mortality is among children. The number of childhood deaths attributable to kernicterus is uncertain, but must be in the thousands.

The patients classed as having severe malaria were so classed based on the WHO criteria. All patients in the study received IV quinine, treatment recommended by the WHO. Supportive therapy was provided as required. Diazepam and phenobarbital were given to control seizures. Exchange transfusions and hemodialysis were performed in cases of severe renal failure, and multiple organ failure.

Results showed that the median age of children in this study was 8 years, with a range of 0–12 years old. The distribution of cases based on age was consistent. An overall mortality rate of 12% was recorded, with the highest rate (16.3%) being in the youngest group – 0–2 years of age. The most commonly seen symptoms were coma, prostration, seizures, anemia and impaired consciousness (drowsy/stupor). There was an overlap in these symptom groups. For example, 25 patients had multiple organ dysfunction and coma, and 15 patients had respiratory distress and other organ dysfunction.

Jaundice was found in 20% of the 374 patients classified as severe malaria. Statistically, jaundice was not a predictor of death. Jaundice ranged from 3 to 38 mg/100 ml in this group of patients. This was associated with a mortality rate of 13.5%. This level of mortality rose to 27% with concomitant cerebral malaria and renal failure.

The authors compare and contrast data from their study and other studies of African children with severe malaria. The median age in the present study was 8 years compared to 26 months in a study of African children with severe malaria (Marsh). In Africa, mortality above the age of 4 years was rare, whereas study in India revealed mortality was about equal in frequency in age groups until age 12. In African children, predictors of death were impaired consciousness, jaundice, respiratory distress, and hypoglycemia. In India, predictors were respiratory failure, cerebral malaria, multiple organ failure, and hyperparasitemia. There were significant numbers of patients with overlapping symptoms.

The authors state that other factors such as malnutrition, seizures, shock, bleeding, intravascular coagulation, and impaired consciousness all need to be considered. African studies suggest that malnutrition is an important contributing factor, but the Indian study did not confirm that finding in their series.

The authors note that many cases and deaths occur in villages in developing countries, and thus the complete clinical picture is not available. Results from this study are important from a worldwide perspective, and suggest guidelines in the diagnosis and treatment of children with severe malaria.

In a recent study (Abro, A., et al. 2009), an attempt was made to examine the jaundice associated with *P. falciparum* and associated liver dysfunction. The study was of 105 adult patients positive for *P. falciparum* based on the examination of patient's blood smears. Clinical examination of the patients showed 23% were jaundiced. Serum alanine amino transferase was high in 68%, and in 11% the enzyme was 3 times normal. Bilirubin was elevated in over 80% of patients, and was over 30 mg/100 ml in 23% of patients. The hyperbilirubinemia was both conjugated and unconjugated. In patients with elevated alanine amino transferase, bilirubin levels were also high. Additionally, hyperbilirubinemia patients had thrombocytopenia, anemia, and renal involvement. In all, five patients died, and the mortality rate was higher in patients with increased bilirubin levels.

In a similar study (Srivastava, A., et al. 2009), seven cases of *P. falciparum* were examined who had a rapid onset of disease. Symptoms included fever, jaundice, oliguria, and evidence of cerebral malaria. Bilirubin levels ranged from 1.9 to 30.7 mg/100 ml. Moderate to severe anemia was also present. These seven patients also showed increased liver enzymes (2–4 times normal), and azotemia (creatinine 1.6–7.4 mg/100 ml).

The patients in this study were treated with IV quinine, and supportive therapy. Of the seven patients, three recovered, and four died. Liver histopathology showed Kupffer cell hyperplasia, pigment deposition, areas of necrosis, and in one case, submassive necrosis. The authors comment that health care workers must be alert to malaria associated with acute liver failure. It is important in patients with altered cerebral function, jaundice, anemia, azotemia to look for *P. falciparum*, as early institution of specific therapy may serve to alleviate symptoms, and lead to recovery.

In another study (Shah, S., et al. 2009), 62 adult patients with confirmed *P. falciparum* malaria, and with jaundice were studied. Age of patients was from 13 to 48 years. All were febrile and jaundiced. Serum bilirubin levels ranged from 3 to 24 mg/100 ml. Ten of the patient's bilirubin levels were over 10 mg/100 ml. About one half of the patients had mixed hyperbilirubinemia. In these patients, serum transaminases were 3 times normal. Thirty-five percent of patients showed hepatomegaly, and 25% had hepatosplenomegally. This study confirms that a significant number of patients with *P. falciparum* and jaundice have significant liver involvement which would justify specific therapies used to correct hepatic pathology.

While in this study, the highest bilirubin levels were about 24 mg/100 ml. The authors note that much higher levels have been described (48 mg/100 ml) (Kochar, D., et al., 2003). The predominant elevation of bilirubin is that of conjugated bilirubin, and this is attributed to increased hemolysis and some degree of renal failure. The authors note that hemolysis alone only produces unconjugated bilirubin, which is generally less than 10 mg/100 ml. The hyperbilirubinemia associated with *P. falciparum* is well recognized, especially in South East Asia

The condition called congenital malaria has been described (Emad, S., et al. 2008), although rare, over 150 cases have been recorded. This entity is defined as the presence of malarial parasites in the blood of the newborn. This condition occurs in less than 5% of pregnancies (Fischer, P. 2003). Prenatal and neonatal mortality rates may be as high as 70% (Menendez, C. 1995).

This paper describes a case of a 12–day-old infant with fever, pallor, poor feeding, and jaundice. Examination showed the patient was lethargic, deeply jaundiced, and and anemic. Serum bilirubin levels were 17.5 mg/100 ml, with 3.5 mg/100 ml being direct, the rest unconjugated. Peripheral blood smears confirmed *P. vivax*. The patient's mother also was positive for *P. vivax*.

Treatment consisted of IV glucose and chloroquinine. The patient also was treated with a packed cell transfusion and phototherapy. The patient responded well to this treatment regime, and recovered. The mother was similarly treated with chloroquinine.

The prevention of malaria during pregnancy may be accomplished by treating pregnant women with chloroquinine. The recommended dose is 4× that recommended for non-pregnant adults. That dose may be associated with teratogenicity. Two cases of brain malformations have been reported. However, teratogenicity has not been confirmed in controlled clinical studies. Given the adverse effects of malaria in terms of spontaneous abortions, still births, and premature births, diagnosis and treatment of pregnant women with chloroquinine seems justified.

In a recent perspective study (Mohapatra, M. 2006), the effects of single complication *P. falciparum* vs. patients with multiple complications were compared. This study was mainly devised to compare the two presentations, and to determine the effects on mortality rates in cases with multiple organ involvement.

Results from 608 patients with complicated *P. falciparum* showed that those with only a single complication such as just anemia or hypoglycemia had a 0% mortality rate. This is compared to patients with more than one complication such as cerebral malaria plus renal failure, in whom the mortality rate was 34%. In patients with four complications – cerebral malaria, plus jaundice, renal failure, and anemia, the mortality rate was 67%.

The authors note that this study shows the dire consequences of multiple complications which may develop rapidly within a matter of hours. The study emphasizes the seemingly changing pattern of symptoms over previous decades. The combination of cerebral malaria, jaundice, and renal failure is the most common combination of complications. The progression of complications is rapid usually moving from complication to complication in under 72 h. The mortality rate increases with increasing number of complications. Patients with five to six complications have a mortality rate of 100%.

In a study looking at the changing spectrum of complications of *P. falciparum* in India (Kochar, D. 2006). In this study, complications of *P. falciparum* malaria were compared from results generated in 1994 and again in 2001. Results showed that in 1994, complications were dominated by cerebral malaria (26%), jaundice (11.5%), bleeding (10%), and anemia (6%). Seven years later, the sequence was jaundice (59%), anemia (26%), bleeding (26%), and cerebral malaria (11%).

The significant changes from 1994 to 2001 relate to both jaundice and anemia. While jaundice is probably present in all forms of malaria, it is more severe in *P. falciparum* (Kochar, D). The knowledge of the changing presentations of malaria is critical to rapid diagnosis and treatment, as many other infectious diseases have similar patterns of symptomology.

In a brief comment on jaundice in *P. falciparum* (Mishra, S., Mohapatra, S., and Mohanty, S. 2003), it was noted that jaundice is one of the most common severe complications of *P. falciparum*. Its incidence may be as high as 45% in all cases. Mortality is stated as being higher in complicated cases of malaria associated with jaundice. The authors note that terms such as malarial hepatitis, hepatic dysfunction, and jaundice are frequently interchanged.

Hyperbilirubinemia may be the result of several factors, including hemolysis, hemoglobinopathies, drug-induced hemolysis, hepatic dysfunction with a decrease in conjugation and/or excretion, and glucose-6-phosphade dehydrogenase deficiency. In spite of these possible causes of jaundice, actual levels of total serum bilirubin are usually less than 5 mg/100 ml, but may be over 50 mg/100 ml. Liver enzymes are nearly always greatly increased as well, indicating hepatocyte necrosis. Liver biopsy in malarial patients does not aid in the actual diagnosis. Blood smears and tests are rapid, facilitating quick initiation of treatment.

In a paper (Newton, C., Hien, T., and White, N. 2000), the clinical features of cerebral malaria are discussed. The authors state that *P. falciparum* is the most severe of the malarias, and is responsible for almost all of the deaths and neurological symptoms. When children are affected, ages 0–2 years old usually have severe anemia, whereas older children and adults usually suffer cerebral malaria, as well as other complications described above.

The neuropathological feature most common in cerebral malaria is the presence of capillaries and venules engorged with parasitized red blood cells. At autopsy, the brain may be swollen, and is slate gray. Petechial hemorrhages are visible. It is the sequestration of red blood cells containing parasites which is believed to be responsible for the many symptoms of cerebral malaria (White, N., and Ho, M. 1992). Indeed, the extent of sequestration is a good predictor of the outcome. Cerebral malaria may be thought of as a disturbance of consciousness. In this case, loss of consciousness should last for more than 6 h following seizure activity in order to differentiate it from postictal depression (Warrell, D., Molyneux, M., and Beales, P. 1990).

Clinical features of adult cerebral malaria are generalized as opposed to focal. The patient is febrile and unconscious with a divergent gaze. The relative functional state of other organ systems should be assessed, as the patient is likely jaundiced, anemic, and acidotic. Metabolic acidosis may result from acute renal failure, or lactic acidosis, or a combination of both. Patients may be hypoglycemic, and 20% of children will present with cerebral malaria.

Seizures may occur in as many as 50% of adults with cerebral malaria, but in recent years this frequency may be decreasing (White, N., et al. 1988). The seizures may occur as partial motor seizures, and be quite subtle as in short-term hand movements, and therefore missed by health workers. Status epilepticus is uncommon, multiple seizures are frequent.

The overall mortality rate from cerebral malaria in adults is about 20%. As stated above, the mortality rate is linked directly to the number of complications, and their severity. Mortality is also linked to the availability of good intensive care facilities. Shock may be an important event precipitating death. Most deaths occur within 48 h of admission, but depend on when in the course of the disease medical treatment is sought.

Cerebral malaria in children is usually associated with seizures. Children have a higher rate of seizures than do adults (Newton, C., et al. 1996). African children usually do not develop renal failure or pulmonary edema. Seizures in children are either focal motor or generalized tonic–clonic convulsions. Seizures are associated with a poor prognosis (Brewster, D., Kwiatkowski, D., and White, N. 1990). About 19% of children with cerebral malaria die, and about another 10% are left with neurological sequelae.

Brain swelling is common in African children with cerebral malaria, and many show symptoms of transtentorial herniation. It is clear that increasing intracranial pressure is a feature of cerebral malaria in African children (Newton, C., et al. 1997).

Neurological sequelae in children are associated with protracted seizures and with deep coma. The causes of seqeulae are not clear, but intracranial hypertension, anemia, hypoglycemia, and jaundice all play a role in what must be the resultant structural damage. Toxins produced by the parasites may also contribute to lasting neurological sequelae.

P. falciparum was the most common cause of seizures in children admitted to a hospital in Kenya. While temperature may induce seizures, most *P. falciparum* children had normal temperatures. Eighty-four percent of seizures in children are complex, and usually with a focus. The seizures associated with *P. falciparum* are recurrent.

Management of children with cerebral malaria is similar to that of adults. In children, hypoglycemia and dehydration should be treated quickly and aggressively. Blood transfusions may be needed to correct anemia, acidosis, parasitosis, and jaundice. Specific parenteral antimalaria treatment is the only intervention which will always produce a positive outcome. However, increased resistance to antimalaria drugs placed the successful treatment regimes in jeopardy.

It is apparent that jaundice is a key feature in *P. falciparum* malaria. The hyperbilirubinemia is frequently not particularly high. However, numerous studies show occasional hyperbilirubinemia ranging well over 30 mg/100 ml, to as high as 50+ mg/100 ml. This high level may be overlooked as differences in concentrations as these high levels are not easily seen. These levels of serum bilirubin are clearly associated with kernicterus and liver damage. In many cases of of jaundice in cases of *P. falciparum* malaria in very young children (0–2 years of age), other factors important in the generation of kernicterus may be present. Thus, hypoxia, prematurity, etc., are present and not quantified. Given the enormous number of patients with malaria worldwide, there can be little doubt that thousands of children develop kernicterus each year. And either die, or suffer the ravages of cerebral palsy. As in the case of polio, many die, but many, many more are left with crippling sequelae. These facts tend to be lost due to poor reporting methods, lack of diagnostic tools, lack of familiarity with the development of kernicterus, and even political reasons.

In future years, education of health care workers, and the availability of effective treatment programs utilizing phototherapy will become available. This in turn should act to decrease both the mortality and morbidity of *P. falciparum* and the associated complications of jaundice.

Jaundice in Congenital Hypertrophic Pyloric Stenosis

Congenital hypertrophic pyloric stenosis is usually initially noticed by the onset of vomiting, frequently in the first to third weeks of life. Cases have been reported in which vomiting does not begin until several weeks after birth. The vomiting can be dramatic, and is termed projectile vomiting – vomit may be projected as far as 3 feet. Gastric peristaltic waves may be seen across the abdomen. The waves are most common after feeding, and just before vomiting. Constipation, weight loss, and dehydration are common. Hypertrophy and hyperplasia of the pyloric region contribute to produce a tumor-like mass (see Fig. 1).

The exact pathogenesis of pyloric stenosis is not certain, but embryologically, the gut goes through a stage when the tube is solid cells. Programmed cell death results in the formation of the definitive tubular form of a functional digestive tract. The heart goes through a similar series of steps in which it is a solid mass of cardiac cells. Programmed cell death results in a four-chambered structure, complete with papillary muscles, cordae tendonae,valves, etc. A defective sequence, or absence of programmed cell death, might be imagined to result in pyloric stenosis. These embryological considerations led early investigators to consider pyloric stenosis a birth defect. Jejunal and ileal stenosis have been described, as well as have stenoses throughout the entire digestive tract.

The majority of cases of pyloric stenosis occurs in males, with a ratio of about 5:1 male vs. female. There is a high incidence of both newborns being affected in twins, which favors a congenital origin. The condition is rare in blacks, and has a high incidence in some northern European countries.

From a pathological standpoint, there is hyperplasia and hypertrophy of the lower pyloric antrum and sphincter, such that the pyloris is about 2 times its normal size, and has a hard knot-like consistency. There is a thickening of the inner circular muscle fibers of the muscularis mucosae. The lumen is severely constricted by this thickening to the point of failure of passage of food.

The main clinical manifestations consist of projectile vomiting, abdominal peristaltic waves, constipation, possible dehydration, jaundice, and a palpable mass. The projectile vomiting has been described above. There is apparently no associated nausea since following vomiting, the newborn again will take food. Vomitus may be blood streaked, which could be a warning of a possible massive hemorrhage.

D.W. McCandless, *Kernicterus*, Contemporary Clinical Neuroscience,
DOI 10.1007/978-1-4419-6555-4_13, © Springer Science+Business Media, LLC 2011

Fig. 1 Schematic representation of the stomach showing the location of the pylorus

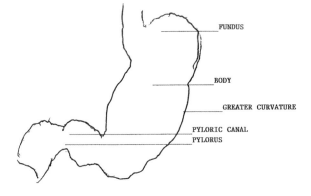

After a few days of vomiting, dehydration may become severe, with a metabolic alkalosis due to loss of gastric hydrochloric acid. Weight loss may be so severe as to result in a weight below that at birth. The pyloric tumor is usually palpable about midway between the umbilicus and the right costal margin. The tumor is usually about 2-cm long, and 1 cm in diameter. Relaxation of the abdominal wall facilitates palpation, and may best be achieved while the newborn is feeding (see Table 1).

From a diagnostic standpoint, if the pyloris is atretic, then vomiting will begin within hours after birth at the initiation of feeding. If the stenosis is lower, in the duodenum or jejunum, then vomitus will contain bile (stenosis distal to the ampula of Vater). Imaging techniques will aid in correct diagnosis, and locate the stenosis. Jaundice must be closely monitored as the predominant bilirubin in serum is

Table 1 Common symptoms of pyloric stenosis

Characteristics	Index cases ($n = 7$)	Historical cases ($n = 40$)	p
Risk factors			
Male sex	6 (86%)	31 (78%)	0.53
Gestational age <38	0	12 (30%)	0.11
Family history of IHPS	0	7 (18%)	0.3
Received erythromycin	7 (100%)	0	<0.001
Received other antibiotics	0	4 (10%)	0.51
Received simethicone, cisapride, or rantidine before IHPS admission	1	7 (18%)	0.66
Maternal history of illicit drug use	0	2 (5%)	0.72
Diagnostic features of IHPS			
Projectile vomiting	2 (29%)	30 (75%)	0.03
Vomiting with feeding	5 (71%)	31 (78%)	0.53
Billous in vomiting	0	1 (3%)	0.85
Blood in vomit	1 (14%)	5 (13%)	0.64
Hypochloraemia	2 (29%)	6 (15%)	0.34
Metabolic alkalosis	4 (57%)	10 (25%)	0.11

unconjugated. Levels of indirect-reacting bilirubin can easily exceed 20 mg/100 ml, a dangerous level in a week or 2 old newborn. Phototherapy should be initiated before unconjugated hyperbilirubinemia reaches these high levels.

The treatment of choice for pyloric stenosis is surgical correction. This procedure should be done as soon as the infant can be readied for surgery. Best results are achieved when there is little or no delay. This surgery can be performed orthoscopically, without the need for more invasive surgical methods. The procedure is to simply split the hypertrophied pyloric muscles without interference with the mucous membrane. Following surgery, feeding is gradually initiated, and a regular schedule is reached after around 4 days post operation.

The jaundice associated with pyloric stenosis, was first described in the medical literature by Martin and Siebenthal (Martin, J., and Siebenthal, B. 1955). Since that time there have been scores of papers examining hypertrophic congenital pyloric stenosis, and its association with jaundice. Pyloric stenosis is readily amenable to surgical correction, but the clinician needs to be mindful of the sometimes rapidly rising jaundice. Unconjugated hyperbilirubinemia can quickly reach dangerous levels, risking the development of kernicterus if unchecked.

In an early review and case presentation (Arias, I., Schorr, J., and Fraad, L. 1959), a couple of cases of pyloric stenosis were presented in which jaundice was a feature. The diagnosis of pyloric stenosis was clear cut, and explanations for the hyperbilirubinemia were sought. Various possible causes were considered including Gilbert syndrome, galactosemia, hemolytic disease, etc., without success. Unconjugated serum bilirubin levels reached over 17 mg/100 ml.

The patients were treated with pyloromyotomy, and the elevated bilirubin levels returned to normal within a few days. In addition to the drop in serum unconjugated bilirubin, bilirubinuria appeared. The fact that the bilirubin returned to normal following surgery argues against Gilbert syndrome being a factor in this clinical presentation. The finding of nearly all unconjugated hyperbilirubinemia argues against any type of biliary obstruction, which could be expected to produce elevated direct bilirubin in the serum. Speculation centered in this paper, on a possible substance excreted in bile which competes with bilirubin for substrate in conjugation. In both cases presented, it was noted that at surgery, the pyloric tumor was seen as angulated posteriorly and laterally in a way that it could have caused compression on the common bile duct. How that anatomic feature might have elevated unconjugated bilirubin in these cases was not obvious.

In another paper (Chaves-Carballo, E., Harris, L., and Lynn, H. 1968), jaundice associated with pyloric stenosis and small bowel obstruction was examined. The authors presented four cases, two with pyloric stenosis, and two with small intestine obstruction. In these cases, unconjugated bilirubin reached over 24 mg/100 ml. In three of the four cases, following surgery, bilirubin levels had returned to normal within about 5 days time.

In another study (Levine, G., et al. 1973), liver biopsy tissue was studied using electron microscopy. The diagnosis of pyloric stenosis was made in five cases using upper gastrointestinal X-ray examination. Open needle biopsy of the liver was made at the time of pyloromyotomy, and prepared for EM.

The results showed that liver function tests were basically normal except for elevated indirect bilirubin. The highest unconjugated serum bilirubin level in these five cases was 8.0 mg/100 ml. Following surgery, all infants' bilirubin levels returned to normal within 5 days.

Results from electron microscopy showed well-preserved architecture except for the presence of dilated endoplasmic reticulum in four of the five cases. These changes were noted in hepatocytes, but not in fibrocytes. There were no alterations seen in the mitochondria, nor were there any signs of cholestasis. Other cell organelles were normal in appearance.

The authors state that they do not believe the dilation of endoplasmic reticulum was an artifact. They state that the dilation of endoplasmic reticulum was the result of an increased protein synthesis rate. The relation between jaundice and ultrastructural change in these infants was not clear.

In another clinical study, (Woolley, M., et al. 1974), 12 infants with hypertrophic pyloric stenosis were studied with respect to their glucuronyl transferase activities. Both jaundiced and non-jaundiced infants were studied. At the time of pyloromyotomy, a wedge of liver was removed for assay and for light microscopy. Hepatic glucuronal transferase activity was estimated in the liver sample.

Results showed that five patients were jaundiced, and seven were not jaundiced. The jaundiced group had glucuronyl transferase activity of between 60 and 80 μg of bilirubin conjugated/gram liver/hour. The non-jaundiced group had enzyme activity between 520 and 800 μg of bilirubin conjugated/gram liver/hour. The microscopic findings were variable and not remarkable as regards jaundice and glucuronyl transferase activity. All patients recovered from surgery and were discharged.

Although the results are from a small group, the authors note that the jaundiced group clearly had lower activity of the conjugating enzyme glucuronyl transferase. The authors state that the reason for the decrease in enzyme in jaundiced pyloric stenosis patients is unclear.

In another case examining jaundice and glucuronyl transferase in pyloric stenosis (Bonnet, J. 1995), a possible relation to Gilbert disease is noted. In this case, the serum bilirubin level was 7 mg/100 ml. A complete blood count was normal. All the usual symptoms of pyloric stenosis (projectile vomiting, etc.) were present. Hypertrophic pyloric stenosis was diagnosed by ultrasonography, and a pyloromyotomy was performed.

A wedge of liver was taken at the time of surgery, and glucuronyl transferase measured. This assay resulted in a value of 2.5 nmoles of unconjugated bilirubin conjugated/gram liver/hour, as compared to a normal value of about 30–40 nmoles/gram liver/hour. Four years later the patient was readmitted for an unrelated G-I problem. At that time fasting serum bilirubin was normal. The liver glucuronyl transferase activity assay was repeated on a fresh sample, and found to be similar to the result 4 years previously.

The authors note that actual estimates of the incidence of jaundice in cases of hypertrophic pyloric stenosis range from about 2 to 8% (Woolley, M., et al. 1974). Other studies examining glucuronyl transferase activity in cases of pyloric stenosis with jaundice have also found low levels of enzyme in liver samples (Labrune, P.,

Mayara, A., and Huguet, P. 1989). Cases of hypertrophic pyloric stenosis in the absence of jaundice show normal levels of activity of glucuronyl transferase. The resolution of the hyperbilirubinemia following surgery has been ascribed to the maturation of the enzyme in newborn infants. The decrease in activity has been explained by coloric deprivation, severe vomiting, dehydration, and possibly to breast feeding (Mitchell, L., and Risch, N. 1993).

Previous studies show a full maturation of glucuronyl transferase activity by about 3 months of age (Onishi, S., Kawade, N., and Itoh, S. 1979). Results from the present study showing quite low enzyme activity 4 years after pyloromyotomy may indicate a permanent deficiency in glucuronyl transferase activity. The authors suggest that hyperbilirubinemia in hypertrophic pyloric stenosis could represent an early symptom of Gilbert syndrome. They are not suggesting that all cases of jaundice in pyloric stenosis represent Gilbert syndrome, only that some Gilbert patients may present as newborns with hypertrophic pyloric stenosis and jaundice. It is also clear that the unconjugated hyperbilirubinemia must be carefully monitored so that proper phototherapy can be initiated when needed (Black, M., et al. 1973).

In another paper on jaundice in hypertrophic pyloric stenosis, an association is made between the jaundice and the possibility of the presence of Gilbert syndrome (Trioche, P., et al. 1999). Molecular/genetic methods were used to lend credence to the possible association between the two.

Three cases were studied, Two were newborn infants, the third was a 30-year-old woman who had a pyloric stenosis surgically corrected at 4 weeks of age. She was noted to be jaundiced at that time. The two newborn infants, both males, were admitted because of projectile vomiting, and had serum unconjugated bilirubin levels of 8.0 and 9.5 mg/100 ml. In each newborn case, pyloric stenosis was diagnosed and a pyloromyotomy performed. The jaundice cleared in 2 days in one case, and in 3 days in the other case, and both were discharged.

In the studies described in this paper, genomic DNA was extracted from leukocytes. Ten patients with pyloric stenosis but without jaundice served as controls. Five exons and the promotor of the bilirubin glucuronyl transferase gene were PCR amplified. The amplified DNA fragments were then sequenced. In the control subjects, only the promoters of the bilirubin glucuronyl transferase gene was PCR amplified and sequenced.

Results showed that one infant and the 30-year-old woman were homozygous for (TA)7 TAA, and the other was heterozygous for (TA)7 TAA. All ten controls were homozygous for (TA)6 TAA. All other features examined (coding sequences, etc.) were normal in all subjects.

The authors note that two of the three patients were homozygous for the (TA)7 TAA polymorphism associated with Gilbert disease, while the third patient was heterozygous for the gene. None of the controls had the polymorphism in the promotor of the bilirubin glucuronyl transferase enzyme. Gilbert disease patients similarly have seven repeats of the TA polymorphism instead of the normal six (Monaghan, G., et al. 1996).

While Gilbert disease is a non-life-threatening syndrome, it may increase the levels of newborn jaundice, and also increase the duration of newborn physiological jaundice (Monaghan, G., et al. 1999). The (TA)7 TAA polymorphism is found in nearly all patients with Gilbert syndrome. The finding of this polymorphism in the three patients in the present study surely indicates that they also have Gilbert newborns with pyloric stenosis disease in addition to jaundice/pyloric stenosis. These data do not say that all jaundiced/pyloric stenosis patients have Gilbert syndrome.

Some investigators have suggested that the (TA)7 TAA homozygosity was a key feature of Gilbert disease, but not the only requirement for diagnosis (Bosma, P., et al. 1995). Other conditions such as a defect in hepatocyte uptake of bilirubin may be essential for having Gilbert disease. The authors also suggest that there is likely a difference between homozygous and heterozygous patients with promotor polymorphism of the bilirubin glucuronyl transferase gene.

Several studies including a recent one (Cooper, W., et al. 2002) have demonstrated a correlation between early exposure of newborn infants to erythromycin and hypertrophic pyloric stenosis. This was a retrospective study examining the incidence of pyloric stenosis in the state of Tennessee over a 12-year period. Records from 314,000 births to parents on Medicaid in Tennessee were examined. Of this group, 804 newborn infants (2.6/1000) had pyloric stenosis and underwent surgical correction.

Results indicated that an early exposure of newborns to erythromycin between days 3 and 13 resulted in an eight fold increased risk of pyloric stenosis. There was no increased risk of pyloric stenosis in infants exposed after 13 days.

This interesting finding had been previously reported (Honein, M., et al. 1999). In this case, a number of newborn pertussis cases occurred, and about 200 neonates in the hospital were administered erythromycin as a prophylactic treatment. A retrospective cohort of infants was used to evaluate erythromycin effects on pyloric stenosis incidence.

Cases of pyloric stenosis associated with erythromycin were compared to a group of pyloric stenosis patients not exposed. Results showed that all cases of pyloric stenosis over a 1-month period had been exposed to erythromycin. This represented a sevenfold increase in incidence. The authors state that the risks and benefits of erythromycin treatment in newborn infants must be considered.

Together, the two papers by Cooper, et al., and Honein, et al. suggest that pyloric stenosis may be an acquired condition. Onset is usually in newborns 3–5 weeks old, although the condition can occur only a few days after birth. Onset is usually characterized by projectile vomiting, while vomiting of a non-projectile nature may proceed the dramatic type by several hours to days. It has been noted that the pyloric tumor may not be palpable immediately after birth in newborns who later are diagnosed with pyloric stenosis.

This then leads to the question of how exactly does pyloric stenosis develop and what is the pathogenesis. The two papers above look at one possible mechanism – erythromycin treatment. These two papers show a clear relationship between pyloric stenosis and erythromycin which certainly flies in the face of the long-standing embryological concept that pyloric stenosis is a birth defect which likely results from a failure in programmed cell death to recanalize the digestive tract in the region of the pyloris.

The studies do not prove why pyloric stenosis develops in erythromycin-treated newborns. The authors state that the pyloric muscle of newborn infants with pyloric stenosis is microscopically similar to that of non-affected newborn infants.

Another hypothesis advanced is that erythromycin has an effect on motilin receptors. Erythromycin is a motilin agonist (Peeters, T. 1993), which acts to induce activity of migrating motor complexes in the stomach. Erythromycin might induce strong contractile activity of the pyloric musculature, leading to hypertrophy and ultimate stenosis. Clearly, the exact cause of pyloric stenosis is speculative, but results do suggest it may be acquired rather than a birth defect.

A recent paper describes the occurrence of hypertrophic pyloric stenosis in identical twins (Yen, J., et al. 2003). In this report, female twins were born at 36 weeks gestation. Laboratory blood data revealed a high serum level of aspartate amino transferase, and low sodium levels in the first twin, and similar results in the second twin. Bilirubin levels were not elevated. Ultrasonography showed a hypoechoic target in the right epigastric area of both twins. Following hydration, both twins had a pyloromyotomy. They were discharged 4 days following surgery.

The authors note that no reported cases in the literature demonstrated liver enzyme alterations or evidence of liver dysfunction except for hyperbilirubinemia. The authors state that the pathogenesis of pyloric stenosis is still somewhat unclear, with some authors suggesting it is acquired (Rollins, M., et al. 1989). Nitric oxide synthetase has been reported to be deficient in nerve fibers in the pyloric sphincter musculature, and this may play a role in the development of pyloric stenosis (Vanderwinden, J., et al. 1992). Another hypothesis is that prostaglandin therapy for maintaining the patency of the ductus arteriosus may play a role in the development of pyloric stenosis (Peled, N., et al. 1992).

As regards treatment, a paper appeared in which a non-surgical approach to hypertrophic pyloric stenosis was successfully tried (Kawahara, H., et al. 2005). In this study, IV atropine therapy was used, and the results were compared to surgical correction.

In the study, the non-surgical approach was utilized in 52 of 85 patients with hypertrophic pyloric stenosis. Atropine was given 6 times per day before feeding. After the cessation of vomiting, and as food volume increased, the atropine was administered orally.

Final results showed that 45 of 52 patients initially placed on atropine therapy had no significant complications, recovered from pyloric stenosis, and their mean hospital was 13 days. Seven of the 52 required surgery to correct the defect. Of a total of 40 who had surgery, four developed infections, and one with hemophilia had postoperative hemorrhagic shock.

The authors conclude that given the longer hospital stay, and possible need for extended atropine therapy, the success rate in this study indicates atropine therapy can be an effective alternative to surgery.

Phototherapy for Hyperbilirubinemia

The advent of phototherapy and its implementation in many countries of the world and its impact on hyperbilirubinemia treatment can hardly be overestimated. Cremer (Cremer, R., Perryman, P., and Richards, D. 1958) and others noted that jaundiced newborn infants exposed to sunlight seemed to appear less jaundiced than those not exposed to the light. Broughton, and especially Jerold Lucey, among others were instrumental in pushing the concept that proper wave lengths of light were capable of lowering serum levels of unconjugated bilirubin (see Figs. 1 and 2).

While phototherapy appeared to decrease serum unconjugated bilirubin levels, there was still the question of whether the photo decomposed breakdown products were potentially neurotoxic. Studies examining these questions were carried out in newborn guinea pigs (Diamond, I., and Schmid, R. 1968).

In these studies, ^{14}C-bilirubin was biosynthesized using bile duct cannulated rats. Radiolabeled photodecomposition products were made by exposing ^{14}C-bilirubin to a mercury lamp. After several hours, the ^{14}C-bilirubin solution was no longer yellow. This preparation was mixed with guinea pig serum and saline prior to infusion into newborn guinea pigs. Some newborns were also infused with ^{14}C-bilirubin, both conjugated and unconjugated.

Results showed that unconjugated ^{14}C-labeled bilirubin, at a dose which significantly exceeded the albumin-binding capacity, left the serum and entered the brain. Serum levels were on the order of 19–26 mg/100 ml. In another group of newborn guinea pigs infused with ^{14}C-bilirubin tightly bound to human albumin, brain entry by the pigment was low. These animals remained neurologically intact, whereas those infused with unconjugated bilirubin which exceeded the albumin-binding capacity developed severe neurological symptoms.

In newborn guinea pigs infused with a mixture of radiolabeled photodegraded bilirubin products, plasma levels were about one half, and brain levels were 7 times less than when 14C-bilirubin was infused. In the group infused with photodegraded bilirubin by-products, there were no discernable neurological signs or symptoms. The photodegraded by-products could have had a higher binding affinity for albumin or some other serum proteins, but the label appeared in urine, and decomposed pigment was more readily dialyzed. This suggests that the photodegraded pigment was water soluble.

D.W. McCandless, *Kernicterus*, Contemporary Clinical Neuroscience,
DOI 10.1007/978-1-4419-6555-4_14, © Springer Science+Business Media, LLC 2011

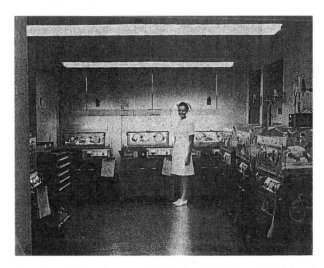

Fig. 1 Picture from an early intensive care unit showing phototherapy administration. Permission from American Academy of Pediatrics, Giunta, F., Pediatr., 47:123–125

Fig. 2 A modern phototherapy unit. Permission from American Academy of Pediatrics, Lucey, J., Ferreiro, M., and Hewitt, J. (1968). See appendix for source

The authors state that previous studies have shown photodegraded bilirubin is converted to diazo-negative compounds which have a higher water solubility than unconjugated bilirubin (Ostrow, J. 1967). This would therefore imply that the water-soluble photodegraded by-products of unconjugated bilirubin would have little access to cerebral tissue. The current study bears this out in that the concentration in newborn guinea pig brain of ^{14}C-bilirubin photodegraded by-products was sevenfold lower than the brain concentration of ^{14}C-bilirubin alone. These studies lend support to the notion that illumination of hyperbilirubinemic newborn infants is a safe and effective means to decrease hyperbilirubinemia, and to reduce the potentially toxic effects of unconjugated bilirubin.

Another study looking at the efficacy and safety of phototherapy involved a total of 111 low birth weight newborn infants who were hyperbilirubinemic. In this study (Lucey, J., Ferreiro, M., and Hewitt, J. 1968), the newborn infants were divided into two groups:a light treatment group and a control (no light) group. Blood was drawn for bilirubin measurement on the first, second, fourth, and sixth days. Newborns exposed to light were placed in the light chamber usually within 12 h or less after birth. Exposure was continued until they were 6 days old. Controls were in similar chambers, but were clothed. The light source was 10 G.E. #20 W daylight bulbs on a wooden cradle which was over the incubator. The newborns' eyes were covered. The infants were under the light continuously except for feeding and obtaining blood. The lights did not alter the incubator temperature. Control and light-exposed infants were on similar schedules as regards feeding, etc. Furthermore, the groups were similar with regards to birth weight, gestation, sex, infection, Apgar scores, and blood group incompatibilities.

Results showed that there was a statistically significant difference between the groups as regards serum bilirubin levels on the fourth and sixth days. The light-treated group had bilirubin levels about one half that of the controls on day 4 (5.6 mg/100 ml vs. 11.7 mg/100 ml), and over one third lower at day 6 (5.3 mg/100 mlvs. 8.2 mg/100 ml).

The authors state that no adverse effects were seen in the light treated group of newborn infants This is in keeping with studies in vitro and in vivo showing no toxic effects of the photodegraded by-product of bilirubin. The present study is important since it was a controlled study in which other factors which might have altered bilirubin levels were equal between the two groups.

The authors note that none of the 53 infants treated with light therapy had a bilirubin level over 12 mg/100 ml. The simplicity of phototherapy means it can be utilized in newborn infants who are sick from other causes such as respiratory distress, sepsis, etc. While more data are needed, the authors state that the light therapy could serve to greatly reduce the need for transfusions, and maintain bilirubin levels low enough to reduce or eliminate the neurological sequelae of kernicterus.

This paper went a long way toward convincing a skeptical group in the USA that there was a probable benefit to phototherapy in hyperbilirubinemic infants. Light therapy had been in use in Europe for many years without any significant adverse effects.

A couple years later, there was an important symposium held to examine phototherapy for hyperbilirubinemia (Behrman, R., and Hsia, D. 1969). Areas of agreement on the part of the participants of the symposium included the effects of phototherapy on bilirubin. They agreed that the effects of light as compared to no light treatment could be distinguished in both Gunn rats and human infants. It was noted that blue violet (440 nm) and yellow green light was most effective in vitro in changing unconjugated bilirubin to water-soluble by-products.

One concern was that since phototherapy seemed to also affect albumin, photo oxidation might decrease the capacity to bind bilirubin thereby acting to move the pigment from serum into brain. In vivo observations on the other hand seemed to indicate that the in vivo capacity of serum to bind bilirubin is not changed by phototherapy. Previous data (Diamond, I., and Schmid, R. 1968) had shown that radioabeled degradation by-products from phototherapy, when injected into Gunn rats did not gain entry into brain.

At the time of the symposium, no ill effects of phototherapy had been described. The argument was made that rather than waiting 5–10 years for controlled prospective studies, phototherapy should be initiated to those who had high levels of unconjugated bilirubin (over 12 mg/100 ml) and not in every jaundiced infant. It was agreed at the symposium that the risks of hyperbilirubinemia vs. possible risks (short and long term) must be weighed on an individual basis. It was further agreed that the etiology of hyperbilirubinemia should be determined, and that the measurement of serum unconjugated bilirubin should be used to assess the progress of treatment.

In another important paper (Ostrow, J.D. 1971), the disposition of photodegraded labeled bilirubin in Gunn rats was examined. ^{14}C-bilirubin was biologically prepared and infused into adult Gunn rats. Following this preparation, the rats were exposed to illumination by six 15 W daylight fluorescent lights. The rats torso was previously shaved to eliminate the masking effect of hair. Urine, feces, and bile were collected and assayed for radioactivity.

Results showed that initially, before the institution of phototherapy, the bile was pale yellow. When the rats were exposed to light, the bile quickly turned a golden brown color. This was coupled with an increase in radiolabel in the bile. Bile flow rate was unchanged. When the lights (phototherapy) were turned off, variables returned to prelight values. These results indicated that phototherapy accelerates bilirubin catabolism and augments biliary excretion of unconjugated bilirubin. In control bile, the ^{14}C- unconjugated bilirubin accounted for less than 5% of the label, whereas in light-treated rats it accounted for 50%.

Spectral and chromatographic analysis of bile showed that the radiolabel was about half- and- half unconjugated bilirubin and water-soluble photodegraded components. In some rats, although light induced an increase in bilirubin breakdown, serum bilirubin levels remained high.

In later studies, Gunn rats were examined as regards the effects of blue light photo therapy on bile and serum components. Rats exposed to light were shaved and exposed for 3.5 h to blue fluorescent lights. Controls were not shaved and not exposed to phototherapy (McDonagh, A., Palma, L., and Lightner, D. 1980).

Results indicated that exposure of Gunn rats to blue light produced a compound which stimulates excretion of unconjugated bilirubin and/or is itself excreted. Thus irradiated rats contain bilirubin and "photo bilirubin," a compound generated by the phototherapy. The same thing seems to occur in vitro (Lightner, D., Wooldridge, T., and McDonagh, A. 1979). The authors speculate that photo bilirubin is not a single component, but a mixture of compounds. It probably contains at least bilirubin 4Z, 15E IX alpha and bilirubin 4E, 15Z IX alpha. These and other bilirubin isomers have overlapping absorption bands. These isomers do not have such strong hydrogen bonds, and therefore are more polar and more hydrophilic.

The authors therefore speculate that with phototherapy near the skin surface, bilirubin is converted to photo bilirubin, a compound with less hydrogen bonding. This new water-soluble substance gains access to the hepatocytes and is excreted into bile without conjugation. The first step, formation of photo bilirubin, is thought by the authors to be a reversible step. The finding that excretion of endogenous photo bilirubin is slower than injected photo bilirubin suggests that photo bilirubin is formed in extravascular spaces, then reenters the circulatory system. This process would be expected to take slightly longer for excretion than if the photo bilirubin was formed in capillaries. The second step, following formation, probably occurs via passive diffusion. The third step, uptake by hepatocytes occurs rapidly. The final step, excretion into bile, is also rapid since conjugation with glucuronic acid is not required. Questions as to why very small amounts of unconjugated bilirubin do appear in bile remain speculative, but may have to do with hydrogen binding.

In another paper (Stoll, M., Zenone, E., and Ostrow, J. 1981), the major mechanism of bilirubin breakdown resulting from phototherapy in the Gunn rat was shown to be photoisomerization. In this study, Gunn rats' femoral veins and bile ducts were cannulated. In phototherapy-treated Gunn rats, six daylight fluorescent lamps emitting 3.2 mW/cm^2 of radiation at 400–500 nM were utilized. Both endogenous and exogenous photo bilirubin were studied. Cannulas were painted black, and bile was collected into dark receptacles.

Results showed that when irradiated ^{14}C-bilirubin was injected into Gunn rats, 76–96% of the label which was non-bilirubin photo products were recovered in a 5-h period in bile. About 20% of this had been reconverted to bilirubin. Unlabeled photo bilirubin, when infused into Gunn rats, resulted in a 5.7 times increase in diazo-reactive compounds, and a nearly 150 times increase in bilirubin excretion. When the two major types of photobilirubins, called photobilirubin 1A and 1B were infused separately, results showed that during preparation for chromatography, most of the photobilirubins had reverted to ^{14}C-bilirubin. By contrast, photobilirubin 2 when infused, was recovered 6 h later as photobilirubin 2, and only a trivial amount had reverted to ^{14}C-bilirubin.

The authors comment that they and others (Onishi, S., et al. 1979) have produced and isolated two unstable photobilirubins (1A and 1B), and two stable photobilirubins. All of these photobilirubin isomers are more polar and water soluble as compared to "regular" unconjugated bilirubin. They are all easily reconverted to bilirubin.

These data, taken with previous results, suggest that photobilirubins are geometric (Z,E) isomers of bilirubin IXa. Their formation accounts for the photodegradation of unconjugated bilirubin both in Gunn rats and in newborn infants. These photoisomers are rapidly formed subcutaneously when exposed to blue light. The photoisomers are subsequently taken up by hepatocytes and excreted into the biliary system. Being unstable, photobilirubins 1A and 1B revert to bilirubin. This accounts for the apparent increased excretion of bilirubin. The more stable photobilirubin 2 remains unchanged after excretion. Taken together, these photodegradation products – bilirubin 1 and photo bilirubin 2 plus minor secondary components account for over 80% of the increased bilirubin catabolism seen during phototherapy of jaundiced Gunn rats and jaundiced newborn infants.

A paper published in Pediatrics (Atkinson, L., et al. 2003) examined the compliance of clinicians to the published guidelines for treatment of hyperbilirubinemia. The guidelines were published by the American Academy of Pediatrics, and were age-specific recommendations for the administration of phototherapy to term newborn infants. This study was a retrospective one in which records from 11 hospitals in the Kaiser program were examined for infants who met criteria for phototherapy. From these records, the numbers of phototherapy treated newborns were determined.

Over 47,000 newborn infants born between 1 January 1995 and 31 December 1996 were examined. Infants of 36 weeks gestation or less than 2,000 g birth weight were excluded. Newborn infants who fell into the classification of no phototherapy, consider phototherapy, or recommend phototherapy, based on serum bilirubin levels were analyzed as to whether phototherapy had been used as treatment.

Results showed that of the total number of infants (47,000), 1086 had received phototherapy (2.3%). Of the 1,086, 59% were in the recommend group, 32% were in the consider group, and 9% were in the no phototherapy recommended group. In the phototherapy recommended group of 1,194 infants, only 54% actually received phototherapy. Increasing serum bilirubin levels was a predictor of phototherapy. There was significant hospital variation in terms of treatment of jaundiced infants with phototherapy. In the group of infants falling into the consider phototherapy classification group, only 16% received phototherapy. Birth weight, gestational age, and hospital were all statistical predictors of whether phototherapy was utilized. Finally, in the group for which phototherapy was not recommended, only 0.2% actually received phototherapy.

The single most significant statistical predictor of use of phototherapy in the recommended group was the hospital of birth. The range for compliance in the recommended group was 27–75%. The authors raise the question of why almost one half of pediatricians did not adhere to the phototherapy recommendations. In one study (Cabana, M., et al. 1999), several barriers to compliance were identified. These included lack of awareness, familiarity, outcome expectancy, self-efficacy, and inertia of previous practice. In this study, the authors speculate that lack of awareness and familiarity are possible since only a couple years had elapsed between the publication of phototherapy guidelines and the study period. Earlier guidelines used lower cut-off points, so lack of awareness of the 1994 guidelines should have resulted in over use of phototherapy. The incidence of kernicterus in the USA has certainly decreased since the reduction of blood incompatability problems, therefore many pediatricians have never seen a case, and so their level of concern might not be high. In addition, some level of hyperbilirubinemia has been judged beneficial because of its supposed antioxidant activity.

The authors note that some pediatricians might believe that if phototherapy is not initiated at the times suggested, it might abate without treatment, or that if it continues, treatment can be started later. Also there is the concept that phototherapy might interfere with feeding, and with the "bonding" between mother and child. Phototherapy and bilirubin assessment might also interfere with hospital release.

As another example of the difficulty in increasing adherence, a Canadian study was quoted in which prior to Canadian phototherapy guidelines were published, 10% of newborns received phototherapy (Escobar, G., et al. 1999). After publication of their guidelines, compliance rose to 17%, and after an educational type interview, compliance reached only 28–31%.

Increased adherence to guidelines depends in part on the belief on the part of the care giver that the treatment is without risk, and is beneficial. With the decrease in kernicterus cases in the USA, and the question of neurodevelopment in hyperbilirubinemic infants, the rationale for phototherapy may not be clear to many practitioners. More research is needed to define long-term sequelae of hyperbilirubinemia, and to try to increase awareness of phototherapy benefits.

In 2004, the Clinical Practice Guidelines for the management of hyperbilrubinemia in the newborn was published (Maisels, M., et al. 2004). This paper represents a thorough description of newborn jaundice, and a framework for the management of hyperbilirubinemia, and for phototherapy and the prevention of bilirubin toxicity in the CNS. While kernicterus is a rare occurrence, it continues to be occurring, and is preventable. Certainly, as stated above, the decrease in ABO and Rh blood incompatibilities has lowered the risk for kernicterus in the USA.

The paper recommends time intervals for bilirubin assessment in newborn infants, and makes further recommendations for newborn infants with other complicating conditions. It recommends follow-up conditions and frequency of discharged hyperbilirubinemic infants. Exchange transfusions are discussed, and recommendations are made for this treatment. In total, this is a very important document which should be read by anyone involved in treating hyperbilirubinemia.

A recent review of hyperbilirubinemia examines early studies on the potential toxicity of unconjugated bilirubin (Wennberg, R., et al. 2006). The authors refer to the 1994 American Academy of Pediatrics guidelines which were laid out for treatment intervention in cases of newborn hyperbilirubinemia. It is noted that the occurrence of kernicterus in the USA is now rare (due to elimination of blood incompatibilities and phototherapy), and also stated that the risks of phototherapy probably outweighed the kernicterus risk when bilirubin levels were below 30 mg/100 ml. Following the 1994 position statement, there has been an upturn in cases of kernicterus, and studies have shown that total serum bilirubin is a poor discriminator of risk for possible free unconjugated bilirubin toxicity.

Current guidelines (2004) stress the importance of parent education about jaundice and their recognizing increasing levels of bilirubin in their newborn infants following discharge from the hospital. In spite of increasing awareness, the 2004 notation states that the risk for toxicity remains low, and continues to recommend total serum bilirubin levels as the benchmark for assessing kernicterus risk. Free, that is unbound unconjugated bilirubin is most likely that which gains entry into the CNS with deleterious results. This fraction of bilirubin is not evaluated when total bilirubin is measured, and several variables influence its level in serum (see chapter "Hyperbilirubinemia Revisited").

The authors state that the goal is to have a mechanism to evaluate the risk for kernicterus without over intervening. If a serum level such as 20 mg/100 ml were used, a large number of newborn infants might be exposed to phototherapy unnecessarily. Esstimates are that about 5,000 newborn infants will exceed the cut off of 25 mg/100 ml and require aggressive phototherapy and/or exchange transfusion.

The authors state that if free unconjugated bilirubin was assessed, this knowledge would permit a considerable reduction in the numbers of newborn infants who would require aggressive treatment, thereby decreasing morbidity and mortality from treatment regimes. This would reduce costs.

The authors note that what is needed is a systematic scientific evaluation of free unconjugated bilirubin. Another factor to be evaluated would be the day-to-day variation in free bilirubin. Current thinking is that the actual levels of free bilirubin are rather stable. While controlled randomized studies cannot be carried out in a pediatric setting because of ethical considerations, retrospective studies might eventually assess the predictive abilities of total bilirubin vs. free unconjugated bilirubin. The auditory brainstem-evoked response (ABR) may currently be the best indicator of bilirubin neurotoxicity (Gupta, A., and Mann, S. 1998; Amin, S., et al. 2001). This method could at least indicate newborn infants in need of follow up. Indications are that sequelae from hyperbilirubinemia are quite rare (Newman, T., et al. 2004) (see the chapter "Auditory Brainstem Responses," this volume).

The authors state that before the above can be achieved, there needs to be a national reporting system in place for infants with very high bilirubin levels (greater than 25 mg/100 ml) and the ability to estimate serum-free unconjugated bilirubin levels. This may mean developing commercial laboratory methodology which would be readily accessible.

The issue of aggressive as apposed to conservative phototherapy in cases of hyperbilirubinemia has been examined (Jangaard, K., Vincer, M., and Allen, A. 2007). In this study, newborn infants weighing less than 1,500 g were studied. This was a randomized study controlled study in which one group received aggressive phototherapy by 12 h of age, whereas in the other group, conservative phototherapy was started when serum bilirubin reached 8.7 mg/100 ml.

The study consisted of 95 newborn infants; 49 were treated conservatively and 46 were in the aggressive treatment group. As regards bilirubin levels, infants weighing less than 1,000 g had lower bilirubin levels when treated aggressively (9.9 mg/100 ml conservative, 8.1 mg/100 ml aggressive). This difference was statistically significant. Other assessed factors were not different between the two groups. Thus, duration of phototherapy, short-term adverse effects, neurological sequelae, and death were all stastically similar between conservative and aggressive phototherapy.

The authors conclude that in newborn infants weighing under 1,000 g, aggressive phototherapy resulted in lower serum bilirubin levels. There were more cases of cerebral palsy in the conservative group, but the difference missed statistical significance. There was also a difference between aggressive and conservative groups as regards verbal development index, with the conservative group having more cases.

This difference just caught stastical significance (0.43). The possible effect on this by faulty hearing was not mentioned.

The authors state that these data are cause for some concern. They state that there appears to be a trend for conservative treatment to result in less favorable neurodevelopmental outcomes. They state that their patient numbers preclude definitive conclusions, and that larger studies should be performed. While the risks for aggressive phototherapy in full-term infants are small, in low birth weight infants, they may be significant and include dehydration, temperature variations, altered electrolytes, etc.

A case study and review of current practice regarding phototherapy was recently published (Maisels, M., and McDonagh, A. 2008). The case centered on a 3,400 g infant with type A Rh-positive blood, same as the mother. All other laboratory tests were normal. At discharge bilirubin was about 7.5 mg/100 ml. Two days later, bilirubin had reached 19.5 mg/100 ml.

A bilirubin level of 19.5 mg/100 ml, in the absence of hemolytic disease or any other complicating criteria, meets the American Academy of Pediatrics guidelines for hospital readmission and aggressive phototherapy. Such treatment should result in a 30–40% drop in serum bilirubin within 24 h. Treatment should be maintained until bilirubin levels reach 13–14 mg/100 ml. Other problems should be treated symptomatically, such as IV treatment for electrolyte replenishment.

In the brief review of phototherapy, the authors point to the 2004 (A.A.P. Subcommittee on hyperbilirubinemia. 2004). The guidelines deal with confounding problems in prescribing phototherapy such as hemolytic diseases, glucose-6-phosphate dehydrogenase deficiency, low birth weiht, gestational age, hypoxia, etc.

The effectiveness of phototherapy depends on the energy output of the light source, and should be monitored with a radiometer. The type of light source also influences effectiveness, but a light with blue emission is most effective. Since light can damage the immature retina, the eyes should always be covered with opaque eye patches. During active hemolysis when bilirubin production is high, bilirubin levels may not drop as fast. Conversely, higher serum bilirubin levels equals more subcutaneous and cutaneous bilirubin available to the blue light. Quite rare toxicity conditions may result from phototherapy, including purpura and bullous eruptions, and complications in patients with cholestasis. Congenital porphyria or use of photosensitizing drugs are contraindicated in phototherapy infants. The authors point out that because of the efficacy of phototherapy for hyperbilirubinemia, exchange transfusions are rarely required.

The above paper drew some criticism from a couple of investigators (to the editor: Kapoor, R. 2008; Csoma, Z., Kemeny, L., and Olah, J. 2008; Ostrow, J., Wennberg, R., and Tiribelli, C. 2008). The critical comments included those that state that the distance of the phototherapy lamp to the newborn's skin should be carefully monitored to prevent skin burns. The second comment stated that phototherapy might be a stimulator of the development of dysplastic nevi. In fact studies in twins have shown that exposure of one twin to phototherapy was associated with a higher number of dysplastic nevi than the twin who did not receive phototherapy for hyperbilirubinemia.

The third comment centers on the use of the total serum bilirubin levels as a predictor of the need for phototherapy. The authors state that it is the free unbound (to albumin) unconjugated bilirubin that is likely to enter the CNS, and produce devastating toxic effects. The authors of this commentary cite increasing evidence and data supporting this contention, and note that guidelines regarding free unbound bilirubin serum levels would improve the selection process for phototherapy. This would reduce the number of infants needing treatment, hospital readmission, and cost.

The authors Maisels and McDonagh respond saying that the distance between the lamp and newborn infant's skin should be closely watched, and the recommendations of the manufacturers closely followed. They also state that dysplastic nevi are a cause for concern. More controlled studies are needed for accurate conclusions. It is noted that Crigler–Najjar infants do not seem to have nevi formation, and these patients receive phototherapy for many years.

Maisels and McDonagh comment that although total bilirubin concentrations are a poor predictor of possible kernicterus, encephalopathy risk increases with increasing serum bilirubin levels. They state that the free unbound unconjugated bilirubin concept awaits validation, and that measurement of this bilirubin moiety is not offered by clinical laboratories in the USA. They say that photoisomers of bilirubin might complicate the interpretation of free unbound bilirubin data in newborn infants. They note that albumin's binding capacity for bilirubin, and bilirubin:albumin molar ratios might be easier to measure (Maisels, M., and McDonagh, A, 2008).

The effects of photo isomers on the measurement of free unbound bilirubin has been investigated (McDonagh, A., et al. 2008). In an attempt to determine the reliability of the peroxidase method for measuring free unbound bilirubin, this recent study focused on the possible influence of photoisomers on the assay. Free unbound bilirubin exists in very small amounts in serum, but it is the free unbound (to albumin) form which may cross into brain, producing toxicity (see the chapter "Hyperbilirubinemia Revisited," this volume).

4Z, 15E bilirubin, a photoisomer of 4Z, 15Z IX was generated by irradiating the non isomer form. This was done in a way in which the total levels of bilirubin were kept constant. About 19–25% of the total bilirubin was in the isomer form following irradiation. Free unbound unconjugated bilirubin was measured by the peroxidase method in a solution of pure 4Z, 15Z bilirubin, and in a solution containing photoisomers.

Both bovine serum albumin solutions as well as human albumin solutions were evaluated. Results showed that different photo isomers were produced following irradiation of solutions of bovine albumin vs. those of human albumin. In addition, the change in peroxidase measureable free unbound bilirubin levels was changed in inconstant ways depending on the photoisomers present.

The authors speculate that the measurement of free unbound bilirubin by the peroxidase method may lead to conflicting results depending on the levels and nature of photoisomers instead of actual levels of free unbound unconjugated bilirubin. Thus, the authors suggest that these impurities in bilirubin may alter experimental results using potentially impure bilirubin solutions. These impurities could also alter results looking at effects of phototherapy on bilirubin levels in newborns.

A recent paper describing use of Kaiser records, examined numbers of newborns, and the categories of jaundiced babies were evaluated (Newman, T., et al. 2009). The purpose was to look at the efficacy of phototherapy. In this study, 281,898 infants with birth weights over 2,000 g, born at 35 weeks gestation or more were included. This represented births at 12 Kaiser hospitals over a 10-year period. From this group 22,547 infants had bilirubin levels within 3 mg/100 ml of the value suggested by the 2004 pediatrics guidelines as qualifying for phototherapy.

Of the 22,547 eligible jaundiced infants, 5,251 (23%) received phototherapy within 8 h of reaching a qualifying level of hyperbilirubinemia. From this group, 354 infants exceeded the guideline for blood exchange transfusion. About 187 newborn infants from this group had rapidly raising serum bilirubin levels such that the threshold was reached by 48 h or less. Hospital phototherapy was judged to be highly effective. In infants with a positive direct antiglobulin test, hospital phototherapy was less effective than in the rest of the group. There was a significant difference in the need to treat with phototherapy between male and female newborns. The female population had 52% more patients needing phototherapy than male infants.

The authors point out that phototherapy is highly effective in the hospital setting. The groups needing phototherapies varies according to which subgroup of infants is being considered. For a review of optical features of blue light phototherapy units, see McCandless, J., and McCandless, D. 2009.

Non-phototherapy Treatment

Several treatment regimens for hyperbilirubinemia not involving phototherapy have evolved over many years. Phototherapy-based treatment is described in the next chapter. Of the non phototherapy-based treatments, some have been directed at stimulating or increasing conjugation of bilirubin, while others are attempts to reduce the levels of circulating unconjugated bilirubin. In either case, the goal is to reduce and maintain unconjugated bilirubin levels below that which puts the patient at risk for bilirubin entry Into the brain and subsequent neuropathological consequences.

An early attempt to reduce serum levels of unconjugated bilirubin involved administration of phenobarbital. In one early controlled study (Stern, L., et al. 1970), 84 newborn infants were included in the investigation. Phenobarbital (8 mg/kg/day) was given to 20 full-term newborn infants. These infants showed significantly lower serum unconjugated bilirubin levels than did 20 non-treated controls. This result occurred after only 4 days of treatment, on the fifth day of life.

A second group of 10 newborn infants received the same dosage from the fifth to ninth day, also had significantly lower serum levels of unconjugated bilirubin as compared to ten control infants. In the case of the second group, hyperbilirubinemia was present on the fifth day, at the onset of phenobarbital treatment.

In yet another group of newborns, the effect of phenobarbital on the in vivo conjugation of salicylamide was studied. Results showed that in ten infants pretreated with phenobarbital, their ability to conjugate the substrate (salicylamide) was greater than in a similar group of 14 newborn infants not pretreated with phenobarbital. The authors state that taken together, their data suggest that the effect of phenobarbital is one of stimulating the enzymatic reaction in which glucuronic acid is attached to bilirubin forming bilirubin diglucuronide, readily excreted into bile by the liver.

Other early studies had shown similar results in which phenobarbital produced an effect on total hyperbilirubinemia in newborn infants (Maurer, H., et al. 1968; Trolle, D. 1968). Several subsequent reports (Crigler, J., and Gold, N. 1969) on the efficacy of phenobarbital on glucuronyl transferase have been described elsewhere in this volume. phenobarbital was also shown to increase the hepatic clearance of bromosulphalein (Yeung, C., and Yu, V. 1971). Although phenobarbital acts to stimulate enzymatic conjugating mechanisms (glucuronyl transferase), it may also act to increase bilirubin transport into the hepatocyte (Black, M., Fevery, J., and Parker, D. 1974).

D.W. McCandless, *Kernicterus*, Contemporary Clinical Neuroscience,
DOI 10.1007/978-1-4419-6555-4_15, © Springer Science+Business Media, LLC 2011

In a paper describing phenobarbital therapy for hyperbilirubinemia in anemia, the small-term effects of the therapy were examined. In this case study, a newborn boy was jaundiced after birth, the levels increasing to a maximum of 6.1 mg/100 ml, then decreased to 1.0 mg/100 ml by day 5. His hemoglobin was low, and by 2 weeks of age had reached 5.1 gm/100 ml, necessitating a transfusion (Schroter, W. 1980).

At the age of 5 months, the patient was diagnosed as having erythrocyte glucose phosphate isomerase deficiency. At this time, his anemia was moderate (hemoglobin: 8.5 g/100 ml). Unconjugated hyperbilirubinemia was 9.7 mg/100 ml, total bilirubin was 10.2 mg/100 ml. The patient required occasional transfusions for severe anemia and jaundice. The jaundice rarely exceeded 11.0 mg/100 ml. At the age of seven, the patient had a splenectomy, after which no further transfusions were required, and hemoglobin stabilized between about 10.5 and 11.5 g/100 ml. Hyperbilirubinemia remained elevated at levels of 8–9 mg/100 ml.

At 6 years of age, the patient had an attack of colicky abdominal pain, resulting from a large gallstone in the cystic duct. At this time phenobarbital was started for hyperbilirubinemia. A low phenobarbital dose of 2 mg/kg/day resulted in a decrease in bilirubin from 9.8 to 1.8 mg/100 ml in just a few days.

There were no more peaks in bilirubin levels, which remained around 2.0 mg/100 ml. Phenobarbital treatment was continued for over 3 years without any problems. A pigmented calcium stone was removed without any problems surgically at age eight.

The authors state that phenobarbital treatment in this patient resulted in the decrease in hyperbilirubinemia in only a few days. At one point, therapy was discontinued for 3 weeks, and bilirubin levels rose. With reinstitution of phenobarbital, bilirubin levels again dropped. These studies demonstrated that a low dose of phenobarbital can be sustained for as long as 3 years with no apparent adverse effects.

In another paper (Cohen, A., et al. 1985), the effects of phenobarbital combined with phototherapy were examined experimentally. Homozygous jaundiced Gunn rats were administered phenobarbital at a dose of 60 mg/kg/day for 7–10 days. Plasma bilirubin levels decreased from 25 to 35% as compared to controls. When ^{14}C-bilirubin was injected as a tracer, total bilirubin was unchanged, but there was a shift in the pool location into the liver. This amounted to a 50% increase in liver, and a 27% decrease in the cutaneous pool of bilirubin.

Subsequent studies showed that a combination of phenobarbital and phototherapy in jaundiced Gunn rats resulted in a greater decline in hyperbilirubinemia than either treatment alone. The authors state that their studies on Gunn rats show that phenobarbital shifts bilirubin into the liver, presumably for conjugation. The studies also suggest that a combination of phenobarbital and phototherapy could be efficacious in patients with Crigler–Najjar syndrome.

In another study (Okuda, H., et al. 1989), the effects of phenobarbital were examined on liver glutathione-S-transferase activity, and on ligand levels in Sprague-Dawley rats. Liver cytosol ($100,000 \times g$) was utilized in the studies. Phenobarbital was administered at various doses from 1 to 125 mg/kg/day for 6 days.

Results showed that the ligandin levels increased in a dose-dependent fashion. Ultimately, levels of ligandin reached a threefold increase over non-treated controls. Glutathione-S-transferase activity against 1-chloro-2,4-dinitrobenzene also was shown to increase as compared to controls regardless of the phenobarbital dose administered. These enzymatic activities were correlated with immunoreactive ligand levels. The cytosolic glutathione-S-transferase activity, however, did not correlate with actual ligandin concentrations. The authors note that these studies suggest that phenobarbital (in vitro) is capable of inducing immunoreactive ligandin concentrations and associated enzyme activities at very low doses. These doses which were successful in induction were as low as 5% of doses previously thought necessary.

In another rather interesting paper (Yin, J., Miller, M., and Wennberg, R. 1991), the induction of bilirubin-metabolizing enzymes by phenobarbital was compared to induction by a Chinese medicine called *yin zhi huang*. In these studies, phenobarbital was administered at a dose of 60 mg/kg/day to adult rats, and *yin zhi huang* was administered at a dose of 30–60 ml/kg. All rats were treated for a period of 5 days. Animals were sacrificed and liver was harvested and prepared for analysis.

Results showed that both phenobarbital and *yin zhi huang* increased the clearance of unconjugated bilirubin from serum, and stimulated conjugation. The biochemical effects between the two treatments was, however, different. Phenobarbital acted to increase cytochrome P-450, while *yin zhi huang* had no such effect. With bilirubin as substrate, both phenobarbital and *yin zhi huang* acted to increase activity of glucuronyl transferase. Bilirubin conjugation in liver microsomes activated by uridine diphosphate-*N*-acetyl glucoseamine was greater in *yin zhi huang*-treated rats than in rats treated with phenobarbital. *Yin zhi huang* was also more effective than phenobarbital in increasing glutathione peroxidase activity. The authors comment that *yin zhi huang* seems to be an effective inducer of bilirubin-metabolizing enzymes.

In another study examining the effects of phenobarbital administration to jaundiced rats, the ability of brain to metabolize bilirubin was compared in treated and control animals (Hansen, T., and Tommarello, S. 1998). The study was undertaken to see if brain mitochondrial membranes might be contributing significantly to bilirubin reduction. Clearance as well as oxidation of bilirubin by brain mitochondrial membranes was examined.

Animals were treated with phenobarbital for 7–8 days, then anesthetized and infused with 50 mg/kg of unconjugated bilirubin. Sixty minutes later, rats were sacrificed. Bilirubin was measured and mitochondrial membranes prepared by differential centrifugation. The mitochondrial membrane suspension was assayed for its ability to oxidize bilirubin.

Results showed that bilirubin levels were slightly lower in phenobarbital-treated rats. Brain levels of bilirubin were also lower in treated rats, but again, not by much. The oxidation of bilirubin in brain was actually significantly decreased in phenobarbital-treated rats as compared to controls. The authors conclude that phenobarbital overall had no effect on brain bilirubin clearance rate. The authors speculate, based on these experiments that phenobarbital was not effective in increasing bilirubin metabolism in rat brain, and that it also was not harmful.

The implication is that phenobarbital treatment of increased bilirubin levels in patients may not have any deleterious effect in the brain tissue of patients.

In another study (Seppen, J., et al. 1996), the ability of Gunn rat fibroblasts to conjugate bilirubin was examined. The authors say that the only defect in Crigler–Najjar syndrome is an absence of glucuronyl transferase activity. The idea (not new) was that cells (fibroblasts) might be transplantable such that, if the enzymatic machinery was present, bilirubin metabolism could proceed, thereby correcting the deficiency.

These investigators isolated Gunn rat fibroblasts and transduced them with a recombinant retrovirus. A cell line was developed which expressed bilirubin UDP-glucuronosyl transferase comparable to that in normal control hepatocytes. In tissue culture, at levels of bilirubin of 5–10 µM., the prepared fibroblasts and hepatocytes conjugated and excreted conjugated bilirubin at comparable rates. When bilirubin levels were higher (20–80 µM), the hepatocytes were more efficient at metabolizing bilirubin.

The authors conclude saying that glucuronosyl transferase can be expressed in fibroblasts, and after that, the cells can completely metabolize unconjugated bilirubin, including uptake and excretion. This suggests that whatever factors are necessary for uptake and excretion, the factors are expressed by the fibroblasts.

The specific activity of UDP-glucose dehydrogenase is lower in fibroblasts than in hepatocytes, and that in itself, may explain the ability of hepatocytes to "handle" higher concentrations of bilirubin than can fibroblasts. Another feature from this study was that the activity of glutathione-S-transferase was lower in transduced Gunn rat fibroblasts than in hepatocytes. Nevertheless, glutathione-S-transferase may bind bilirubin, thereby reducing its toxicity. While its presence in fibroblasts is lower than in hepatocytes, it may also contribute to the ability of the fibroblasts to lower unconjugated bilirubin levels. In total, these results suggest that the transplantation of transduced fibroblasts back into the donor might lower serum bilirubin, and could suggest possibilities for extrahepatic gene therapy(see chapter on Gene Therapy).

In a recent overview of various questions concerning hyperbilirubinemia (Stanley, I., et al. 2004), the issues and risks of blood exchange transfusions for elevated bilirubin levels are examined. This was a retrospective study in which 15 previously published studies relating to transfusion were included. This review is extensive and well done. Several questions regarding hyperbilirubinemia are addressed, and the reader is encouraged to obtain a copy of this paper.

As regards transfusions, patients were divided into three categories. Category 1 contained patients who were term births without evidence of disease other than jaundice. Category 2 contained 10–50% patients that had Category 1 characteristics. Category 3 consisted of patients (90%) who who were over preterm or term and whose preexchange clinical condition was not stable and/or who had other clinical problems besides jaundice. One feature noted was that transfusion is no longer a mainstay in treatment, so its use now seems to be increasingly for preterm, very low birth weight at risk newborn infants.

The total number of transfused patients reviewed from the quoted 15 published studies was over 7,000. The overall mortality rate for all patients ranged from 0 to 7%. The reviewers of these 15 studies note that attributing mortality to blood exchange transfusions is difficult because many infants had serious underlying medical problems including kernicterus, and death could have been due to factors other than transfusion. Infants dying within 6 h following transfusions are often thought of as dying as a result of the transfusion per se, and using that criteria, the 6-h mortality ranged from 3 to 19/1000 cases transfused. If hemolytic diseased infants were excluded, the "6 h mortality" rate for blood exchange transfusions drops to 3–4/1,000 cases. These cases of mortality were from older literature, and the patients were born before 1970. This reflects changes in treatment patterns.

In terms of morbidity, many complications can be attributed exchange transfusion. Again, however, these are infants who in many cases are acutely sick, and may have other major underlying medical issues in addition to hyperbilirubinemia. Complications associated with blood exchange transfusions include infections, hemolysis, acidosis, cardiorespiratory reactions, serum potassium and sodium alterations, apnea, cardiac arrest, etc. In many cases the morbidity was minor, for example, transfusion site infections. In studies examining morbidity rate, it varied between about 5–7% in all infants studied.

In a recent study (after phototherapy was established), the morbidity rate was 3%. This morbidity rate was confined to exchange transfusion patients who were defined as "sick" as opposed to healthy. In the same study (Jackson, J. 1997), the mortality rate was 4.7%. Thus, in the sick group of 25 patients, the combined mortality/morbidity rate was as high as 32%.

Again, the authors point out that phototherapy is improving, and the result is that blood exchange transfusions are less and less required. Procedure such as transfusions should be performed by a team thoroughly familiar with the procedure. Since it is now infrequently performed, that alone may account for an increase in mortality and morbidity. On the other hand, more modern monitoring devices to assess potential problems may counter the relative unfamiliarity with blood exchange transfusions.

Another issue with blood exchange transfusions is whether to use single or double volume exchange transfusion in cases of hyperbilirubinemia. The author (Milligan, T., 2006) say that usually twice the blood volume of the newborn infant is exchanged with donor blood. Assuming that the morbidity/mortality rates of transfusion increase as the volume exchanged increases, this paper examines adverse effects of double vs. single exchange transfusions. The idea was to see if single exchange was as effective as double exchange in lowering hyperbilirubinemia, and at the same time reduce risk.

This was a retrospective study, examining literature results. Only one study fulfilled the criteria of being a randomized control trial. In this study (Amato, M., et al. 1988), 20 full-term newborn infants who were candidates for blood exchange transfusions due to ABO incompatability were randomly divided into two groups. The groups were those that received a single volume transfusion, and those receiving a double volume transfusion. The volume of newborn blood is taken as 80–90 ml/kg.

A single exchange is defined as replacing about 80–90 ml of blood, whereas a double exchange transfusion is taken as replacing 160–180 ml of blood. Base line data such as birth weight, bilirubin levels, etc., were similar between groups.

Results showed no difference in total bilirubin levels immediately following blood exchange transfusion between the two groups. The authors conclude that, based on this controlled study, there is no clear evidence to indicate suggesting that the single volume technique is safer or more effective in lowering hyperbilirubinemia than is the double volume transfusion technique.

The author comments that exchange transfusions are commonly used in blood incompatability cases in newborn infant nonimmune anemias, and disorders such as glucose-6-phosphate dehydrogenase deficiency also qualify as conditions possibly needing exchange transfusion. Early studies showed that before the development of blood exchange transfusion, the mortality rate was over one third, and there was a 90% risk of neurological sequelae. Now, with aggressive therapy, including phototherapy for hyperbilirubinemia, blood exchange transfusions in the USA are uncommon.

In Third World countries, where medical care is rudimentary, and knowledge of the perils of hyperbilirubinemia are not widely recognized, referral may to medical care is likely to be late. Once kernicterus has developed, exchange transfusion may not be in time to prevent permanent neurological damage. Entry of unconjugated bilirubin into brain may occur at lower serum levels in premature infants, or newborn infants with other problems such as hypoxia. These confounding problems occur in Third World countries at a high incidence.

The author suggests that the mortality rate from blood exchange transfusions (1%) is frequently associated with cardiac arrest, and/or air embolism directly from the process of transfusing the patient. It would seem that the time required for a double blood exchange transfusion would increase the risk for problems. Since improved treatments other than blood exchange transfusions have improved the outcome for jaundiced newborns, the risk/benefit of blood exchange transfusions should be reexamined.

In another paper looking at various methods for reducing hyperbilirubinemia in newborn infants, blood exchange transfusions were discussed (Akobeng, A., 2005). The author of this paper states he could not find any randomized controlled trials on blood exchange transfusions vs. either no treatment, or phototherapy. It is difficult, given the efficacy of phototherapy, and the patient population, to imagine any such study now. The author states that the general opinion regarding blood transfusions is that it is an effective means of lowering serum hyperbilirubinemia, and the effect is of course rapid. The rapidity of lowering serum bilirubin levels may be beneficial from the standpoint of neurological and developmental sequelae. If serum unconjugated bilirubin levels (or free bilirubin levels) are such that neuropathological damage is being actively done, then stopping that process immediately is critical. Phototherapy can be fast, but never as fast as blood exchange transfusion.

In a more recent paper (Newman, T., et al. 2006), outcomes as regards neurological and developmental sequelae were examined in a group of 140 infants who had total serum bilirubin levels of over 25 mg/100 ml. Both phototherapy and blood exchange transfusions were employed as treatment in all 140 patients. The breakdown was that 135 newborns received phototherapy and 5 received exchange transfusions. In all cases, outcomes were assessed using hospital records, interviews, questionnaires, and various testing modalities, including the Wechsler intelligence scale and the Beery- Buktenica developmental test.

The major thrust of this paper was to evaluate possible intellectual/developmental impairments from hyperbilirubinemia, and this will be examined in another chapter (see the chapter "Neurobehavioral Teratology," this volume). At the same time, a comparison was made between phototherapy and blood exchange transfusion. Results between the two treatment groups showed no measurable differences. Bilirubin levels in the transfusion group were greater than 25 mg/100 ml. Phototherapy was also used as the only treatment modality in infants with bilirubin levels over 25 mg/100 ml. There were no cases of kernicterus in the 140 newborn infants with hyperbilirubinemia, and the authors note that even with bilirubin levels at or over 25 mg/100 ml, treatments – either phototherapy or transfusions – were successful, and there was no demonstrable morbidity.

In a brief review of the above paper (Watchko, J. 2006) several features are emphasized. First, in the Newman study, newborns were carefully monitored, such that the duration of hyperbilirubinemia above 25 mg/100 ml was less than 6 h in 75% of newborns, and exposure to levels over 20 mg/100 ml was less than 24 h. in one half of patients. In addition, most of the newborn infants were term babies. All infants in this study (140) whether treated with phototherapy or blood transfusions ultimately did well. These findings therefore, lend support to the recommendation of the American Academy of Pediatrics. that hyperbilirubinemia should be kept below 25 mg/100 ml.

In an overview of treatment for hyperbilirubinemia (Springer, S., and Annibale, D. 2008) several methods are discussed. The method of double blood exchange transfusion is recommended when hyperbilirubinemia exceeds 25 mg/100 ml, or is rising rapidly, or continues to rise in spite of other treatments (phototherapy). Increase in serum unconjugated bilirubin of more than 0.5–1.0 mg/100 ml per hour is a signal that transfusion may be needed.

The risks associated with blood exchange transfusion are significant, and not to be taken lightly. These risks include infection – both from the transfused blood itself, and at the transfusion site, anemia, apnea, cardiac arrest, air embolism, etc. Usually two transfusions are required. The mortality rate from the transfusion alone may be 3/1000, and the morbidity rate 5/100. Critically ill newborn infants carry a higher risk factor.

Several other treatments have been tried in hyperbilirubinemia in addition to blood exchange transfusions and phototherapy. These include phenobarbital administration to possibly increase conjugation and excretion of unconjugated bilirubin. This has been discussed above in this chapter. Albumin would appear to be a possible treatment modality since its binding to bilirubin prevents unconjugated bilirubin from entering brain. This method of treatment is not recommended because of possible side effects. It should be noted, however, that Waters (Waters, W. 1961) showed that albumin prevented the development of kernicterus in newborn puppies without any adverse effects from the treatment. The data showed albumin protected the puppies from bilirubin toxicity.

The parenteral administration of immunoglobulin G may act to decrease the need for transfusions in blood incompatibility cases. Accelerated meconium evacuation with glycerin suppositories has been tried with somewhat equivocal results. This may possibly work by decreasing transit time, thereby reducing the opportunity for enterohepatic circulation of bilirubin. Another possible treatment involves the administration of Sn-mesoporphyrin, which acts to inhibit bilirubin production. It is thought to act by blocking heme oxygenase thereby decreasing the catabolism of hemoglobin. This treatment modality needs further evaluation.

Blood exchange transfusion, although associated with clear risks, can prove highly beneficial in cases of hyperbilirubinemia which do not respond satisfactorily to phototherapy. The risks are lowered if done in settings with extensive familiarity with the procedure. An additional benefit of exchange transfusion is that additional albumin, ready for binding to bilirubin is made available. New treatments are being developed, and after evaluation, may prove beneficial. Given the nature of the patient and ethical considerations, controlled clinical trials of existing treatment modalities are scanty.

Hyperbilirubinemia Revisited

In recent years there have been numerous papers published reexamining various concepts regarding unconjugated bilirubin and its apparent toxicity. Overall, results of these studies have shown that measurement of serum levels of unconjugated, conjugated, and/or total bilirubin may not be adequate for assessing the risk for kernicterus in newborn jaundiced infants. Especially measuring total bilirubin does not assess the contribution of unconjugated and conjugated bilirubin. It is assumed that in newborn infants the majority is unconjugated. Also not evaluated by either method is the contribution of "free" unbound to albumin bilirubin. It is this moiety of bilirubin, free and unbound to albumin which can move easily and rapidly into brain tissue.

Earlier studies had shown the importance of using a highly purified form of bilirubin in any study of unconjugated hyperbilirubinemia (Ostrow, J., Hammaker, L., and Schmid, R. 1961). Using such highly purified unconjugated ^{3}H labeled unconjugated bilirubin, the ATP-dependent transport was studied in rat liver canalicular plasma membrane vesicles (Pascolo, L., et al. 1998).

It was found that purified free unconjugated bilirubin transport was highly ATP dependent. ADP, AMP, GTP, etc., did not stimulate bilirubin transport. At a very low unbound bilirubin level (less than 70 nM unconjugated bilirubin) the ATP-dependent transport showed saturation kinetics. Labeled unconjugated bilirubin uptake was inhibited by unlabeled bilirubin, showing that the species being transported was indeed unconjugated bilirubin. When ATPase was inhibited, the stimulatory effect of ATP was inhibited as much as 100%.

Other studies described in this paper showed that neither mdr-1 nor the canalicular bile acid transporter play a role in the transport of free unconjugated bilirubin. Uptake of labeled bilirubin in TR rats was similar to that seen in Wistar rats, suggesting the canalicular multispecific organic anion transporter was not involved in bilirubin transport by plasma membrane vesicles. Taken together, the results of this study suggest that unconjugated bilirubin transport across the canalicular membrane in rat liver is highly ATP dependent; however, the exact nature of the transporter is unclear.

Another study emphasizing the importance of measuring unbound (to albumin) levels of unconjugated bilirubin in jaundiced newborn infants was performed (Weisiger, R., et al. 2001). In this paper, the concept that the binding constant of

D.W. McCandless, *Kernicterus*, Contemporary Clinical Neuroscience,
DOI 10.1007/978-1-4419-6555-4_16, © Springer Science+Business Media, LLC 2011

albumin/bilirubin in serum is independent of the actual concentration of albumin was examined.

In terms of materials/methods, radiolabeled bilirubin was collected by a bile duct cannula following infusion of ^{14}C-aminolevulinic acid. As much as was possible, all steps were carried out in the dark. Any illumination was provided by red lamps with no emission below 600 nm. This process produces no photodegradation of bilirubin when in solution with albumin, maintaining purity of the ^{14}C labeled sample. Purified bilirubin was stored in sealed vials in the dark until use.

Results from this study showed a very small level of impurities of bilirubin (0.13%). Knowledge of these values allowed estimation of the true binding affinity over a wide range of albumin concentrations. The authors state that they show the albumin binding of bilirubin decreases as albumin levels rise. The estimation of the degree of reduction of true binding affinity in this study is attributed to use of purified albumin, correction for poorly bound impurities, and differences in buffer concentrations. The binding affinities of albumin is reported previously to vary with its concentration for other metabolites and drugs (Fenerty, C., and Lindup, W., 1989; Clegg, L., and Lindup, W., 1984).

The reason for decreased albumin binding may relate to the self aggregation of albumin at increasing concentrations, thereby decreasing binding capacity for bilirubin. Binding was decreased in the presence of KCl as opposed to the presence of sucrose. The range of KCl which decreases bilirubin binding has not been determined.

The authors point out that the increase in binding of albumin to free bilirubin required correction of the binding constant for impurities in bilirubin. In spite of prevention, there were some impurities formed by photoxidation and photoisomerization. Serial ultrafiltration was utilized to remove impurities, and this proved to be effective. Ultrafiltration membranes have improved such that almost no leakage of bilirubin bound to albumin across the membrane occurred which might compromise the data.

The clinical implications of the results of this study are that the constant for albumin binding to free unconjugated bilirubin are less than one half that previously reported. The presence of fatty acids act to increase the affinity of albumin for bilirubin, and these may have been present in previous estimates, giving falsely elevated constants for binding. Still unclear are the effects of variable binding constants. The assumption had been that the constant remained unchanged; however, this study shows there may be a variation due, for example, to changes in chloride, which could increase free unbound bilirubin levels.

In another paper (Mukerjee, P., Ostrow, J., and Tiribelli, C. 2002), the saturation status of unconjugated bilirubin was reexamined. Previous studies had shown a pKa value ranging from a low of 4.2 and 4.9 to values of 8.1 and 8.4. These variations could be the result of experimental differences. In the present paper, results are reported which examine whether some previous systems were supersaturated with mesobilirubin-13a (a close analog of bilirubin 9a).

Results showed that while supersaturated aqueous systems of unconjugated bilirubin are optically clear before centrifugation, there may be significant precipitation (sedimentation) upon centrifugation. In addition, there may be small fine colloids to small to sediment. In terms of pH, it was found that the lowest unconjugated bilirubin level at which sedimentation did not occur at $100,000 \times g$ was 100 nM at pH 7.4, higher than previously reported (62 nM) (). Increasing pH resulted in an increase in unconjugated bilirubin to levels of 17 μmolar at pH 8.05, and 34 μmolar at pH 8.2.

These data, and other results from these studies, support the previously reported values for aqueous pKa of 8.12 and 8.44 derived from partition studies (Hahm, J., et al. 1992). Previous reports of low pKa were limited by improper pH measurements, or assurance that mesobilirubin was below saturation and not self-associated. The authors state that these are important data (pKa) from the standpoint of the neurotoxicity of unconjugated bilirubin in jaundiced newborn infants.

In 2003, a splendid review article was published by members of the group responsible for many of the new concepts regarding bilirubin metabolism (Ostrow, J., et al. 2003). The purpose of the review was to outline some of the novel ideas suggested by more recent experiments using better methodology, and attempting to eliminate possible impurities in assay technique.

One concept mentioned several times in the review is the potential reversibility of neurological symptoms of bilirubin encephalopathy if recognized early. This is the overriding significant concept regarding bilirubin encephalopathy. Nearly all of the metabolic encephalopathies are defined as biochemical alterations which if treated aggressively and in a timely fashion, can be reversed. The reversible nature of these disorders is important in experimental studies in that neurochemical changes can be reassessed when the symptoms revert toward normal. Untreated bilirubin encephalopathy becomes untreatable kernicterus.

In the review, various factors which may result in elevated bilirubin are enumerated, such as hemolysis, enterohepatic circulation of bilirubin, hereditary defects, premature birth, and other causes, all of which are described in other chapters in this volume. Symptoms, such as opisthotonus, hypertonia, hearing loss, seizures, stupor, coma, and death, may all occur in severe cases of kernicterus.

One important aspect reviewed is that of the free unbounded unconjugated bilirubin. This refers to the small fraction of unconjugated bilirubin not bound to the carrier protein albumin. Results from studies (Ostrow, J., Pascolo, L., and Tiribelli, C. 2003) show that levels of free bilirubin of from 80 to 800 nM may produce toxicity. This level of free unconjugated bilirubin can permit binding of the bilirubin to plasma membranes, changing normal membrane function, and binding to mitochondrial membranes. This results in an effect on oxidative phosphorylation and changes in energy metabolites, and may result in apoptosis. These data are in support of Gunn rat studies performed many years earlier by many investigators showing an effect both in vitro and in vivo of the mechanisms of toxicity of unconjugated bilirubin.

In this review, the authors point out the significance of the remarkable animal model of bilirubin encephalopathy/kernicterus, the Gunn rat. This animal model lacks the bilirubin-conjugating enzyme glucuronyl transferase in the homozygous (bb) animal. Littermate non-jaundiced animals serve as ideal controls. An almost exact phenotype in humans exists, and is the Crigler–Najjar syndrome type 1. Many neurological symptoms of these above two conditions are identical, as well as the neuroanatomical site of bilirubin deposition in the brain.

Unconjugated free bilirubin in the diacid form may diffuse across all cell membranes. It is the diacid form of unconjugated bilirubin thought to be responsible for kernicterus. Hepatocytes and trophoblasts can also take up free unconjugated bilirubin by carrier-mediated facilitated diffusion. Whichever mechanism is used, only free bilirubin crosses; bilirubin bound to albumin does not cross the blood–brain barrier.

In humans, it has been shown that the affinity for unconjugated bilirubin decreases as the concentration of serum albumin increases (see above). Due to the techniques used, it was determined that early studies using less accurate methodologies have significantly overestimated the actual amount of unconjugated bilirubin bound to albumin in vivo. This means that free unconjugated bilirubin is able to gain access to highly sensitive cells, such as those in the brain, at much lower concentrations than heretofore thought. In fact, the levels of free bilirubin that cross plasma membranes are at least an order of magnitude lower than previously thought. The binding of unconjugated bilirubin may actually be even lower in fetal/newborn plasma. This is because the affinity of bilirubin is lower for alpha fetoprotein than is the affinity of bilirubin for adult serum albumin. All of this plus the strong possibilities that other factors which would lower the affinities of bilirubin and albumin, such as drugs used to treat newborn infants for a variety of newborn problems. In addition, it has been shown that nearly 10% of unconjugated bilirubin is bound to apolipoprotein D. This protein has a lower affinity for unconjugated bilirubin than does albumin.

The authors of this review point out that the free unconjugated bilirubin exceeds its water solubility at a concentration of 70 nM, resulting in self-aggregation. This occurs when total unconjugated bilirubin is about 5 mg/100 ml (85 μmolar). One cannot always be sure that two sites on the albumin molecule are always available. It is noted that pH plays a critical role in binding decreases as pH drops. This in turn increases the possibility of bilirubin entry into brain. With so many unpredictable and difficult to measure variables, it is easy to see that no clear-cut formula can be derived to predict jaundice outcomes in any single case. Risk assessment for bilirubin encephalopathy and kernicterus based on one clinical test such as total serum bilirubin levels is highly uncertain.

In terms of entry of molecules into the brain parenchyma, it depends in part on the operation of the blood–brain barrier, and the operation of the blood–cerebrospinal fluid (CSF) barrier. The structural component of the blood–brain barrier is the tight junctions with no fenestra of the capillary endothelial cells. This anatomical barrier acts to reduce or eliminate entry into brain cells of various compounds and drugs, but maybe not bilirubin. The endothelial cells of capillaries in the choroids plexus of the blood–cerebrospinal fluid barrier have no tight junctions, and do have fenestrations. This translates to a more readily crossed barrier than the blood–brain barrier. Total protein levels in the CSF are less than 1% that found in plasma.

Energy requiring transport in both the blood–brain barrier and the blood–CSF barrier consist of two classes of ATP-binding cassette (ABC) transporters. The transporters are localized in capillary endothelium. Multidrug resistance-associated proteins (MRP) and multidrug resistant P-glycoproteins (MDR/PGP) are two such proteins. Significant data exists showing RNA expression MRP in newborn and adult rat brain, including neurons and astrocytes. Studies, including in knockout mice show these transporter play a major role in limiting drug buildup in the brain. MDR1 functions to export toxic compounds from epithelial cells, and are expressed in brain. Evidence exists for a role for MDR1 in bilirubin export from brain.

Other mechanisms exist to manage increasing levels of bilirubin if transporters are not enough. Binding of free unconjugated bilirubin by cytosolic glutathione-S-transferase acts to keep the free pigment at low levels. The glutathione-S-transferase isozymes have different binding characteristics, and different concentrations in various brain regions. Some studies question the actual role of the transferases in protecting brain from bilirubin, but they have been shown to exist in the cerebellum, a site of bilirubin deposition in kernicterus.

The authors of this review point out that, unlike serum unconjugated bilirubin levels, estimation of free unconjugated bilirubin levels did correlate with neuropathology in both Gunn rats and in premature newborn infants. Both free unconjugated bilirubin levels and brainstem auditory evoked potentials can predict the presence of reversible neuropathological impairment. This has been demonstrated in Gunn rats and newborn infants. Treatments might include inducing transporters in an attempt to move pigment out of astrocytes and neurons.

The authors note that evidence up to the date of this review (2003) should suggest the complexities and uncertainties of unconjugated hyperbilirubinemia. The evidence supporting the role of free unconjugated bilirubin in producing symptoms and lesions in newborn infant brain is undeniable. The dependence by neonatologists and pediatricians on serum levels of total or unconjugated bilirubin is frought with danger. Even worse is the estimation of jaundice by comparing skin color to a graded yellow card, then reading a range of possible serum bilirubin levels. Further, even with proper bilirubin measurement, the guidelines for treatment of jaundiced newborn infants may be high. That is to say that cerebral damage, given the right circumstances and levels of free unconjugated bilirubin in serum, may occur at serum levels well below that previously proposed.

In a more detailed assessment of free unconjugated bilirubin and its deleterious effects, previously published data was recalculated taking into consideration new concepts (Ostrow, J., Pascolo, L., and Tiribelli, C. 2003). While low levels of unconjugated bilirubin may be neuroprotective in neonatal infants due to its antioxidant properties, higher levels of bilirubin are toxic. Initially, free unconjugated bilirubin produces relatively modest neurological symptoms, and at this stage, may be reversible. Later as levels rise, precipitation of bilirubin in brain neurons and astrocytes produces much more severe neurological symptoms which are probably not reversible. This stage is overt kernicterus.

A key concept is that the bilirubin entering the brain cells is free – that is unbound to albumin unconjugated bilirubin. Also, as human serum albumin increases, its affinity for unconjugated bilirubin rapidly decreases. This serves to overestimate the affinity for albumin, thereby underestimating the levels of free unconjugated bilirubin ready to enter the brain.

By recalculating several previously published papers' values, a better assessment of the ability of free unconjugated bilirubin to enter the brain could be evaluated. Comparison of data was achieved using only manuscripts reporting total unconjugated bilirubin, serum albumin, and chloride concentrations.

Results showed that when the albumin affinity constants are applied, free unconjugated bilirubin exceeds the level of unconjugated bilirubin solubility in water below which albumin binding is saturated. At normal albumin levels this occurs at a total unconjugated bilirubin level of about 5 mg/100 ml. This is rarely seen in adults except in those with Crigler–Najjar syndrome. This means with normal albumin levels, only newborn infants would ever be exposed to free unconjugated bilirubin levels which were higher than aqueous solubility levels.

In recalculating free unconjugated bilirubin levels, neurotoxic effects of bilirubin might be expected at levels only just above the aqueous solubility level. Impurities and slight variations in content of human albumin, and Dulbecco's medium can account for some variations in results. Significantly, these results show that measurable toxicity in CNS cells can occur at levels before supersaturation and precipitation of unconjugated bilirubin occurs. This disproves the early concept that in order for toxicity to occur, there must be coarse aggregation of unconjugated bilirubin. The protective effect of very low levels of free unconjugated bilirubin is lost as levels of the pigment exceed the aqueous saturation level.

At the levels of free unconjugated bilirubin at which toxicity begins, astrocytes and neurons showed impairment of mitochondrial functions and apoptosis. The toxic effects may explain structural features associated with toxicity. These changes really represent the metabolic (biochemical) changes of early bilirubin encephalopathy before the irreversible features of advanced kernicterus occur. Astrocyte membranes are also affected as reflected by an increased release of lactic acid dehydrogenase, and a decreased uptake of glutamate. Another investigator (Ahlfors, C. 2000) obtained similar results in reassessing free unconjugated bilirubin, and concluded that bilirubin encephalopathy/kernicterus probably only occurred at a level over 60 nM (70 nM equals aqueous saturation). These results are in agreement with those described in the Ostrow paper, given the nature of possible minor methodological variations described above.

Many cells of the body are able to protect themselves from potentially toxic compounds, such as bilirubin, by transporting them out of the cell. This is achieved by the multidrug resistance-associated protein (MRP) transporters. These proteins are a subclass of ATP-binding cassette transporters (ABC transporters). Many of these transporters are expressed in animal CNS cells. One such transporter, MRP1 may be able to transport unconjugated bilirubin. In one study (Gennuso, F., et al. 2004), experiments were performed examining the ability of astrocytes to intracellularly mobilize and redistribute MRP1 when exposed to a level of unconjugated bilirubin (40 nM) below the aqueous saturation level.

In this study, cells were labeled by immunofluoresence and visualized by confocal laser scanning microscopy. The effects of inhibiting MRP1 activity was on the unconjugated bilirubin toxicity, and on apoptosis.

Initial results showed that the localization of MRP1 in astrocytes was perinuclear, and asymmetrically placed in the cytoplasm. This localization was within the Golgi apparatus. The exposure of astrocytes to purified unconjugated bilirubin resulted in a redistribution of MRP1. The redistribution was both dose and time dependent. After 30 min of exposure to unconjugated bilirubin, the MRP1 signal had increased, and moved throughout the astrocytic cytoplasm. Fusion images suggested the MRP1 had moved along the astrocyte cytoskeleton.

When a higher dose of unconjugated bilirubin was used, results were different. In this case, at 5 min of exposure, the intensity of the MRP1 signal was intensified as before, but remained localized to the perinuclear region occupied by the Golgi apparatus. At 10 min, the immunofluoresence had spread throughout the cell and was further intensified, but after 30 min the immunofluoresence had again returned to the Golgi site. When MRP1 was inhibited by the inhibiting MK571 the movement of MRP1 did not occur, and the astrocytes had an increased vulnerability to super saturated concentrations of free unconjugated bilirubin.

The high dose of unconjugated bilirubin resulted in astrocyte toxic effects including nuclear staining by propidium iodide, mitochondrial activity, and apoptosis. Astrocytes appeared shriveled. Addition of MK571 alone had no effect, but further impaired cell function in the presence of the high dose of unconjugated bilirubin.

The authors note that results from this study demonstrate that MRP1 expression is stimulated by unconjugated bilirubin. This consists of an increased immunoflu-ourescence and movement of MRP1 from a perinuclear region to the vicinity of the plasma membrane. This migration is both dose and time dependent. The MRP1 inhibitor MK571 blocks the protective MRP1 effects.

Unconjugated bilirubin has the ability to upregulate and redistribute MRP1 from the perinuclear located Golgi apparatus region to the plasma membrane, where neurotoxins such as unconjugated bilirubin can be pumped out of the cytoplasm. These mechanisms can also serve to remove toxins from inside the nucleus. Thus, the authors suggest that MRP1 is a critical defense mechanism against the deleterious effects of free unconjugated bilirubin. This mechanism may exist in many cells, including those of the CNS. This finding is not unique in that MRP1 also serves to extrude other toxins from brain cells. Demonstration of this mechanism in astrocytes is noteworthy given the important role of astrocytes in the protection of neurons from toxic agents now including free unconjugated bilirubin.

Shortly after the discovery of MRP1 (Cole, S., et al. 1992), it was demonstrated that the transporter was capable of moving unconjugated bilirubin out of cells (Jedlitschky, G., et al. 1997). MRP1 and MDR1 (another transporter with similar characteristics) are expressed in the cells of the blood–brain barrier, and in cells of the blood–cerebral spinal fluid barrier. The purpose of this paper (Rigato, I., et al. 2004) was to directly demonstrate that MRP1 could transfer free unconjugated bilirubin across plasma membranes of Madin-Darby canine kidneyII cells.

[3]H labeled unconjugated bilirubin was prepared in bile duct canulated rats, then highly purified (Bayon, J., et al. 2001). Madin-Darby canine kidney II cells and clones were utilized, and cultured using standardized methodology. Membrane vesicles were subsequently obtained from the canine kidney cells, with modification of previously described techniques. These involved differential centrifugation using a sucrose gradient. Western blotting was used to produce bands which could be visualized and quantified.

The transport of labeled unconjugated bilirubin was estimated by incubating vesicles with unconjugated bilirubin at a concentration of about 17–28 nM. These experiments were done under a dim light in order to try to eliminate photodegradation. To measure transport, the incubation was stopped, and samples vacuum filtered through a nitrocellulose membrane, then radioactivity on the filter was counted by scintillation spectroscopy. The contribution of MRP1 was taken as the difference between transport in MRP1 cells vs. those wild-type cells not transfected with MRP1.

Results from these experiments showed that transport of unconjugated bilirubin from the canine kidney ll cells which were overexpressing MRP1, demonstrates that unconjugated bilirubin is a compound which stimulates MRP1. It also demonstrates that both ATP and glutathione are required in order for the transport to occur. These results regarding GSH have been observed by others for MRP1-mediated transport (Mao, Q., Deeley, R., and Cole, S. 2000). Low levels of GSH would be conducive to less unconjugated bilirubin, favoring an antioxidant action of bilirubin. As levels continued to increase due to further low GSH concentrations, toxicity would occur.

When the MRP inhibitor MK571 was introduced, ATP-dependent unconjugated bilirubin transport was inhibited. Since MK571 inhibits other transporters besides MRP, this cannot be taken as definitive proof that MRP1 is a transporter of unconjugated bilirubin. Nevertheless data from this study support the concept that MRP1 (but not MRP2) is active in the ATP-dependent transport of unconjugated bilirubin. The presence of MRP1 transporter in many cells of the body suggest it could be involved in the transportation out of the cell of toxic unconjugated bilirubin.

It should be remembered that low intracellular levels of unconjugated bilirubin are protective in that they are potent antioxidants (Ostrow, J. and Tiribelli, C. 2003). The "changeover" from a beneficial antioxidant to a highly toxic agent is about at a concentration of 70 nM. The presence of a transporter system working at these nanomolar levels of free unconjugated bilirubin lends credence to the concepts that neurotoxicity can exist at levels of free bilirubin above about 70 nM, and that MRP1 transporters are capable of keeping free unconjugated bilirubin at low beneficial levels. It may be shown that MRP1 is not the only transporter which is able to "work" on free unconjugated bilirubin. The authors suggest that further studies on other transporters such as MRP4 need to be done. And it must be remembered to use clinically relevant levels of purified free unconjugated bilirubin in such studies in order for them to be relevant.

In a recent paper (Rigato, I., Ostrow, J., and Tiribelli, C. 2005), the cytoprotective effects of unconjugated bilirubin are discussed. The calculated pKa value of unconjugated bilirubin is initially discussed as it has significant importance as regards the bilirubin levels at which cytoprotective and cytotoxic effects occur. When pKa values are below 5.0, the unconjugated bilirubin dianion is predominant. If pKa values are 8.0 or above, then the unconjugated bilirubin is in the diacid form. The diacid form is that which is believed to readily pass through cell membranes, thereby, at high levels (over 70 nM), is capable of disrupting normal cellular function such as oxidative phosphorylation.

At levels below 70 nM, unconjugated bilirubin may offer some protection to cells because of its antioxidant effects. The protection may be available to all tissues of the body, and independent of whether bound or unbound to albumin. This role for unconjugated bilirubin serves to decrease concentrations of the free form, as well as protecting cells from reactive oxygen species. In this process, unconjugated bilirubin is converted to biliverdin.

A large variety of disease states have been suggested as recipients of the beneficial antioxidant properities of free unconjugated bilirubin. One of these is atheromatous disease where unconjugated bilirubin acts to inhibit oxidation of low-density lipoprotein. Unconjugated bilirubin even when bound to albumin is a good scavenger of lipid peroxyl, thereby exerting an effect on plaque formation (Wu, T., et al. 1994).

Some populations of patients who have mild hyperbilirubinemia show a decreased incidence in other potentially harmful conditions. For example, in a study of Gilbert disease, patients described in another chapter in this book, none of a group of 50 cases ever were shown to have developed ischemic heart disease. This was compared to a group of over 300 non-Gilberts syndrome patients in which the incidence of ischemic heart disease was about 2.5%.

Coronary artery disease is another clinical entity for which there is an apparent positive effect of unconjugated bilirubin. This was a study (Hopkins, P., et al. 1996) in which 161 patients and 155 controls were compared as to serum bilirubin concentrations and coronary artery disease. Results showed that the incidence of coronary artery disease was lower in patients with higher levels of unconjugated bilirubin. It was suggested that the elevated serum bilirubin in the patients with lower artery disease as compared to patients with low unconjugated hyperbilirubinemia was an independent protective factor favoring a decrease in incidence of coronary artery disease.

In another category of demyelinating diseases a protective correlation has been found for unconjugated bilirubin. For example, oxidative stress may interfere with oxidative phosphorylation in cerebral cells in patients with multiple sclerosis, and also in animal models of multiple sclerosis. Free radicals are suspected in the pathogenesis of multiple sclerosis. In the animal model of multiple sclerosis (experimental allergic encephalomyelitis), administration of unconjugated bilirubin served to prevent the induction of both acute and chronic symptoms (Liu, Y., et al. 2003). Unconjugated bilirubin also stopped the progression of established disease.

One other interesting observation (Muller, N., et al. 1991) was that in schizophrenic patients there was a frequency of 25% with elevated bilirubin levels as compared to an incidence of elevated bilirubin in non-schizophrenic patients of only 7.3%. It also seemed that the symptoms in the schizophrenic patients were worse than those schizophrenic patients who had normal bilirubin levels. While a cause/effect relationship for bilirubin and schizophrenia has not been established, these results certainly bear further investigation.

The effects of unconjugated hyperbilirubinemia seem far reaching. Unconjugated bilirubin may play roles in cellular protection by its antioxidant properties. It may also act to enhance various disease states by furthering symptoms in both incidence and severity. Many disease conditions are influenced by unconjugated hyperbilirubinemia, including coronary artery disease, plaque formation, Gilbert syndrome, cardiac failure, T-cell metabolism, breast cancer, GI cancers, multiple sclerosis, amyotrophic lateral sclerosis, schizophrenia, depression, and others. It will take many controlled studies to sort out the effects, both good and bad, on the above examples of disorders already thought to be influenced by hyperbilirubinemia.

In late 2004, there was a conference sponsored by EASL on the molecular basis of bilirubin encephalopathy and bilirubin toxicity. The results of this symposium were published in the form of short synopses from various investigators who presented at the meeting. The following are brief summaries of some of the presentations:

Steven Zucker, University of Cincinnati, and Richard Wennberg, University of Washington, Seatle, both presented regarding the question of how bilirubin crosses membranes and enters cells. The issue of the nature of a transporter of unconjugated bilirubin across liver cells' membranes is unclear. Passive diffusion is possible in many tissues of the body, suggested by their diffuse yellow color. The biochemical characteristics of unconjugated bilirubin are conducive for its crossing phospholipids membranes.

Richard Wennberg confirms that the unconjugated bilirubin diacid diffuses rapidly across phospholipid membranes. Liver uptake is more rapid when bilirubin is bound to bovine serum albumin. This probably represents dissociation limited diffusion in which diffusion is a representation of the concentration of donor and acceptor molecules. These results suggest that the blood–brain barrier may not be much of a barrier to the entry of unconjugated bilirubin into neural cells. Passive diffusion of bilirubin into brain is offset by subsequent metabolism and/or active secretion of the pigment back into plasma. When unconjugated bilirubin is high in plasma, it probably does not reflect any change in the blood–brain barrier.

Jean-Francois Ghersi-Egea, Lyon France states that entry into brain parenchyma is well regulated by the blood–brain barrier, and blood–cerebral spinal fluid barrier. Tight junctions associated with the blood–brain barrier regulate its function, whereas the blood–cerebral spinal fluid barrier has no such tight junctions. Endothelial cells of each have specific efflux transporters such as MDR1 described above, which effectively remove potential toxins from inside the cells. The intrinsic ability of neurons to metabolize drugs and toxins is low.

Jon F. Watchko, University of Pittsburgh, reported that passage of unconjugated bilirubin across the blood–brain barrier and into neurons is possibly dependent upon the function of MDR1 transporters. Significant evidence exists to support this role for MDR1. These transporters are expressed by both cells of the blood–brain barrier and by some brain neurons. This activity has been shown to possibly decrease unconjugated bilirubin apoptosis.

There were two reports on possible new therapies for unconjugated hyperbiliru-binemia. One from Anja Hafkamp, Groningen, The Netherlands suggested that plasma unconjugated bilirubin levels were decreased by the administration of the pancreatic lipase inhibitor, orlistat. Over 50% of Gunn rats had lowered plasma unconjugated bilirubin with a combination of orlistat and phototherapy as compared with 31% decreased bilirubin levels with phototherapy alone.

Libor Vitek, Prague, Czech Republic stated that administration of zinc methacry-late which acts to trap unconjugated bilirubin in the intestine, was able to reduce serum unconjugated bilirubin levels by 25% in Gunn rats. This decrease in bilirubin level was not associated with elevated serum zinc.

For more summaries, the reader is referred to the J. Hepatology 43:156–166. 2005.

In a somewhat mordant commentary, the question of "free" unconjugated biliru-bin and its toxicity and importance is examined from a different perspective (McDonagh, A., and Maisels, M. 2006). The authors note that for many years, unconjugated bilirubin was recognized as being potentially toxic. Later it was deter-mined that bilirubin, unconjugated and not bound to albumin (free) was the producer of toxicity as it could readily cross biological membranes.

While the authors recognize that unbound unconjugated bilirubin is the frac-tion to measure in order to assess risk of kernicterus, they state that (1) kernicterus is rare, and usually does not occur until bilirubin levels exceed 30 mg/100 ml and (2) application of the American Academy of Pediatrics guidelines, and early use of improved phototherapy should virtually eliminate most potential cases of encephalopathy without need for additional measurements. They quote a study in which only 1/10,000 newborn infants from a hospital required blood transfusions. Therefore, the authors state, if newer phototherapy is used, and serum bilirubin lev-els are maintained below 25 mg/100 ml, the amount of free unconjugated bilirubin is inconsequential.

It is noted, however, that in premature infants the risk for brain damage exists. The authors quote a paper (Sugama, S., Soeda, A., and Eto, Y. 2001) in which are described three infants with kernicterus, showing signs of hypotonia and choreoa-thetosis, with MRI evidence of brain damage. Bilirubin levels were less than 15 mg/100 ml. In some premature (25–29 weeks) brain damage was detected at even lower bilirubin levels, and serum albumin was especially low.

The authors of this commentary state that assessment of free unbound uncon-jugated bilirubin levels should be viewed cautiously because its measurement is difficult for technical reasons. Second, there is no readily available instrument to measure free bilirubin. Third, the free bilirubin levels vary with the attention to detail of measurement, so inconsistent results, difficult to assess, may occur. Total biliru-bin levels are easy to measure. Fourth, elevation of free bilirubin may be fleeting and difficult to "catch" and/or easily missed.

In addition, the authors of the commentary call into question the validity of the peroxidase method of measuring free unconjugated bilirubin. This is based on data from high-pressure liquid chromatography showing at least two isomers of free bilirubin. One, a photoisomer of the biosynthetic bilirubin isomer, may comprise 10–20% of the total bilirubin present in serum. This photoisomer of bilirubin (and possibly others) would lead to spurious results obtained by the peroxidase method.

The authors of the commentary state that if the peroxidase method is shown to be correct, and made easy to use, then it might have value. The authors continue by saying that all of the questions regarding free bilirubin need resolution. They say that the "poorly understood and confusing field of bilirubin toxicity" needs consistent terminology. They take exception, for example, to the term "free bilirubin" which might be confused with the term "unconjugated bilirubin."

In a different vein, the true pKa values of unconjugated bilirubin are very important as they affect the actual distribution of unconjugated bilirubin. These values have a bearing on the potential interactions of unconjugated bilirubin, and on the cerebral damage in hyperbilirubinemic newborn infants. The authors of this manuscript (Ostrow, J., and Mukerjee, P. 2007) had previously measured the pKa values for unconjugated bilirubin and found values over 8.0 (8.1 and 8.4). Other investigators found pKa levels lower, even below 5.0. If low pKa numbers (below 5.0) are used, the ratio of diacid/dianion at pH 7.4 would be 4×10^{-6} as opposed to 0.58 (pKa over 8.0). The solubility of unconjugated bilirubin diacid would change from 5×10^{-8} M to less than 10^{-14} M. This difference is highly significant.

Accordingly, the authors have reevaluated their findings of pKa values. In this study, ^{14}C-unconjugated bilirubin was biosynthesized in bile fistula-prepared rats using highly purified bilirubin. ^{14}C-unconjugated bilirubin was isolated from bile by previously described methods. Further purification steps were performed. The purification of labeled bilirubin was measured. Partition methods employed in this study have been used successfully to determine with high precision, pKa values in previous studies.

The authors note that while a wide range of values of pKa have been reported (4.4– 8.3), solutions for study must avoid supersaturation to prevent aggregation and degradation of unconjugated bilirubin must be avoided by conducting experiments quickly, under anaerobic conditions, and in the dark, or under low-light conditions. The unconjugated bilirubin as stated before, must be highly purified.

Results from the present study, taking into account all the circumstances regarding purity of the bilirubin, etc., confirmed the previous pKa values of 8.12 and 8.44 previously published (Hahm, J., et al. 1992). The authors state that studies by others in which there were errors in pH measurements, and long duration of ^{13}C-NMR analysis (overnight) render results "not interpretable." The high pKa values result from three internal hydrogen bonds in unconjugated bilirubin. Effects of these H bonds include donation of an H bond from the $-$OH moiety of the $-$COOH group, hindered solvation of the $-$COOH group, and restricted rotation of the previously mentioned two groups, which lead to a reduced solvation of the two groups.

The authors explain that mechanisms by which bond breaking during ionization might raise pKa of unconjugated bilirubin. The presence of two hydrogen bonds which anchor the $>$C=O group of dissociating $-$COOH groups acts to position molecules to donate a hydrogen bond. The donation of the hydrogen bond serves to raise pKa values in unconjugated bilirubin. The degree of the raising of pKa is not clear, but two full units are possible. Other workers suggest that hydrogen binding may lower pKa, but the authors dismiss this because they think others are using inappropriate model systems.

Another feature of unconjugated bilirubin which might raise the pKa involves the restricted rotation of the −COOH and −COO⁻ groups. In unconjugated bilirubin molecules, the two hydrogen bonds to the >C=O group would restrict free rotation of the −COOH group. The −COO⁻ group would also be restricted. One investigator (McCoy, L. 1967) suggests that crowded microenvironments with concomitant rotation, possibly combined with impaired solvation can serve to increase pKa values by several units. It is thought likely that such a microenvironment is possible in unconjugated bilirubin, effectively increasing its pKa value.

The hydrogen bonding associated with the −COOH and −COO⁻ groups of unconjugated bilirubin molecules restrict the environment of the molecule. This leads to a situation in which water solvation needs to be inhibited. Steric inhibition of solvation would likely increase the energy in charged molecules. The resultant sterically hindered unconjugated bilirubin −COOH and −COO⁻ groups might have higher pKa groups. The exact increase in pKa due to this phenomenon is unclear, but might be in the range of 1.0–2.0 pKa units.

A comment or two are in order regarding the investigations described in this chapter. Most of the studies summarized are from one productive group. What is noteworthy is the absolute attention to detail, and in making sure that the technical methods used were as impeccable as possible. Every step was examined for potential errors, and corrected beforehand. An example is performing experiments quickly and in darkened areas, or under dim red light to prevent breakdown of labeled free bilirubin.

I am a fan of trying to do experiments correctly, regardless of time or effort. As I have said elsewhere in this book, for example, what can be learned by grinding up whole brain to try to answer questions regarding bilirubin toxicity in the hippocampus or cochlear nucleus? These small areas represent a small fraction of whole brain. Similarly, the authors of most of the manuscripts quoted in this chapter state that other studies in which precautions on purity of bilirubin, for example, were not taken, yield data which are not interpretable. I could not agree more. What is gained by doing a study incorrectly which produces data which are not interpretable?

Auditory Brainstem Response

Auditory brainstem response to auditory stimulation is a non-invasive technique used to test the functional integrity of the brainstem connections and nuclei involved with hearing This method has been used in evaluating potential early brain damage from high serum levels of unconjugated bilirubin. The concept is that small discreet brain regions such as the cochlear (eighth) nucleus is uniquely sensitive to toxic effects of unconjugated bilirubin. Since function of this nucleus and its brainstem connections can be tested without invasive techniques (see Fig. 1), it is an ideal target to determine if unconjugated bilirubin brain damage has occurred in newborn infants.

It is well known that unconjugated bilirubin enters the brain when serum levels reach a point in which albumin can no longer bind the pigment. Unconjugated bilirubin has a predilection for certain selective punctate areas, and one of these areas is the nucleus of the eighth cranial nerve. Obviously, decreased hearing or total loss of hearing has a significant impact on the development of speech, and on the development of intellect. Because of concerns about hearing loss, there has been an increase in awareness of the importance of testing newborn infants for hearing, and screening programs have evolved in many regions. This was possible in part due to the development of cost-effective equipment and to technological advances in auditory brainstem response (ABR) testing methods. These advances have shown that speech-evoked ABRs are different between children with normal hearing and many children with learning problems (see Figs. 2, 3, and 4).

From a physiological standpoint, ABR waves are the resultant data which are recorded. Wave I represents the auditory nerve action potential in the distal portion of the eighth cranial nerve. Wave II is generated from the proximal portion of the eighth nerve as it enters the brainstem. Wave III arises from second-order nerve activity near or in the cochlear nerve at the level of the auditory pons. Wave IV arises from pontine third-order neurons in the superior olive. Wave V probably originates from the inferior colliculus, and is the wave most often examined in auditory brainstem responses. These technologies have been in use for over 30 years,

This Chapter is authored by Laura Muncie and David McCandless.

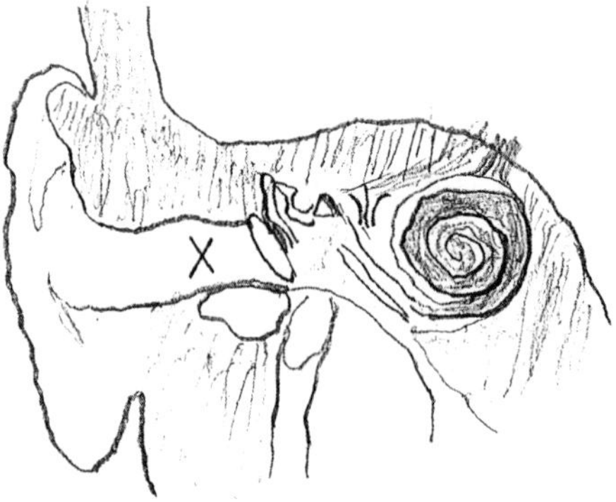

Fig. 1 Schematic representation of the location (x) of the ear insert responsible for delivering click stimulation to the auditory pathway. See appendix for source

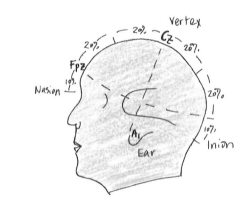

Fig. 2 Schematic drawing of the ABR recording electrode placement (A1, Cz, and Fp2) according to the 10–20 system. See appendix for source

Fig. 3 Drawing of the ABR recording electrode placement viewes from the top (A1 and A2, Cz, and Fp2) according to the 10–20 system

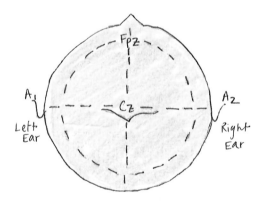

and have been utilized in investigations of hyperbilirubinemia and its consequences (see Fig. 5).

In an early study of ABR in hyperbilirubinemic newborn infants (Perlman, M., et al. 1983), auditory brainstem responses were obtained and studied from 24 infants

Fig. 4 Cartoon showing the
central auditory pathway.
Adapted from M.S. Hill,
USW, edu. au

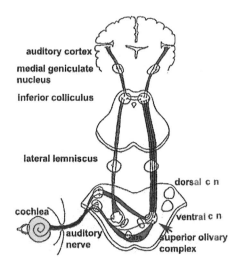

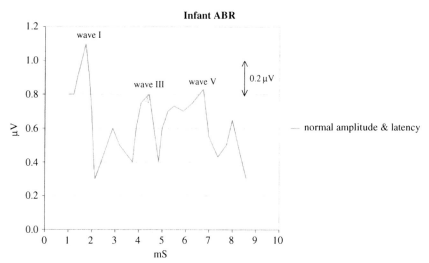

Fig. 5 Schematic representation of an ABR recording of a normal non jaundiced newborn. Waves I and III, V are shown-note latency and amplitude

with serum bilirubin levels between 15 and 25 mg/100 ml. These results were compared to those of 19 non-jaundiced newborn infants who were otherwise matched for gestational age, and age at time of testing. Results showed that the wave complex IV–V was absent in at least one recording in 10 out of the 24 hyperbilirubinemic infants, but present in all 19 controls. Newborns with hyperbilirubinemia also had a longer brainstem transmission time, indicating an increased latency at both upper and lower brainstem levels. There were statistically significant transient alterations in ABRs, implying a transient brainstem bilirubin-induced encephalopathy (Lenhardt, M., McArtor, R., and Bryant, B. 1984).

The authors state that entry of unconjugated bilirubin into the brainstem occurs in a transient fashon, and at these moderate bilirubin levels is novel. It may reflect a brainstem encephalopathy occurring at bilirubin levels thought to be "safe," and raises questions about current concepts regarding transfer of bilirubin into cerebral tissue.

A similar paper (Chin, K., Taylor, M., and Perlman, M. 1985) looked at auditory brainstem responses and visual-evoked potentials, and demonstrated improvement following exchange transfusions. This paper describes two cases of hyperbilirubinemia in newborn infants whose bilirubin levels had risen to 12–15 mg/100 ml.

Results showed changes in wave I (eighth cranial nerve), wave III (superior olive), and wave V (inferior colliculus). The two newborn hyperbilirubinemic infants clearly show a shortening of the wave I–III and I–V latencies (brainstem transmission times) following exchange transfusions. These improvements following transfusion had been reported before, but in newborn infants with much higher bilirubin levels (Wennberg, R., et al. 1982). The authors postulate that the latency changes in the above two newborn infants were related to the decrease in bilirubin concentrations following exchange transfusions (see Fig. 6).

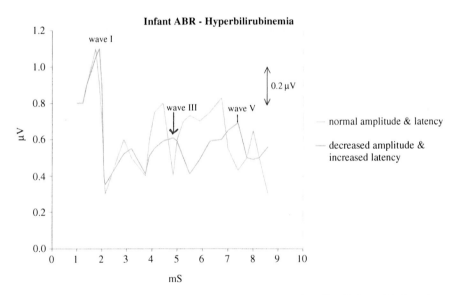

Fig. 6 Schematic graph comparing an ABR recording in a non-jaundiced infant with that of an early kernicteric infant. Note deficiencies between the latency and amplitude of waves

The authors note that other birth factors such as asphyxia and acidosis may influence ABRs, but the two infants in this study did not show evidence of either confounding condition. The authors also raise concerns about the safety of moderate hyperbilirubinemia. The lack of evidence of hearing loss in some long-term follow-up studies may suggest that the changes seen in this study were reversed as the hyperbilirubinemia dropped.

In another study (Shapiro, S. 1988), hyperbilirubinemic Gunn rats were used to evaluate the timing of changes in brainstem auditory-evoked responses. In this experiment, jaundiced Gunn rats were used on postnatal day 18. They were anesthetized, then sulfonamide was injected IP to displace bilirubin from its albumin-binding site. Brainstem responses from auditory stimulation were recorded at various intervals after sulfonamide administration.

The recording of brainstem activity was achieved by placing electrodes at the cranial vertex and right and left mastoid processes. Scalp electrical activity was amplified and filtered, then averaged. The stimuli were 100 μs clicks at 75 dB by a Sony Walkman 4LIS speaker placed over the right external acoustic meatus. Each auditory brainstem response, a result of averaging, was plotted separately. Temperature of the Gunn rats was maintained at 37°C. Bilirubin levels in serum were measured by clinical laboratory methods.

Results showed that the responses to auditory stimulation were similar to those seen in hyperbilirubinemic humans, and this further validates the Gunn rat as an excellent model for the Crigler–Najjar syndrome. Injection of sulfonamide into infant Gunn rats lowered serum bilirubin from 8.0 l to 2.8 mg/100 ml, a drop of 65%. This indicates that the sulfonamide succeeded in pushing bilirubin off its albumin-binding sites and into brain. The molar ratio of bilirubin to albumin was also decreased.

The changes of Gunn rat acoustic brainstem responses to stimuli after sulfon-amide treatment consisted of an increase in I–II IWI responses, and a decrease in amplitude of waves II and III. These changes in amplitude suggest anatomical local-ization of auditory dysfunction. The changes in I and II IWI waves and amplitude abnormalities in wave II serve to localize bilirubin damage to the cochlear nuclei. These changes in brainstem auditory-evoked potentials were seen to occur in as lit-tle as 2 h after sulfonamide injection. Histological evidence suggests the cochlear nuclei suffer significant damage by bilirubin (Jew, J., and Williams, T. 1977; and Dublin, W. 1951). Wave III changes indicate either dysfunction of the ascending auditory pathways or abnormal input from the cochlear nuclei, or both. Wave IV was not affected in this study.

In the discussion, the author mentions that while his study did not focus on changes occurring in less than 2 h, changes were in fact seen in as little as 15 min after sulfonamide injection. This is in agreement with other results (Ahlfors, C., et al. 1986) showing change in responses in monkeys in as little as 12 h. Control animals injected with sulfonamides showed no abnormalities in brainstem auditory-evoked potentials.

The author comments that this study is the first to show a reversal of the changes in brainstem auditory-evoked potentials. It has been shown that reversibility is pos-sible following exchange transfusion (Wennberg, R., et al. 1982). Treatment aimed at lowering serum bilirubin levels in a timely manner are most likely to succeed in reversing central nervous system damage. Since the recording of auditory responses does not involve sacrifice of the experimental animal, variables such as time of exposure and of reversibility can be conveniently studied.

Another study (Karplus, M., et al. 1988) aimed to correlate changes in auditory brainstem responses to stimuli with actual intracerebral bilirubin concentrations in the brainstem. To achieve this, adult Sprague-Dawley rats weighing 300–450 g were utilized. Animal were divided into the following groups: high bilirubin, low bilirubin, and controls. Bilirubin and sulfisoxazole were administered by IV infusion. The low group received 50 mg/kg, and the high group received 100 mg/kg. Blood samples were taken from the tail vein, and at the end of the experiment, the brainstem and cerebellum were taken for bilirubin analysis.

Results showed that serum bilirubin levels were 13.6 mg/100 ml 60 min after infusion in the low bilirubin group, and 36 mg/100 ml in the high bilirubin group. Brainstem/cerebellum bilirubin concentrations were 0.93 μg/gram tissue in the low group, and 3.2 μg/gram in the high bilirubin group. Results of auditory brainstem-evoked responses showed a reduction in amplitude of wave I and III in the high bilirubin animals. Wave IV was absent in three of the seven high bilirubin animals. There was an increase in the reduction of wave amplitude as brain bilirubin levels increased.

The authors note that previous studies have shown similar amplitude alterations in both animal models and in humans exposed to elevated serum bilirubin levels. The present study showed similar results but correlated brainstem changes with levels of bilirubin. There was a correlation with bilirubin in that as the brainstem/cerebellum level of bilirubin increased, so increased the aberrations in the auditory brainstem-evoked responses.

The authors note that the case could be stronger if autoradiographic techniques were used to further localize the changes to foe example, the cochlear nuclei. The authors also suggest that the peripheral nerve effects involving hearing might precede effects of bilirubin on central nuclei.

Another study (Esbjorner, E., et al. 1991) studied auditory brainstem responses in nine hyperbilirubinemic neonatal infants. This study was undertaken to examine the reserve albumin concentration for monoacetyldiaminodiphenyl sulphone (MADDS), which is a deputy ligand for bilirubin found in jaundiced neonates; and its relation to the auditory brainstem-evoked responses.

The subjects were nine hyperbilirubinemic infants admitted to the intensive care unit with high serum bilirubin levels. Phototherapy was immediately initiated, and acoustic recordings taken as soon as possible after starting therapy as was possible, usually within 2–24 h. One infant had ABO incompatibility, all other infants were healthy. Serum reserve albumin concentrations were determined by the MADDS technique. Acoustic brainstem responses were recorded without sedation in a soundproof room after feeding.

Results showed that there was a prolongation of latencies and absence of waves in two of thenine jaundiced infants. This represents the first time a correlation has been shown between a neurophysiologic event and concentrations of serum albumin, according to the authors. Stated differently, this shows the importance of albumin's binding characteristics for bilirubin. This phenomenon may play an important role in bilirubin's access to brain tissue. The authors found the lowest reserve albumin concentration in the two infants who had pathological auditory brainstem responses.

In the case of all nine children, there was a negative correlation between wave latencies and the reserve albumin concentration as estimated by the MADDS method
proved to be a better estimator of potential bilirubin toxicity than was actual serum
bilirubin concentrations.

Another paper (Shapiro, S., and Conlee, J. 1991) attempted to refine the anatomical localization in the brainstem of Gunn rats from whom brainstem-evoked auditory
responses were recorded.

Results showed differences in auditory brainstem responses in jaundiced Gunn
rats that were mild to severe; the most severe were recorded from Gunn rats
which received sulfadimethoxine There was a significant variability in the jaun-
diced/sulfadimethoxine group which ranged from normal to complete disappearence
of waves I–IV. The volume of the cochlear nucleus examined histologically, correlated with wave I latency, I and II interwave interval, and wave I, II, and 4IV
amplitudes (see Fig. 7).

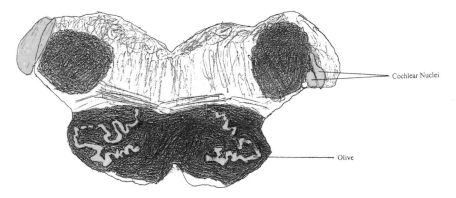

Fig. 7 Drawing showing bilirubin staining in the cochlear nuclei and olives

The authors note that this is the first study to compare histological features of
brainstem auditory areas with brainstem auditory-evoked responses in Gunn rats.
The volume of brainstem nuclei and cellular size as estimated histologically correlated with auditory brainstem responses. Most auditory brainstem responses were
similar to those previously described by these authors and others (see above).
Previous studies of newborn human infants have shown that in kernicterus, there
is involvement of brainstem auditory structures from a physiological standpoint, but
structural changes may not be obvious.

The authors note that in their study, the animals were severely affected due to
the administration of sulfadimethoxine. In addition, the time of the experiment was
long as compared to other studies.

Within the cochlear nucleus, the authors note, the areas of spherical cells were
decreased as compared to the globular cells, which were unaffected by hyperbilirubinemia. This suggests that perhaps specific classes of cells are affected in jaundiced
Gunn rats. These data indicate that specific sites in the brainstem cochlear nucleus
are sensitive indicators of structural abnormalities in jaundiced Gunn rats.

The authors further suggest that since there were no gross degenerative changes at the stages studied, perhaps subcellular and/or biochemical alterations should be looked for in the brainstem at this early stage. These early (biochemical) changes may predate the development of gross degenerative changes, and represent a stage where reversibility is still possible. This concept fits well with the notion that early metabolic encephalopathy is potentially reversible.

In a study of auditory brainstem responses in hyperbilirubinemia in primates, six newborn Rhesus monkeys were infused with unconjugated bilirubin (Wennberg, R., et al. 1993). Variations in arterial pH were produced in inspired CO_2. When unconjugated bilirubin levels reached 20 mg/100 ml, CO_2 exposure resulted in a decrease in arterial pH to about 7.0.

When bilirubin levels reached 22–33 mg/100 ml, there was a lengthening of the wave II and IV inter peak latencies and decreased amplitudes. This developed in hyperbilirubinemic monkeys 2–4 h after CO_2 administration. The correction of respiratory acidosis served to return auditory brainstem responses toward normal within 3–20 min. Reexposure of the primates to CO_2 immediately reproduced the auditory brainstem response alterations seen earlier. Results showed that when brainstem bilirubin levels were near the toxic point, changes in CO_2 produced alterations in the auditory response. The authors suggest that in spite of hyperbilirubinemia, the adverse effect of bilirubin in monkeys can be controlled to some extent by manipulating the arterial pH.

Another paper examining the reversibility of auditory brainstem responses has been published (Shapiro, S. 1993). In this study, 17–20-day-old Gunn rats were injected with sulfadimethoxine at a dose of 100 mg/kg which acts to competitively displace bilirubin from albumin-binding sites. This effectively pushes bilirubin into tissues, including brain, producing toxicity.

Results showed that the baseline values in jaundiced Gunn rats included a prolongation of the wave I–II and I–III interwave intervals, and a decrease in amplitude of auditory brainstem responses in waves II and III. These changes were noted at 4 and 8 h post-sulfadimethoxine administration.

These studies showed that therapeutic intervention as late as 8 h with human serum albumin caused a recovery of the altered brainstem auditory-evoked responses. The degree of recovery between Gunn rats treated with albumin at 8 h was not significantly different than the recovery seen when treatment started at 2 h. These data indicate that the time after acute bilirubin toxicity occurs, during which recovery is possible, is at least 8 h.

These data again stress the concept of bilirubin toxicity being initially a metabolic encephalopathy. In terms of auditory brainstem responses to auditory stimuli, the initial lesion seems to be a biochemical one. After a period of time (8+ hours at least), bilirubin produces permanent damage by either causing irreversible structural damage, or becoming bound to brain tissue such that it cannot be withdrawn by treatment .

Another study examined auditory-evoked brainstem responses in newborn infants with hyperbilirubinemia (Agrawal, V., et al. 1998). The study was a prospective cohort study in which 30 hyperbilirubinemic newborn infants were compared to 25 normal control infants. The hyperbilirubinemic infants had serum bilirubin levels at or above 15 mg/100 ml. Brainstem acoustic responses were recorded at the peak of bilirubin levels, before treatment started, after therapy, and at age 2–4 months.

The brainstem acoustic response is an effective technique for the assessment of the integrity of the auditory pathway in human newborns, and is also an optimal method for assessing possible bilirubin damage since the cochlear nucleus seems to be especially sensitive to bilirubin toxicity. Damage at this site may foretell damage to other select cerebral sites such as the basal ganglia and cerebellum. Deafness is a frequent sequelae to hyperbilirubinemia. In the present study, 17 (57%) of the 30 jaundiced newborn infants showed abnormalities in brainstem responses to auditory stimulation. Increased latencies of waves and interwave intervals were the results of hyperbilirubinemia. When patients were followed up, only three had persistent abnormal brainstem auditory response on stimulation.

The authors conclude that auditory brainstem responses may detect substantial evidence of bilirubin encephalopathy at a stage before any signs or symptoms. The alterations in brainstem auditory-evoked responses were transient, and largely absent in nearly all patients followed up at 2–4 months. The changes were positively correlated with serum bilirubin levels. The advantages of measuring brainstem auditory responses are that the method is non-invasive, and cost-effective.

In a somewhat amusing paper in which the term vigintiphobia (Fear of 20) is used in the title, the concept is advanced that it is the levels of serum unbound (unconjugated) bilirubin which is a meaningful indicator of kernicterus, not total bilirubin (Funato, M., et al. 1994). In this study, 72 newborn infants with total bilirubin levels over 20 mg/100 ml were studied. Auditory brainstem responses were measured in 37 of these patients just before treatment with phototherapy or exchange transfusion. The patients were all term infants.

Infants were divided into three groups according to their acoustic brainstem responses: group A = normal ABR, group B = prolonged wave I latency, and group C = prolonged I– III and/or I–V interpeak latency and/or poor amplitude. Results showed that group A and C differed as regards total bilirubin levels, whereas A/B and B/C differences were not statistically significant. The peak unbound bilirubin concentrations were higher in groups B and C as compared to group A. In a logistic regression model, abnormal ABR findings were significantly associated with unbound (unconjugated) bilirubin levels.

The authors state that the auditory pathway is uniquely vulnerable to unconjugated bilirubin toxicity. The findings of a prolongation of latency of wave I (peripheral conduction time), and changes in interpeak latencies of wave I–III and/or I–V (central conduction time) reflect early brain damage from bilirubin. This may well be the single earliest indicator of brain damage, a very important finding. The authors further note that, because of the correlation of ABR deficiencies with the measurement of unbound unconjugated hyperbilirubinemia, this measure, and not total serum bilirubin should be used whenever possible to evaluate any newborn infant at risk of bilirubin encephalopathy/kernicterus.

The ability of auditory brainstem response to predict brain damage was called into question by a study (Yilmaz, Y., et al. 2001) in which auditory brainstem responses were measured in 22 hyperbilirubinemic newborn infants. The newborns were followed up until 1 year of age. Results showed that in two cases, there were altered auditory brainstem responses, but no other neurological sequelae not related to auditory problems. Two other patients had normal auditory responses, but one was cerebral palsied, and the other hypotonic. The findings of patients with neurological sequelae and normal auditory brainstem responses, and visa versa, cast doubt on the method as being absolute as a predictor of outcomes.

In another Gunn rat paper (Ahlfors, C., and Shapiro, S. 2001), the purpose of the study was to try to correlate bilirubin and auditory brainstem responses as regards the physical state of the bilirubin. The authors note in the introduction that the usual laboratory test for jaundiced newborns is total serum bilirubin, which actually correlates poorly with bilirubin encephalopathy/kernicterus. The levels of unconjugated unbound (to albumin) bilirubin are those which correlate best with bilirubin toxicity (see the chapter "Hyperbilirubinemia Revisited," this volume). It is the unbound unconjugated bilirubin moiety which enters brain.

In this study, 21 jaundiced Gunn rat newborn pups at postnatal day 16, and weighing 27 g were tested for auditory brainstem responses. Results showed an increase in wave I–II and I–III conduction times (corresponding to wave I–III and I–V in humans) in jaundiced Gunn rats as the bilirubin toxicity progressed. Total bilirubin was measured by conventional methods. Unconjugated unbound bilirubin was measured by the peroxidase-diazo method (Ahlfors, C. 2000).

Results showed the mean total serum bilirubin levels were on average 8.4 mg/100 ml, and the mean albumin level was 2.8 g/100 ml. The mean total bilirubin/albumin molar ratio was 0.35. The mean value for unconjugated unbound bilirubin was 5.8 μg/100 ml (0.099 μmol/l). In terms of correlations, only the unconjugated unbound bilirubin levels correlated significantly with changes in interwave intervals and amplitudes of the auditory brainstem responses. There was no significant correlation between altered auditory brainstem responses and total serum bilirubin levels or molar ratios of bilirubin/albumin.

These data, the authors note, suggest that unconjugated unbound bilirubin levels are more useful and predictive with regards to neurological damage and sequelae than are measurements of total bilirubin levels so commonly used. However, the unconjugated unbound bilirubin accounted for only 30% of the alteration in auditory brainstem responses, suggesting other factors such as the blood–brain barrier may be a contributor to the overall risk for bilirubin encephalopathy/kernicterus. While the bilirubin/albumin ratio did not correlate with bilirubin-induced acoustic brainstem responses in jaundiced Gunn rats, it does correlate in newborn jaundiced human infants.

The authors comment that while bilirubin-induced changes in auditory brainstem responses are transient, they are associated with brain lesions in kernicterus. Resolution of altered acoustic brainstem responses in hyperbilirubinemia does not assure that there will be no neurological sequelae. The adoption of the unconjugated unbound bilirubin peroxidase method to clinical laboratory protocols should be a priority.

In another study (Jiang, Z., et al. 2007), 90 term newborn infants who had hyperbilirubinemia were tested for altered acoustic brainstem response. The purpose of the study was to look for differences between auditory brainstem responses and bilirubin levels in a large number of subjects.

Results showed that the response threshold for acoustic responses was elevated in hyperbilirubinemic infants. In terms of acoustic brainstem responses, wave latencies and I–V intervals were increased significantly in hyperbilirubinemic newborns. Twenty-five neonates (28%) had abnormal responses which suggested peripheral auditory changes, while 16 subjects (18%) had alterations implying a central involvement.

The authors noted that while there was a correlation between auditory brainstem response and total serum bilirubin, it was not a close one. Thus, differences were noted between 11–20 mg/100 ml and over 20 mg/100 ml, but not between 11–15 mg/100 ml and 16–20 mg/ml.

In a case report (Smith, C., et al. 2004), a premature infant (33 weeks) was delivered to an ORH-negative blood type mother. The pregnancy was complicated by fetal anemia. Three fetal transfusions were performed. Following birth, severe hemolytic anemia accompanied by edema confirmed a diagnosis of hydrops fetalis secondary to Rh isoimmunization. Neurological exam by day 2 showed diminished tone and reactivity. By day 3, unconjugated bilirubin levels were at 13.3 mg/100 ml.

Even while the unconjugated bilirubin was only 13.3 mg/100 ml, other complicating factors indicated that the patient was a candidate for exchange transfusion. In agreement was the neurological exam suggestive of early bilirubin encephalopathy/kernicterus. Acoustic brainstem response was tested before transfusion. Results from this test showed baseline acoustic brainstem responses characterized by prolonged interpeak latencies of waves. The patient was not sedated for the exam. A double exchange transfusion was performed at 75 h of age. Phototherapy, initiated on day 1, was minimized due to skin bronzing.

The authors comment that this was a complicated case necessitating early exchange transfusion in spite of unconjugated bilirubin levels of only 13.3 mg/100 ml. Other reports suggested changes in wave latencies at unconjugated bilirubin levels at or below 20 mg/100 ml (Wennberg, R., et al. 1982). As suggested by the present study, the correction of the altered acoustic brainstem response through treatments aimed at lowering hyperbilirubinemia demonstrates the clinical importance of acoustic brainstem responses as an indicator of bilirubin toxicity in any given case. This is especially important in cases like the present one in which other confounding problems rendered the infant less "accessible" for clinical evaluation than in otherwise normal newborn infants.

Another study (Ahlfors, C., and Parker, A. 2008) examined the correlation between unbound, unconjugated bilirubin and the auditory brainstem response in newborn infants. Automated auditory brainstem response to stimulus is frequently used to screen newborn infants for hearing deficits. Since at least 60% of newborns experience elevated bilirubin levels during the first few days of life, it seemed appropriate to see if there was a correlation between bilirubin and acoustic brainstem responses. Small changes in the auditory brainstem response have been seen previously in infants with total bilirubin levels of 10 mg/100 ml (Lenhardt, M., McArtor, R., and Bryant, B. 1984). The study was done to evaluate the correlation of both total serum bilirubin, and the levels of unbound unconjugated bilirubin with changes in acoustic brainstem responses.

In this study, charts were reviewed to find newborn infants with gestational age of 34 weeks or greater, Patients with no other problems and who had the auditory screening test within 4 h of bilirubin measurements were chosen. Unbound unconjugated bilirubin was measured by the peroxidase method in a clinical laboratory.

Forty-four infants met the above requirements for inclusion in the study. There were 28 males and 16 female infants. In terms of auditory brainstem responses sufficient to indicate the need to refer the patient for further tests, the correlation with the unbound unconjugated bilirubin (2.63 μg/100 ml) levels was at a significantly higher level of confidence than total serum bilirubin levels. In addition, increasing levels of unbound unconjugated bilirubin correlated well with an increasing severity of abnormal auditory brainstem responses.

The authors comment that unbound unconjugated bilirubin levels are shown in this study to correlate with auditory brainstem responses, and have been shown to correlate better than total bilirubin levels with neurotoxicity (Nakamura, H., et al. 1992). These results have some consequences since auditory brainstem responses sufficient to indicate referral for further testing may result from hyperbilirubinemia and be transient. The authors suggest that their data are sufficient to warrant further studies examining the effects of unbound unconjugated bilirubin on auditory brainstem responses This represents further evidence of the importance of unbound unconjugated bilirubin in determining/predicting bilirubin's toxicity in brain.

In a recent paper, studies are described in which a new model was developed for hyperbilirubinemia which was a hemolytic anemia model. Bilirubin levels in serum are sufficient to result in kernicterus. The idea was to create a model of bilirubin encephalopathy/kernicterus more closely correlated with kernicterus in human infants. The authors suggest that the use of sulfanamides in Gunn rats, which displaces bilirubin from albumin-binding sites, may not be as good of a model as one incorporating hemolysis as the key feature (Rice, A., and Shapiro, S. 2008).

In this study, phenylhydrazine (PHZ), a substance which lyses red blood cells, was administered to newborn Gunn rats. Total bilirubin levels and hematocrit levels were then measured after administration of PHZ. Auditory brainstem responses were obtained in 40 jaundiced and 10 non-jaundiced littermates.

Results showed that a baseline total bilirubin level of 10.8 mg/100 ml in jaundiced 15-day-old Gunn rats increased to 27.3 mg/100 ml with PHZ. This treatment did produce a high mortality rate at the high PHZ treatment level of 75 mg/kg. Clinical observations of jaundiced Gunn rats included abnormal muscle tone in the extremities. Hematocrit levels were lowest at 24 h after PHZ administration.

Acoustic brainstem responses were recorded at 48 h following PHZ injection. The acoustic brainstem response animals were similar to controls in all physiological parameters such as weight, age, etc. Results of the acoustic brainstem responses showed a significant difference between the 75 mg/kg dose in jaundiced Gunn rats and saline controls. There was an increased latency and decreased amplitude of waves II and III (corresponding to waves III and IV in humans).

The authors comment that there are not a lot of animal models for hyperbilirubinemia and associated kernicterus. Animal models in which bilirubin is administered are generally not satisfactory because of a variety of other bilirubin-induced changes. Gunn rats are ideal in many ways since they have a similar enzymatic defect as have newborn humans with Crigler–Najjar syndrome.

In this study, the authors discuss a model in which hemolysis is a key feature. Induced by PHZ, this results in altered auditory brainstem responses in the homozygous hyperbilirubinemic Gunn rats, but not in heterozygous littermates. The heterozygous littermates have elevated bilirubin levels from hemolytic anemia, but not changes in auditory brainstem responses. The high doses of PHZ produced changes in the auditory responses similar to sulfonamide-treated Gunn rats, namely latencies in waves II and III, increased I–II and I–III IWIs, and decreased amplitude of waves II and III.

The authors cannot rule out the possibility that PHZ has some as yet unknown direct effect on brain metabolism, or produce changes in auditory brainstem responses by some mechanism besides through elevated bilirubin levels. The authors note, however, that using PHZ to produce hyperbilirubinemia might prove to be more clinically relevant than using sulfonamides to create bilirubin encephalopathy/kernicterus. This model will allow correlation of total bilirubin, free unbound unconjugated bilirubin, and albumin binding with the highly selective brain damage produced by bilirubin. This model would permit a more direct animal comparison of neurotoxocity to human kernicterus caused by hemolytic anemia.

Progressive Familial Intrahepatic Cholestasis

Progressive familial intrahepatic cholestasis (PFIC) is a progressive disorder that affects children. It is initially characterized by intrahepatic cholestasis which worsens over time. It may present with jaundice in the first weeks of life, or may appear after several months. Byler's disease, another term for one type of PFIC, was coined after an immigrant who brought the disorder to the USA. PFIC is a rare inherited disorder in which at least three subtypes have been identified, and called PFIC-1, PFIC-2, and PFIC-3. These three subtypes carry descriptive names which partially describe them: PFIC-1 is called familial intrahepatic cholestasis, PFIC-2 may be called bile salt export pump deficiency, and PFIC-3 is multidrug resistant-associated protein deficiency. A mild not as severe form of PFIC is called benign recurrent intrahepatic cholestasis.

Early descriptions of intrahepatic cholestasis center around congenital malformations – complete or partia – of bile canaliculi, ducts of Hering, and first-order bile ducts (for review, see Smetana, H., Edlow, J., and Glunz, P. 1965). In many cases of intrahepatic cholestasis due to lack of intrahepatic bile passages, giant cell transformation of hepatic parenchyma has occurred. Whether this giant cell transformation causes failure of patency of bile canaliculi or visa versa, is not clear. The outcome is quite similar to PFIC: elevation of total bilirubin soon after birth, or somewhat later, followed by absence of bile in the duodenum, jaundice, pale stools, pruritis, liver failure, enceph alopathy, and death. In fact, the differential diagnosis of PFIC and intrahepatic biliary atresia or hypoplasia is often difficult to make. The clinical presentation of neonatal hepatitis is also quite similar to the above two clinical entities. Laboratory results in all three, such as serum transaminases, alkaline phosphate, and conjugated hyperbilirubinemia, may be features of any of these three disorders. Liver biopsy is usually helpful in establishing a diagnosis (van Nievwkerk, C., et al. 1996).

In a case report (Khalil, A. 2000) of PFIC, a patient was presented at the age of 2 months with deep jaundice, and severe itching. The parents were first cousins, but there was no history of similar problems in other children. Physical exam showed a deeply jaundiced infant covered with scratch marks. Both liver and spleen were

---

This chapter is dedicated to Isabella (Bella) Hubner, and her parents Doug and Michelle Hubner.

D.W. McCandless, *Kernicterus*, Contemporary Clinical Neuroscience,
DOI 10.1007/978-1-4419-6555-4_18, © Springer Science+Business Media, LLC 2011

palpable below the costal margins. There were no signs of chronic liver disease. Total bilirubin was 20 mg/100 ml, with a direct reading of 8.7 mg/100 ml.Liver biopsy studies showed prominent parenchymal bile stasis, and partially distorted cytoarchitecture, but no signs of a diminished number of intrahepatic bile ducts. These results from the liver biopsy indicate a diagnosis of Byler's disease, or PFIC. The fact that in this case serum gamma glutamyltransferase is elevated may indicate that this case has type 3 PFIC. This type has the poorest prognosis, and the development of liver failure and subsequent encephalopathy is likely. This patient is a clear candidate for a liver transplant (see Figs. 1, 2, and 3).

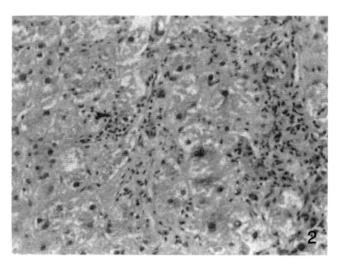

Fig. 1 Liver biopsy specimen from a patient with intrahepatic cholestasis. From Arch. Pathol. Id = 1543-2165-123-3-4-FO2

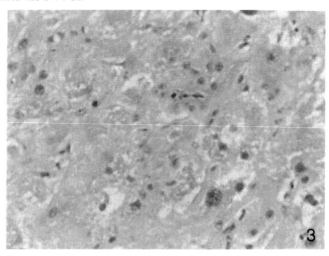

Fig. 2 Higher magnification showing ballooned hepatocytes, and canalicular bile Stains. From Arch. Pathol. Id = 1543-2165-123-3-4-FO2

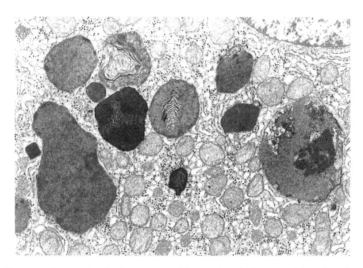

Fig. 3 Electron micrograph of a hepatocyte with cytoplasmic lysosomal lemellar phospholipid inclusions. From Arch. Pathol. Id = 1543 1 = 10 ASICCBA%3E2.0.CO %3B2

A recent paper (de Vree, J., et al. 1998) reports clinical research examining possible genetic defects resulting in type 3 PFIC. In this study, two livers from PFIC transplant patients were utilized for study. In the first case, a newborn boy had recurrent episodes of jaundice from 3 months of age. At that time, he had severe jaundice and pruritis. At 3 years, he had hepatosplenomegaly, elevated liver enzymes, increased gamma glutamyltransferase (gamma gt) activity, and high serum bile acid. Orthotopic liver transplantation was performed at age 3.5 years.

A second patient had jaundice and pruritis since the age of 8 months. By 3 years of age, he had hepatosplenomegally, increased gamma gt activity, and increased bile acids. Liver enzymes were mildly elevated. As in case one, a liver biopsy showed portal inflammation and fibrosis. Treatments with ursodeoxycholate were not successful. Liver transplant was performed at age 9 years.

Results showed that the histological and biochemical findings from the livers of the above two cases classified them both in the third group of PFIC patients. Microscopic examination of the livers of both patients showed extensive proliferation of bile ducts and metaplasia of hepatocytes. Bile salt analyses were normal. Microscopic examination of liver samples from both patients showed a complete lack of canalicular staining using a polyclonal antibody against MDR3 P-gp, alpha REG1. As an internal control, liver sections were stained successfully with other stains for canicular transporters. These data are strongly indicative of an absence of MDR3 P-gp in the two patients.

Rt-PCR results suggested that each patient was homozygous for the mutation producing their MDR3 defect, and disease process. Results from these studies could be used to confirm homozygosity and heterozygosity of parents, and also to demonstrate that PFIC has a recessive inheritance aspect.

The authors state that these studies elucidate a mutation in the MDR gene, and thereby illustrates the functional role for MDR3 P-gp. This defect was responsible for the development of a newer form of PFIC. The intrahepatic cholestasis was relentless, and led to liver failure, corrected by liver transplantation. MDR3 P-gp was not measureable in either patient.

Murine studies using knockout mice have indicated that heterozygous parents of homozygous offspring will not develop liver disorders. The histological features seen in the two patients, and consisting of inflammation, changes in bile duct epithelium and ductular metaplasia can be accounted for by the detergent-like action of bile salts. These bile salts are inactivated by phospholipids in non-diseased people. The lack of phospholipid secretion in homozygous patients accounts for ductal changes.

In the MDR2 mouse model, feeding the bile salt UDCA has resulted in the complete stoppage of the progression of liver disease (Balistreri, W. 1997). In addition, UDCA is effective in several other examples of pediatric hapatobiliary disease. The use of UDCA is less effective in patients with elevated gamma GT levels. The authors speculate that patients not responding to UDCA have a complete defect in phospholipid secretion and that UDCA cannot reduce the bile salt toxicity. Those who do respond may have only a partial defect in bile salts. Both the above two patients had a complete defect, and in fact were unresponsive to UDCA treatment. Successful treatment with UDCA might obviate the need for a later liver transplant due to liver failure and hepatic encephalopathy.

In a paper by Jansen and Muller (Jansen, P., and Muller, M. 2000), defects in the so-called ABC transporter superfamily are briefly reviewed. The ABC transporters are key in most human organs, and therefore play important roles in a variety of disorders including PFIC, eye disorders such as retinitis pigmentosa, familial HDL deficiency, hyperinsulin-induced hypoglycemia, adrenoleukodystrophy, etc.

As stated earlier, PFIC type 1 is one of the three described variants of PFIC. In this variant, children often have recurrent episodes of intrahepatic cholestasis. This progresses to a permanent cholestasis with concomitant fibrosis of the liver, cirrhosis, and liver failure. The decay in liver function results in encephalopathy and death in the absence of liver transplantation. Jaundice and pruritis are severe. The bile canaliculi seem to be filled with a course appearing bile when observed microscopically. Patients may be descendants of Jacob Byler, an Amish man who emigrated to the USA and founded a large Amish population, in whom there were many cases of PFIC (Bylers syndromee). Many patients with PFIC exist who are not related to the Amish family.

In this group of PFIC patients, the genetic defect was traced to the F1C1 locus on chromosome 18q21–q22. This locus has been characterized by mapping and gene scanning to a region which encodes a subfamily of P type ATPases. These ATPases are actually not ABC transporters, but belong to an ion transport pump encoding family. In the liver, the F1C1 is not highly expressed, but in the pancreas, small intestine, and stomach it is highly expressed. Subsequently, the exact relation between F1C1 and the liver is not obvious.

A mild (benign) form of PFIC is similar to the above progressive form, but does not lead to liver failure. Patients are severely jaundiced during attacks, but do not have elevated gamma GT in their serum. Although liver failure is not a feature, sometimes liver transplant is indicated to ease the pressure on social life of patients due to cholestatic episodes.

Studies have suggested that the F1C1 locus was not involved in all patients with PFIC. Other patients had the defect traced to a locus on chromosome 2q24. This was later shown to be the bile salt export pump (BSEP). This gene encodes the canalicular bile salt export pump. This pump is a P-glycoprotein of the ABC transporter superfamily. This protein is specific to the liver and is located in the plasma membrane of the hepatocyte. The authors showed that antibodies for BSEP did not stain canaliculi in type 2 PFIC patients. This indicates an absence of a BSEP.

As seen in type 1 PFIC, serum gamma GT is not elevated, and bile duct proliferation is absent. The difference, however, is that type 2 PFIC starts as a giant cell hepatitis. These patients are severely jaundiced, and the disorder quickly progresses to a stage in which liver transport is required. When examined microscopically, the liver shows inflamination, giant cells, and portal fibrosis. This is more extensive than in type 1 PFIC. PFIC type 2 patients do not respond positively to USDC therapy. Bile acids can be found in low levels in the bile of type 2 PFIC patients.

As described above, the PFIC type 3 subtype is characterized by elevated gamma GT activity. Microscopic examination of liver shows an extensive bile duct proliferation and portal fibrosis (de Vree, J., et al. 1998). The gene mutation in this type is a phospholipids translocator. This gene belongs to the ABC transporter superfamily. The mechanism of toxicity centers on phosphotidylcholine which normally acts to dilute the toxic effects of bile acids. In the absence of a phospholipids transporter, the bile acids are toxic to both bile canaliculi and cholangiocytes. This then represents the primary mode of hepatic toxicity in PFIC type 3.

In type 3 PFIC, patient symptoms appear somewhat later than in type 1 or type 2. Jaundice may not be as severe, and liver failure occurs at a later age. Liver transplantation is usually necessary, although treatment with UDCA therapy may provide some degree of relief from symptoms. This paper concludes with the comment that liver transplantation remains the only cure for these disorders, and will be necessary in almost all cases, due to encephalopathy and liver failure. In the absence of liver transplant, death occurs.

In an interesting report (Ko, J., et al. 2007), neonatal intrahepatic cholestasis related to citrin deficiency is described. Over a 2-year period, 47 patients admitted to the hospital were diagnosed with neonatal cholestasis. There was a diverse range of "sub diagnoses" including Dubin-Johnson syndrome, extrahepatic biliary artresia. PFIC, Alagille syndrome, and of the 47, three cases of citrin deficiency induced intrahepatic cholestasis. The authors term this disorder neonatal intrahepatic cholestasis caused by citrin deficiency (NICCD).

In these three cases, birth weights ranged from 2.7 to 3.1 kg. The patients were all jaundiced, and neonatal hepatitis and biliary atresia was suspected. In all cases, direct serum bilirubin was between 2.7 and 3.2 mg/100 ml. Blood ammonia was elevated in all three cases. Gamma GT and alpha fetoprotein were both also elevated as compared to normal controls. Galactosemia was present in all three patients. Results from liver biopsies showed cholestasis and microvascular fat changes, and moderate periportal fibrosis.

The authors note that citrullinemia is described as having two types: a neonatal onset variety (CTLN1), and an adult onset variety (CTNL2). The neonatal variety is characterized by severe hyperammonemia, lethargy, poor feeding, and tachypnea. The adult variety can occur later in life, and is characterized by frequent bouts with hyperammoninemia, mental problems, encephalopathy, and death within a few years after onset. A gene called SLC25A13, which encodes for citrin, is defective in CTNL2.

Citrin functions in the liver urea cycle. It acts to supply mitochondrial aspartate to the cytoplasm for incorporation into argininosuccinate. Lack of synthesis of argininosuccinate may cause citrulline accumulation (Saheki, T., et al. 2004). Some other patients shown to have a deficiency of the SLC25A13 gene have idiopathic neonatal hepatitis with accompanying cholestasis, hyperaminoacidemia, galactosemia, and hepatic steatosis.

The authors comment that in their study, percutaneous liver biopsy was successful in eliminating biliary atresia as a cause, as well as other causes of PFIC. Both plasma amino acid analysis and serum gamma GT levels were useful in making the diagnosis. Patients with NICCD had hyperammonia and galactosemia. Hypoglycemia was present in one of the three patients. Plasma amino acid analysis showed that citrulline, methionine, threonine, and arginine were elevated. An estimate of the occurrence of NICCD in Japan was 1:34,000 (Shigematsu, Y., et al. 2002).

The management of NICCD is aimed at treating galactosemia and cholestasis. A medium chain triglyceride formula and diet supplementation with fat soluble vitamins may have some effectiveness in preventing the complications of prolonged cholestasis. The authors state that the prognosis of these three patients was good because liver function was normalized before 13 months of age, after which growth accelerated. One patient developed liver failure and required a liver transplant. The prognosis of CTLN2 is poor, however, liver transplantation is effective.

There appears to be a relation between the liver/cholestasis, and thyroid disease. This relationship is due to the fact that the liver is the key organ in thyroid hormone metabolism. A recent paper examines this relation in two cases of hyperthyroidism and intrahepatic cholestasis (Soylu, A., et al. 2008).

The first case presented with jaundice and tachycardia, and a non-palpable thyroid gland. Total bilirubin was 15 mg/100 ml, direct bilirubin was 11 mg/100 ml. Alkaline phosphatase, lactate dehydrogenase, and gamma glutamyl transferase assays for hepatitis were negative. A preliminary diagnosis of intrahepatic cholestasis was made. Thyroid hormone levels were elevated, and the assessment was that the patient had a diffuse hyperplastic thyroid gland. Liver biopsy showed acute intrahepatic cholestasis. The cholestasis was limited to the bile canaliculi. Bilirubin levels continued to rise, reaching a total of 37 mg/100 ml, the direct being 24 mg/100 ml. Pruritis was severe. The authors state that the opinion was that the intrahepatic cholestasis was secondary to hyperthyroidism. Radioactive iodine was administered to this patient for thyroid ablation. Over a course of 3 months, all abnormal blood tests had returned to normal.

In a second case, a patient had been on antithyroid medication for about 1 year. Following discontinuation of therapy, the patient developed pruritis and jaundice. On examination, she was found to have subclinical hyperthyroidism and bilirubin levels 24.5 mg/100 ml total, and 17.5 mg/100 ml of that was direct reacting. A liver biopsy showed degeneration and regeneration of hepatocytes, pigment accumulation in hepatocytes, bile plugs, and dilation of sinusoids. These results were consistent with a diagnosis of intrahepatic cholestasis. Treatment with corticosteroids and ursodeoxycholic acid did not seem to alter jaundice or pruritis, so rifampicin was also administered. This resulted in the gradual resolution of the intrahepatic cholestasis and lowering of serum bilirubin levels.

The authors point out that the occurrence of hyperthyroidism and intrahepatic cholestasis together is rare, and the precise mechanism is unknown. Patients with high levels of T3 and T4 may develop severe hypoxemia, which may adversely affect the pericentral portion of the hepatic acini. The authors emphasize that hyperthroidism should be considered in liver dysfunction.

The development of liver failure is a common evolution of liver disease which starts as progressive familial intrahepatic cholestasis. Liver failure per se may be classed as acute (presenting within 28 days of jaundice), and subacute (presenting between 29 and 72 days of the onset of jaundice) (Gimson, A. 1996). These distinctions are important in suggesting etiologies in hepatic failure. Acute or fulminant hepatic failure is not common, and usually results from hepatitis or drug hepatotoxicity. In the pediatric setting, increasing bilirubin levels, especially direct reacting levels may lead to rapid liver damage and intrahepatic cholestasis. Viral hepatitis is considered the most common cause of acute liver failure. Cannalicular cholestasis is a common feature seen in liver biopsies from affected patients. In fulminant hepatic failure, resultant hepatic encephalopathy is associated with cerebral edema. The higher the grade of encephalopathy, the worse is the mortality rate. Cerebral edema is a major cause of mortality in hepatic encephalopathy, and is present in as many as 80% of grade 4 encephalopathic patients.

Treatment of cholestasis and hyperbilirubinemia in pediatric cases has been described (Tobias, J. 2002). The drug of choice for treating pediatric cases has been ursodeoxycholic acid (UDCA). This study was a retrospective study looking at results from five patients who had been treated in a pediatric intensive care unit with UDCA over a 12-month period.

Results showed that four of the five patients responded favorably to UDCA treatment. The bilirubin levels in the four who responded averaged 5.5 mg/100 ml total, with 4.2 mg/ml being direct reacting. The fifth patient had increasing bilirubin levels to greater than 30 mg/100 ml. This fifth patient did have a brief decrease in bilirubin levels when UDCA was initiated, but soon began to rise in spite of dose increases.

The authors note that regardless of the cause of liver damage, the first concern is to reduce and eliminate treatable causes of cholestasis and hyperbilirubinemia. The goal of the diagnostician is to search for infections (hepatitis), hepatotoxic drugs, and structural biliary problems which can be surgically corrected. The authors were unable to define an etiological factor in any of the five cases, and surmise that intrahepatic cholestasis in children is probably multifactorial. Possible factors include sepsis, shock, total parenteral feeding, and presence of bile acids.

Untreated cholestasis and hyperbilirubinemia may have serious deleterious effects which include hepatocellular injury which may be due to several mechanisms. These include the solubilization of membrane-bound cholesterol and phospholipids from the hepatocyte membrane, an increase in intracellular calcium levels accelerating apoptosis, and finally, the uncoupling of oxidative phosphorylation by bile acids with resultant depletion of ATP (Javitt, J. 1966). Class 1 antigens may be released which in turn may lead to autoimmune injury (Krahenbuhl, S., et al. 1995).

The nature of "protection" by UDCA is not entirely clear, but may act by increasing bile flow, and by having a direct protective effect from deleterious bile acids on hepatocytes. UDCA protects membranes, and mitochondrial injury. Dosage regimens are usually listed as 10–20 mg/kg, with dosages as high as 30–45 mg/kg in cases which do not respond to lower dosages.

The exact incidence of acute hepatic failure has not been established because the International Classification of Diseases does not have a code for acute hepatic failure which limits inclusion of many databases (Calmus, Y., et al. 1990). The mortality rate from acute hepatic failure has dropped recently due to more aggressive therapy, and the advent of orthotopic liver transportation (Sass, D., and Shakil, A. 2005).

Encephalopathy is a critical phenomenon in judging the severity of liver failure, and is described as existing in four stages. Intracranial pressure monitoring due to edema is important in hepatic encephalopathy in stages 3 and 4. Oxygen consumption may be an indicator of cerebral blood flow.

The major successful treatment of acute liver failure is orthotopic liver transplantation. In an effort to increase efficacy, orthotopic liver transplantation should be balanced against potential success. Without transplantation the mortality rate approaches 80%, and it is important to recognize irreversible liver failure early in order to achieve transplantation before complications which might render surgery impossible. Early recognition of reversible cases could prevent unnecessary transplantation.

Criteria viewed as indicating a transplant include patients under 10 years of age. Also included are patients with bilirubin levels greater than 17.5 mg/100 ml lasting longer than 7 days, and a prothrombin time greater than 100 s. Patients with grade 3 or 4 encephalopathy are also candidates. Serum creatinine greater than 3.4 mg/100 ml is also viewed as an indicator for transplantation.

Alternative therapeutic strategies include auxiliary liver transplantation, and hepatocyte transplantation. In auxiliary liver transplant therapy, a partial liver graft is placed heterotopically or orthotopically, leaving part of the host liver in situ. This seems effective in patients who might be expected to recover liver function at a later date. When that happens, immunosuppresion treatment can be stopped, and this occurs in about 70% of patients (Hoofnagle, J., et al. 1995). Regeneration seems to be most successful in younger patients. The other possible alternative treatment consists of hepatocyte transplantation. In this method, hepatocytes from a donor are prepared and infused into the recipient. This method may serve to provide functional hepatocytes with a reasonable life span, but exactly how much liver function recovery is possible is not clear. Controlled clinical studies are needed for the appropriate evaluation of hepatocyte transplantation (Chenard-Neu, M., et al. 1996; Wang, X., and Anderson, R. 1994).

The clinical complexity of hepatic failure and subsequent encephalopathy can be illustrated by the management of seizures in hepatic failure (de Lacerda, G. 2008). Acute or chronic liver failure may lead to hepatic encephalopathy, and in stages 3 and 4, seizures may be an important feature (Anderson, G. 2004). The presence of excitatory neurotransmitters in brains of hepatic encephalopathy patients could be a causative mechanism for seizure propagation. Controlling seizures in patients with hepatic encephalopathy is complicated. The use of antiepileptic drugs such as phenytoin or benzodiapines may not be effective (Bhatia, V., Batra, Y., and Acharya, S. 2004). Barbiturate induced sedation, and hypothermia therapy may be beneficial (Eleftheriadis, N., et al. 2003; Shet, R. 2004).

As a further complication, antiepileptic drug metabolism is usually impaired in PFIC due to decreased liver function capacity. Antiepileptic drugs are metabolized by the liver. Liver enzymes are expected to be influenced by liver disease according to the exact genetic defect in each individual case. An example is that CYP3A4 is affected in hepatocellular dysfunction, whereas CYP2E1 is affected in cholestasis. Hepatic function is very difficult to estimate in liver failure, making antiepileptin drug adjustment much more difficult to determine than it is even in non-liver disease patients.

In a case of idiopathic fulminant hepatic failure (Mohanty, A., and Schiff, E. 2009), the importance of identifying the cause of liver failure if possible, as well as the value of prognostic screening are discussed. The authors describe a case of a young patient most likely exposed to ecstacy (3,4-methylenedioxymethamphetamine). The exposure could not be verified.

The patient had normal vital signs when first examined, and was diffusely jaundiced. Total bilirubin was 31.3 mg/100 ml, the direct contribution was 24.3 mg/100 ml. Imaging studies were negative for biliary dilation or cholelithiasis. Hepatitis screening was negative. Other tests including liver biopsy were negative or inconclusive. Over 4 days the patients condition worsened, and encephalopathy deepened. A liver transplant was performed on the seventh day of the hospital. During surgery the liver was seen as enlarged and cholestatic. The patient recovered without major complications. No cause was ever identified for the fulminant hepatic failure, except for possible exposure to the drug ecstacy.

This study is noteworthy because in spite of no diagnosis as to cause of the hepatic failure, transplantation was successful. The etiology of liver failure is an excellent prognosticator of transplantation success. The authors suggest that other prognostic scoring systems may not always accurately predict outcome, and a liberal dose of clinical judgment must be applied.

The development of animal models of acute liver failure has languished to some extent. This is due to the multiple causes of liver failure, and many complications thereof (Belanger, M., and Butterworth, R. 2005). These authors correctly state that currently liver transplantation holds the most hope for successful treatment of liver failure, but the number of donors is limited as are medical resources. Many patients die while waiting for a transplant, and even after transplant, the mortality rate is too high.

Several models of liver failure are available including hepatotoxic drug models, viral models, surgical models, and combination models. As regards drug models of acute hepatic failure, acetaminophen is a drug which produces acute liver failure in humans who overdose with the drug. Acetaminophen toxicity results from a toxic metabolite *N*-acetyl-p-benzoquinoneimine. A model using acetaminophen and glutathione in combination produced a reasonable and reproducible result (Kelly, J., et al. 1992).

Thioacetamide is another compound administered to animals in order to produce acute liver failure (Zimmerman, H. 1986). This model seems to elevate ammonia and produce encephalopathy, but thioacetamide has been found in brains of animals so treated. Therefore, some results from this model could be due to the drug and not to acute liver failure.

Viral models have had only limited utility in spite of the fact that viral hepatitis is a major cause of acute liver failure. The use of rabbit hemorrhagic disease virus (RHDV) has been described (Farivar, M., et al. 1976) This produces a model similar to acute liver failure in humans, the symptoms of which include elevated bilirubin levels, elevated transaminases, hypoglycemia, stupor, coma, and death. Eighty-five percent of animals die within 36–54 h. One drawback to this model is that only rabbits are susceptible to this virus.

Surgical models include total hepatectomy, partial hepatectomy, liver devascularization, and partial devascularation. The disadvantage of complete hepatectomy models is that the liver is not there to possibly contribute to the progression of liver failure. The use of partial hepatectomy has been somewhat successful. One proposal was remove about 95% of the liver, which resulted in an 80% mortality rate (Roger, V., et al. 1996). This model has not been studied enough that a clear description of biochemical changes or of neurological involvement has been characterized.

Complete hepatic devasculairzation is performed by doing a portacaval anastomosis, then ligating the hepatic artery. This is a reasonable model in that blood ammonia levels are elevated as they are in the human counterpart. Acute hepatic failure is produced, followed by the development hepatic encephalopathy. Death may occur in only hours. Combined surgical and drug methodologies have also been developed. These include a 70% hepatectomy followed by galactosamine administration. This produced a 78% mortality rate and liver failure with encephalopathy in about 5 days (Karrer, F., et al. 1984).

The precise mechanism by which hepatic encephalopathy is produced via liver failure is not completely known. There have been many investigations into changes to various neurochemical substances in the brains of animal models with hepatic encephalopathy. Many of these result in alterations, but which might be primary and which are secondary is hard to determine. The positive feature of at least early hepatic encephalopathy is that the symptoms may be reversible, suggesting as in nearly all metabolic encephalopathies, that the early changes are of a biochemical nature and occur before permanent (structural) changes can develop. Although technically beyond the scope of this book on kernicterus, a brief comment on the pathogenesis of hepatic encephalopathy in this chapter is warranted.

A major feature of hepatic encephalopathy is the elevation of blood ammonia. That this elevation in ammonia levels is a key feature of hepatic encephalopathy and a likely primary toxin in the development of symptoms has been speculated for many years (Zimmerman, H. 1986; Chalmers, T. 1960; Zieve, L. 1966) Several mechanisms of ammonia toxicity have been suggested including a depletion of alpha-ketogluterate, decreased activities of pyruvate and alpha-ketoglutarate decar-boxylases, or a decrease in ATP levels due at least to increased energy consumption from the formation of glutamine. Any of these, or a combination is postulated to lead to a decrease of available ATP necessary for normal cerebral function (Zieve, L. 1966). Further, analysis of symptoms and their anatomical localization in the brain suggest specific cerebral regions one might examine in any neurochemical study. As in other metabolic encephalopathies, the initial neurochemical alterations seem to be limited to specific and highly reproducible brain regions. The reasons for selective vulnerability of various regions to insult in metabolic encephalopathy remain a mystery.

With these caveats in mind, an early in vivo study was undertaken to exam-ine the effects of ammonia on regional cerebral energy metabolism (Schenker, S., et al. 1967). In these studies, ammonium acetate was injected into rats so as to produce symptoms in a few minutes. Animals were sacrificed at various time points such that analyses were performed at precoma stages, coma, and recovery time points. Results were compared to neurologically intact control rats. Energy metabolites were measured in the brainstem and cerebral cortex of ammonia injected rats.

Results showed a time-graded decrease in both phosphocreatine and ATP in the base of the brain (brainstem), but not the cerebral cortex. The decrease stopped, and levels of high-energy phosphates were returning toward normal as the rats were recovering from ammonium acetate-induced coma. The correlation between recovery of consciousness and recovery of energy metabolites is a key observation. Both glucose and glycogen were decreased in comatose rats (38 and 73%, respec-tively), but the decreases were similar in cortex and brainstem. Corollary studies on these rats showed no uncoupling of oxidative phosphorylation in the comatose rats brainstem (see Figs. 4 and 5).

This study shows conclusively that ATP and phosphocreatine were significantly and consistently depleted in the brainstem of rats rendered encephalopathic with ammonium acetate. These changes were anatomically selective to the brainstem, the site which mediates the symptoms produced. Further, with recovery from coma, these high-energy metabolites were recovering from their depletion in accordance with symptomology. Furthermore, given the central position of energy-yielding metabolites, such as ATP and phosphocreatine, it is easy to see how depletion of the key compounds would have profound secondary effects on many other neurochemicals.

Fig. 4 Brainstem
phosphocreatine in ammonia
intoxication. Ammonia was
injected and symptoms
monitored. ** = 0.01 level of
confidence significantly
decreased as compared to
saline injected controls.
Adopted from Schenker, S.,
et al. 1967

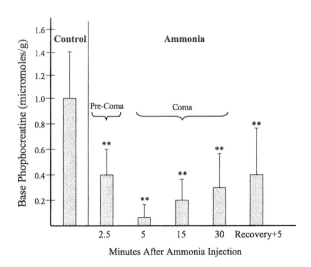

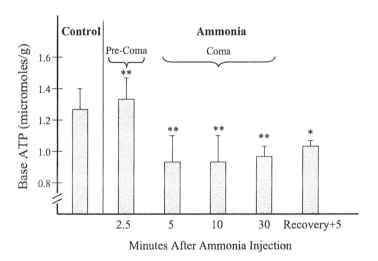

Fig. 5 Brainstem ATP in ammonia intoxication. Ammonia injected, and symptoms were monitored. ** = 0.01 level of confidence, * = 0.05 level of confidence. Adopted from Schenker, S. et al.
1967

A very strong case exists for elevation of blood ammonia to be the key alteration
in liver failure and resultant stupor and coma, and other neurological symptoms.
These symptoms are mediated in the brainstem, site of ATP and phosphocreatine
depletions. This alone is a strong argument for these energy metabolite changes
being the primary change in this metabolic encephalopathy. By primary changes is
meant those alterations which occur initially, and lead/cause secondary changes in
a wide variety of neurochemicals. Adequate energy is required for most cellular

metabolic processes. Depletion of high-energy metabolites could easily lead to a cascade of events finally resulting in massive biochemical alterations in brains of animals made encephalopathic. These changes over time would certainly be expected to have a similar counterpart in humans with hepatic encephalopathy caused by liver failure due to hepatitis, cirrhosis, progressive familial intrahepatic cholestasis, etc.

This may seem a long way from hyperbilirubinemia and kernicterus; however, conjugated hyperbilirubinemia may cause liver damage due to bile acids. This in turn can lead to intrahepatic cholestasis and subsequent liver failure. The increased ammonia resulting from hepatic failure alters brain energy metabolites, and thereby explains the symptoms seen in hepatic encephalopathy, and explains possible resultant death. These concepts await further clarification.

As stated above, a wide variety of neurochemical changes have been studied and described over the past 40 years. These include aquaporin-4, a player in edema formation in hepatic encephalopathy (Rao, K., and Norenberg, M. 2007, MBD), neurosteroids in hepatic encephalopathy (Ahboucha, S., and Butterworth, R. 2007), NMDA receptors (Llansola, M., et al. 2007), cytokine flux (Wright, G., et al. 2007) nitric oxide in hepatic encephalopathy (Rose, C., and Felipo, V. 2005, MBD), phosphate- activated glutaminase in hepatic encephalopathy (Romero-Gomez, M. 2005), and the role of glutamate receptors in hepatic encephalopathy (Cauli, O., et al. 2005).

All of these above neurochemical changes, and many more, in models of hepatic encephalopathy in both animal models and the human disease state may be significant to the ultimate understanding of the pathogenesis of hepatic encephalopathy. Further descriptions are beyond the scope of this chapter. Examination of these alterations must be done in early and late symptom stages, and upon recovery. And it is essential they be done not in whole brain, but in the areas reflected by the symptoms. Inclusion in the sample of contiguous non-affected tissue dilutes the affected regions, thereby masking essential critical changes.

Kernicterus in Older Children and Adults

Kernicterus represents a metabolic encephalopathy afflicting newborns and children with unconjugated hyperbilirubinemia. Also occurring are several types of hyperbilirubinemia with both unconjugated and conjugated present in the serum. These may make an appearance in post newborn years, some in adulthood. The Crigler–Najjar type 2 is an example of an inherited unconjugated hyperbilirubinemia which can appear after birth, and last as long as 50 years (ref 53 in ostrow book). Rarely, type 1 Crigler–Najjar syndrome may have a late appearance. Gilbert disease is another example of mild unconjugated hyperbilirubinemia which may only "appear" in the teens, or later. There may be a genetic relation between Crigler–Najjar type 2 and Gilbert disease.

Conjugated hyperbilirubinemia consists of the Dubin-Johnson syndrome, Rotor syndrome, and various examples of cholestasis. Cholestasis will be discussed in a later chapter. Dubin-Johnson and the Rotor syndromes are characterized by hyperbilirubinemia in which conjugated bilirubin constitutes one half or more of the total bilirubin levels. These are almost always mild elevations, although hyperbilirubinemia over 20 mg/100 ml has been seen. While overt kernicterus may never be seen in these cases, mild neurological complications may occur.

These syndromes – Gilbert syndrome, Dubin-Johnson syndrome, and the Rotor syndrome and their treatments are included because they represent disordered examples of bilirubin metabolism. Knowledge and awareness of these disorders is important, and may aid in a better understanding of less benign forms of hyperbilirubinemia. A possible relation between Crigler–Najjar type 2 syndrome and Gilbert syndrome is suggested by the finding of both disorders in the same extended families (Ostrow, J. 1986). This possible relation, and late onset in some cases of Crigler–Najjar type 2 will be discussed later.

Gilbert disease was first described (Gilbert, A. 1907) in the early 1900's. The disorder was characterized as being a mild hyperbilirubinemia without bilirubinurea or other liver disease. This disorder could be distinguished as not involving severe hyperbilirubinemia. Soon it was realized that many liver syndromes and diseases involved elevated serum bilirubin; this mild form without hemolysis or other liver problems soon came to be called Gilbert syndrome. It was later shown that Gilbert syndrome represents a situation in which bilirubin conjugation capacity is reduced by about one third (Powell, L., et al. 1967).

D.W. McCandless, *Kernicterus*, Contemporary Clinical Neuroscience, 203
DOI 10.1007/978-1-4419-6555-4_19, © Springer Science+Business Media, LLC 2011

The usual presentation of Gilbert syndrome is in teens and young adults who present with a mild unconjugated hyperbilirubinemia. Patients diagnosed with Gilbert syndrome may have symptoms such as fatigue, abdominal pain, muscular weakness, etc. These symptoms do not always correspond to bouts of increased serum bilirubin levels. In some studies, males have a higher incidence than females (Auclair, C., et al. 1976).

Bilirubin levels in Gilbert syndrome are ordinarily in the 2–3 mg/100 ml range, but are occasionally higher. Conditions such as stress, fever, infection, and alcohol abuse may act to elevate bilirubin levels. In terms of blood tests, indirect bilirubin levels are the only abnormality. Other liver function tests are normal.

Liver biopsies examined with light microscopy are generally unremarkable. Results of electron microscopy examination have been equivocal. Some investigators have found no specific changes in ultra structure as compared to controls (Jezequel, A., et al. 1980). Others have found changes in both endoplasmic reticulum and mitochondria (Arias, I., et al. 1969; Berk, P. 1970).

The cause of these diverse results is unclear, but could be related to bilirubin levels, which cannot be constantly monitored. The long-term nature of the syndrome, and frequency of bouts may play a role. There may be two groups of Gilbert syndrome patients – those with normal endoplasmic reticulum, and those with hypertrophied liver endoplasmic reticulum (Berk, P. 1970). The clinical significance of these findings is not clear, but may indicate a heterogenous nature of Gilbert syndrome similar to that seen in the Crigler–Najjar syndrome. In fact it has been suggested by many that Gilbert syndrome may be an extension (mild) of the Crigler–Najjar type 2 syndrome (Goresky, C., et al. 1978).

As regards mechanisms, both elimination of loading doses, and disappearance of labeled bilirubin show that there is a reduction in the hepatic uptake of bilirubin (Felsher, B., et al. 1973). Not all studies have shown a defect in hepatic uptake of bilirubin (Black, M., and Billing, B. 1971).

There also exists the possibility of a reflux of unconjugated bilirubin from liver cells back into the circulation. There is also a partial defect in the conjugating enzyme glucuronyl transferase in all cases of Gilbert disease (Black M., and Billing, B. 1969). The correlation between hyperbilirubinemia and glucuronyl transferase activity in Gilbert disease is not the best, but this may be explained by the finding that 50% of Gilbert patients have a mild compensated hemolytic state (Berk, P., and Blaschke, T. 1972).

Treatment of Gilbert patients with phenobarbital has been shown to be effective (Reyes, H., et al. 1969) in reducing hyperbilirubinemia. The exact mechanism of phenobarbital's ability to lower serum bilirubin levels is not entirely clear. Some investigators speculate phenobarbital acts to increase hepatic uptake of bilirubin (Berk, P., and Blaschke, T. 1972), while other evidence indicates an effect of the drug on the activity of glucuronyl transferase (Owens, D., and Sherlock, S. 1973).

One interesting observation in Gilbert disease patients is that fasting may serve to increase serum bilirubin levels as much as twofold (Bloomer, J., et al. 1971). It seems that the fasting-induced hyperbilirubinemia is caused by a decrease in bilirubin uptake by the hepatocyte This fasting phenomenon can be useful in diagnosing Gilbert disease in older people, and not confusing it with other disorders producing mild hyperbilirubinemia such as alcoholism, heart disease, biliary tract disease, tumors, etc.

In an interesting study (Fevery, J., et al. 1977), investigators studied unconjugated bilirubin and bilirubin conjugates in patients with Crigler–Najjar syndrome, Gilbert syndrome, patients with hemolysis, and controls. Results showed that patients with Gilbert syndrome had a twofold increase over normal of bilirubin monoconjugates in their bile. Moreover, the increase was even more dramatic in children with the Crigler–Najjar syndrome. In addition, unconjugated bilirubin was present in the bile of both Gilbert syndrome and the Crigler–Najjar syndrome patients from 30 to 57%, whereas control levels were less than 1%. Bile composition was also abnormal in patients with hemolytic disease.

In this study, there was a wide variation in glucuronyl transferase activities in Gilbert patients (4–45%of control levels). In Crigler–Najjar patients, values for glu-curonyl transferase activity were zero or very close to zero. Analysis of bile showed that conjugated bilirubin was present in samples from both Gilbert patients and patients with Crigler–Najjar syndrome.

The authors point out that in their study there was not a clear correlation between glucuronyl transferase activity and conjugating capacity in Gilbert syndrome. This suggests possible lack of adequate sensitivity of the enzyme assay technique, or of the presence of an abnormal enzyme capacity in Gilbert syndrome.

Evaluation of bilirubin kinetics in patients with Gilbert syndrome provides use-ful information (Okolicsany, L., et al. 1978). In this study, Gilbert syndrome was diagnosed based on high unconjugated bilirubin levels combined with normal liver function tests and no hemolysis. The subjects ages ranged from 17 to 63 years. In these patients, bromosulphthalin (BSP) kinetics were studied. Bilirubin loading tests were performed by injecting 2 mg/kg of bilirubin IV.

Results showed that, when compared to controls, Gilbert syndrome patients could be divided into two groups. In the first group, bilirubin production rates were 337 μmoles per day average, compared to 409 μmoles/day for controls. In a second group of Gilbert patients the bilirubin production rate was on average 886 μmoles/day. In both groups there was a highly significant reduction in bilirubin clearance of 12 and 23 ml/minute, respectively. This compared to clearance in con-trols of 53 ml/minute. Results from BSP studies were essentially normal, suggesting different mechanisms of excretion for BSP as compared to bilirubin in these cases.

The authors note that they were able to classify Gilbert syndrome into two groups based on the level of daily bilirubin production. The increased production of biliru-bin in 40% of patients is in agreement with other studies (Powell, L., et al. 1967). These different groups of Gilbert patients lends support to the concept of a heteroge-neous condition as regards Gilbert syndrome. These results indicate several variants in expression of variations in genetic defects in Gilbert syndrome.

The Dubin-Johnson syndrome, also called chronic idiopathetic jaundice, is a chronic hyperbilirubinemia in which both unconjugated and conjugated bilirubin are elevated in the serum. This is a case where total bilirubin levels may reach 27 mg/100 ml, but are usually much lower. As in Gilbert syndrome, liver function tests are usually normal. Urine urobilinogen has been noted to be elevated in some cases. Loading studies using both bilirubin and BSP have shown that the primary defect consists of an inability to transport organic anions into bile (Berk, P., Wolkoff, A., and Berlin, N. 1975). BSP test results are altered due to regurgitation of the dye as its glutathione conjugate moves back into plasma from the hepatocyte. Reflux of dyes not conjugated by the liver are not seen.

The Dubin-Johnson syndrome is usually a mild chronic form of hyperbilirubinemia of little clinical consequence. It may first be noted, for example, in pregnant women, or women using oral contraceptives which act to reduce hepatic excretory function.

Although relatively rare, Dubin-Johnson syndrome cases have been reported to be higher than elsewhere in certain groups of people in Israel. There is an animal model of Dubin-Johnson syndrome in the Corriedale sheep This animal model has chronic conjugated hyperbilirubinemia, and liver histopathological features similar to Dubin-Johnson patients.

Liver biopsies from Dubin-Johnson patients show the presence of green-black to black pigment in lysosomes in liver cells as shown by light microscopy. These dark lysosomes give an overall green-black color to the surface of the liver from Dubin-Johnson patients. This syndrome is considered a disorder of direct-reacting bilirubin, which undergoes a significant regurgitation into the serum. This might account for the green-black appearance of the liver. It is important to measure serum indirect and direct bilirubin levels in a way that accounts for albumin-bound pigment, which in turn may answer why there is a substantial amount of conjugated bilirubin. Some methods fail to assess albumin-bound bilirubin, thereby underestimating the true value (Weiss, J., et al. 1983).

The Rotor syndrome is described as a benign, chronic, inherited conjugated hyperbilirubinemia. Bilirubin levels are under 10 mg/100 ml in most cases, but various factors as noted above may cause short-term spikes in levels of bilirubin (Levels may be as high as 25 mg/100 ml. The levels of unconjugated vs. conjugated bilirubin are usually about equal. Liver function tests are normal, and unlike Dubin-Johnson syndrome, the liver does not show green-black staining under light microscopy, and other histological features are normal. BSP studies are different from Dubin-Johnson syndrome in that they indicate a defect in uptake by the hepatocyte (Delage, Y., et al. 1977). The hepatocyte storage capacity may also be altered (Schoenfield, L., et al. 1963). The Rotor syndrome is essentially asymptomatic. Another difference between the Dubin-Johnson syndrome and Rotor syndrome is the finding that administration of Tc HIDA does not permit visualization of the liver, biliary tract, or small intestine as it does in the Dubin-Johnson syndrome.

BSP studies in the Rotor syndrome indicate a modest reduction in transport of the dye, but an almost complete obliteration of any storage capacity. The reduced storage capacity in hepatocytes in the Rotor syndrome has suggested that there could be a reduction of intracellular binding proteins, which could be then primary defect.

When family heterozygotes of patients with Rotor syndrome were tested for BSP uptake and clearance, results were about halfway between that of homozygous patients and normal controls . In similar studies in possible candidates for Dubin–Johnson syndrome heterozygotes, BSP metabolism does not suggest the existence of obligate heterozygote.

As was pointed out at the beginning of this chapter, Crigler–Najjar syndrome may have a late onset. This onset may appear in the late teens or early adulthood. While it is known that Crigler–Najjar type 2 occurs well after the newborn period, most frequently Crigler–Najjar type 1 occurs soon after birth Rare cases, however, are described in which neurological symptoms of Crigler–Najjar type 1 occur late. The following case report is of such a case (Blaschke, T., et al. 1974).

This case was of a normal term female who developed jaundice shortly after birth. Her unconjugated bilirubin levels averaged 25 mg/100 ml. Shortly after birth, the family declined further medical help. The patient passed the usual milestones, and passed into adolescence. She contracted childhood diseases with no apparent adverse effects, and made proper progress in school. A cousin with a similar history suddenly developed neurological symptoms suggestive of kernicterus, and died at age 16. This brought the family back for more medical help.

At that time she was admitted to the clinical center at NIH for a period of 10 weeks. She had extensive physical and neurological exams, and all were normal except for the presence of jaundice. Plasma bilirubin concentrations ranged from 18 to 30 mg/100 ml, nearly all were unconjugated. EEG results showed a diffuse and slow background, with occasional paroxysmal bursts.

Treatment attempts included phenobarbital and glutethimide, but both were ineffective in reducing bilirubin concentrations. A 4-day exposure to sunlight for 6–8 h per day only lowered serum bilirubin by 0.5 mg/100 ml, and the drop was transient. Subsequently she was given genetic counseling, she was married, and then lost for follow up for 6 months.

When she was next seen, her family gave a history of bizarre behavior and withdrawal, followed by speech problems, ataxia, and seizures. She was readmitted to the NIH clinical center, and her physical exam was unchanged, but she suffered significant neurological decline. She displayed incoordination, ataxia, decreased awareness, athetoid movements, and motor seizures. She seemed to have a significant intellectual decline, and the EEG showed epileptiform activity.

Based on reports of beneficial effects of exchange transfusion and plasmapheresis on both bilirubin levels and neurological signs and symptoms, these procedures were done. She was treated with plasmaphoresis twice as well as phototherapy. Bilirubin levels reached a low of 6 mg/100 ml, but the following morning after treatment, bilirubin levels were 13.2 mg/100 ml. A special phototherapy device with eight blue lights was placed close to the patient in bed and phototherapy administered 12 h per day. Over 2 weeks, bilirubin levels gradually rose to 20 mg/100 ml.

A neurological exam now showed bradykinesia suggesting basal ganglia involvement, hyperreflexia indicating pyramidal system damage, rapid alternating movements indicating cerebellar involvement, and myoclonic jerks. A liver biopsy was normal. Additional phototherapy was ineffective in reducing the serum bilirubin concentration, which remained about 24–27 mg/100 ml. The patient was discharged on the 128th day, and she was followed for some time. Her bilirubin levels fluctuated between 25 and 34 mg/100 ml and correlate more or less with her neurological status. She continued to show seizure activity and all of the neurological deficits previously described.

The authors comment that the reason for such a late onset of Crigler–Najjar syndrome type 1 is unclear. The only speculation was that the patient had a spike in unconjugated bilirubin coincidently coupled with a drop in albumin, thereby pushing the molar ratio of bilirubin to albumin over 1.0, thereby "spilling" bilirubin into the brain.

The authors also note that after plasmophoresis and removal of a large quantity of bilirubin, the patient still had hyperbilirubinemia, and her jaundice was still quite evident. This indicates a large amount of bilirubin deposited in body tissues. Albumin may be able to draw some bilirubin out of tissue deposits. These deposits have been noted even upon autopsy.

The authors further speculate that perfusion of a Crigler–Najjar patient's blood over an albumin-conjugated agarose column could be a highly efficient method for the removal of unconjugated bilirubin from plasma. Studies on this method in Gunn rats have shown that in this model of Crigler–Najjar syndrome, a high degree of success was reached. The authors further speculate that liver transplantation seems to hold promise for future treatment of intractable hyperbilirubinemia.

One feature of this case not commented on by the authors is the observation that as the bilirubin levels drop, the neurological signs and symptoms abate, and visa versa. This was noted even late in the course, even after weeks of very high serum bilirubin levels. In one sense this flies in the face of the concept that much earlier entry of bilirubin into key cerebral areas such as the basal ganglia and cerebellum, would have caused irreversible structural damage to neurons and mitochondria. In favor of that concept is that newborns with hyperbilirubinemia can develop neurological signs and symptoms, and die within only hours to a couple days. Autopsy reports show significant structural damage.

On the other hand this finding of seeming correlation between bilirubin levels and neurological findings emphasizes the "metabolic encephalopathy" concept for kernicterus. Most importantly it may hold some hope for some level of amelioration with treatment of the devastating neurological aspect of this disorder. There can be no doubt that the development and wide use of phototherapy in infants has prevented neurological symptoms and kernicterus in many infants.

Treatment of Gilbert disease, Dubin-Johnson syndrome, and the Rotor syndrome is usually unnecessary. Bilirubin levels in these disorders rarely exceed 10 mg/100 ml, and the disease processes are benign. The prognosis for one lifelong absence of any sequelae are excellent. Once diagnosis is established (by exclusion), the patient should be reassured that nothing else is wrong. Certain therapies such as phenobarbital or glutethimide might help reduce bilirubin, although such therapy is not recommended merely for cosmetic reasons. Counseling is best for patients who are concerned about the cosmetic effects of hyperbilirubinemia.

Cerebral Palsy and Counseling

Cerebral palsy is a disorder/symptom which can result from several underlying causes. Conditions from which cerebral palsy may occur can arise before birth, at birth, or after birth. The usual classification of cerebral palsy is based on symptoms/location of cerebral damage. Cerebral palsy affects muscle coordination and bodily movement. Cerebral palsy can be mild and hardly recognizable or so severe that any semblance of normality is not present. There is no cure once cerebral palsy is acquired, but treatment such as physical therapy may ameliorate some symptoms.

The incidence of cerebral palsy varies with the study. Most place the incidence at about 2–3 cases per 1000 births. There is some evidence that the incidence has increased in recent years. Since cerebral palsy is a result of brain damage, trends such as wearing helmuts while riding bicycles lowers the incidence of brain injury from falls. This in turn lowers the incidence of cerebral palsy. Other fashions such as skate boarding serve to increase incidence of brain injury.

Classifications of cerebral palsy are usually based on the nature of the movement disorder, which in turn reflects the damaged brain region. In this sequence, the classification includes spastic cerebral palsy, athetoid cerebral palsy, ataxic cerebral palsy, mixed cerebral palsy, etc. Other classifications are topographic, such as monoplegic, paraplegic, hemiplegic, quadraplegic, etc. These are based on the number and site of the limbs affected. Another classification system is based on the timing of the brain damage resulting in cerebral palsy. This encompasses prenatal birth problems, and postnatal incidences. Examples of this classification include prenatal anoxia and genetic factors in the prenatal group. Hyperbilirubinemia-induced kernicterus and subsequent cerebral palsy are in the postnatal group.

Other classification systems are based on the functional capacity of the patient, or on therapeutic considerations. The early diagnosis and classification of the child's brain injury is important from the standpoint of counseling and treatment aimed at minimizing physical disability. The brain damage, once it has occurred, is usually stable and not progressive. The actual prognosis is less than favorable and the social and psychological effects are sometimes difficult to treat. The majority of patients do not reach an independent state.

Treatment is generally based on trying to teach the patient as much self sufficiency as is possible. This involves an accurate assessment of the disabilities, and

D.W. McCandless, *Kernicterus*, Contemporary Clinical Neuroscience, 209
DOI 10.1007/978-1-4419-6555-4_20, © Springer Science+Business Media, LLC 2011

the setting of realistic goals. The patients need physical therapy in order to use the full compliment of residual physical assets. Many affected children have excellent mental capacities, and these patients should be encouraged. It is critical to pay attention to the psychological and emotional needs of cerebral palsy children in order for them to achieve as much independence and satisfaction as is possible. For this reason, counseling for both patient and parents should be started early, and continued.

Counseling goals will be discussed in greater detail in the second part of this chapter. Physical therapy treatments are best carried out in short frequent sessions rather than long tedious sessions. The goal should be to prevent the development of contractures. This is achieved by movement against the direction of the contracture in commonly affected joints such as ankles, knees, wrists, and fingers.

Spastic cerebral palsy is characterized by increased activity of deep tendon reflexes and increased stretch reflexes. This form of cerebral palsy is the most common, affecting 70–80% of cerebral palsy patients. These patients are characterized by having spasticity in groups of muscles, acting to prevent movement of an affected limb. While the progression of spastic cerebral palsy due to brain injury does not occur, the spasticity at the muscle level can increase over time. Spastic cerebral palsy may alter growth patterns in children.

Ataxic cerebral palsy is the least common form of cerebral palsy, affecting 5–10% of patients. Ataxic cerebral palsy is most frequently caused by damage to the cerebellum. Ataxic cerebral palsy usually affects both arms and legs, and may be associated with hypotonia. Ataxic cerebral palsy is associated with intention tremor. Some treatments such as benzodiazepines may act to alleviate some of the symptoms of ataxic cerebral palsy. Some patients with ataxic cerebral palsy have been treated with temporary cooling of an affected limb. For example, cooling of an arm may result in a period of disappearance of intention tremor sufficient for a meal to be consumed.

Athetoid cerebral palsy is the type most frequently associated with brain damage from hyperbilirubinemia. Athetosis is characterized by involuntary uncoordinated motions with constantly varying levels of muscle tension. In these cases the initial symptom may be hypotonia which can last several months. Then in the late first year or second year appears the wondering movements of fingers, hands, arms, and feet which characterizes athetoid cerebral palsy. This writhing movement may also affect the face and tongue. This level of loss of muscular control often affects the ability to hold posture. These movements reflect damage to the basal ganglia. This represents the site of damage in essentially all cases of kernicterus. Movement of facial muscles can have a profound effect on the ability to eat, talk, and even affect breathing. Athetoid cerebral palsy accounts for 20–25% of all cerebral palsy cases.

There has long been an association between hyperbilirubinemia and cerebral palsy. In one study (Hyman, C., et al. 1969), 405 infants with hyperbilirubinemia were followed for a minimum of 4 years, assessing central nervous system abnormalities. The initial treatment of these jaundiced infants was varied as none were born at the hospital where the study was performed. The goal of the authors was to keep bilirubin levels at or below 20 mg/100 ml. Three hundred and fifty one of these infants had exchange transfusions. Careful documentation of neurological states, muscle tone, moro reflex, etc., was kept. All patients had follow up examinations until 4 years of age. A variety of classical features were examined including hearing, verbal communication, EEG, and neurological/psychological evaluation.

Results showed that 396 infants had hemolytic disease, and 9 had hyperbilirubinemia from other causes. Of the patients followed for 4 years, 346 had no evidence either from histories, or other examinations of any central nervous system involvement. Fifty-nine patients showed one or more of the following symptoms: hearing loss, athetosis, seizures, strabismus, or other minor disabilities.

Of the 59 with central nervous system involvement, 3 had athetoid cerebral palsy. In these three patients, bilirubin levels were 33 mg/100 ml or greater. In these cases, neurological involvement was diagnosed at the time of hospital admission because of the degree of central nervous system involvement. One of the three patients with athetosis had marked retardation. The other two had lesser degrees of mental retardation. The authors note that while 59 of 405 children had at least one CNS abnormality, the numbers in most instances were too low to permit statistical analysis, and further studies are warranted.

In a later study (Scheidt, P., et al. 1991), patients enrolled in a phototherapy trial were further evaluated for neurological complications. In this study, 1,339 newborn infants at six separate facilities were randomly assigned to phototherapy or control groups. Patients were selected for inclusion based on probable need for treatment for hyperbilirubinemia (low birth weight, bilirubin over 13 mg/100 ml within the first 96 h, etc.). Patients receiving phototherapy received this treatment continuously for 96 h. Exchange transfusion was done in both groups if bilirubin levels reached as high as 20 mg/100 ml., or at lower levels if there was a risk factor such as low birth weight. Patients were followed up yearly.

The use of exchange transfusion resulted in bilirubin being kept at a low level in both groups. Results showed that of the 1,339 newborn infants, none had evidence of athetoid cerebral palsy of any type. Three infants did have bilirubin levels over 20 mg/100 ml, and these three had no evidence of cerebral palsy. There were cases of cerebral palsy in the two groups, but their bilirubin levels were not different than those without cerebral palsy. Furthemore, no other evidence suggested any central nervous system toxicity from bilirubin in these infants.

The authors note that this study did not allow bilirubin levels to rise over 20 mg/100 ml to any appreciable extent in that exchange transfusion was initiated at that level regardless of whether patients were in the control or phototherapy groups. Thus, the failure to find any significant central nervous system impairment which could be attributable to bilirubin may have been related to the fact that very high levels of unconjugated bilirubin never occurred. Still, upto 20 mg/100 ml. There was no evidence of brain damage.

In another study (Graziani, L. 1992), 249 newborn infants who were of less than 34 weeks gestation were examined in terms of bilirubin levels, cranial ultrasonographic abnormalities, and severe neurological sequelae. From this group, 45 had some form of spastic cerebral palsy, while in 204 infants there was no evidence of cerebral palsy, but there was abnormal developmental tests. All except seven of those with cerebral palsy had evidence of intracranial hemorrhage. Ultrasonographic changes did not correlate with serum bilirubin levels.

The authors further note that there was no evidence to indicate that hyperbilirubinemia upto 22.5 mg/100 ml was related to cerebral palsy or any other neurological indicator in this study. Other problems such as intracranial hemorrhage, severe periventricular echodensity, and bronchiopulmonary dysplasia were correlated with cerebral palsy, but not hyperbilirubinemia. Again, however, bilirubin levels were not permitted to exceed 22.5 mg/100 ml.

In a short paper (Okumura, A., et al. 2001), two patients are described who had hyperbilirubinemia and cerebral palsy. The first case was a 2,078 g newborn who was of 33 weeks gestation. Total bilirubin reached 13 mg/100 ml on day 3, but no further bilirubin levels were measured. The patient returned to the hospital on day 48 with bilirubin levels of 19.7 mg/100 ml. Phototherapy reduced this high bilirubin level. A second case was a 994 g newborn delivered by C-section at 26 weeks gestation. This patient was on phototherapy from birth. The bilirubin levels in this case peaked at 8.8 mg/100 ml on day 7.

The patients had no history of blood incompatibilities, and a metabolic screen was negative. No signs of kernicterus were seen in the immediate neonatal period, but by 3 years of age, the patients could not sit, and psychomotor development was severely retarded. Dystonic posture was noted, and mild rigidity was present in the extremities. MRI of these patients showed abnormal high intensities in the globi pallidi bilaterally. The brainstem auditory-evoked response showed elevated thresholds, and altered interwave separation.

The authors state that the findings in these two patients are entirely consistent with a diagnosis of athetoid cerebral palsy. This occurred in the absence of any clear-cut signs/symptoms of kernicterus in the newborn period. Other cases of kernicterus have been documented after the newborn period who also did not show symptoms early. In addition, the second case did not demonstrate any severe elevation of serum bilirubin. This phenomenon has been previously reported (Gartner, L., et al. 1970).

Both of these two patients had MRI data which were consistent with kernicterus, and included abnormal high intensity areas on T2 weighted images in the globe pallidi bilaterally. In addition, the auditory-evoked response was abnormal, a feature commonly seen in kernicteric infants (see Figs. 1 and 2).

In another paper (Sugama, S., Soeda, A., and Eto, Y. 2001), the occurrence of MRI images consistent with a diagnosis of kernicterus are presented, but in patients without the usual symptoms. These three cases were in preterm infants who had relatively low levels of hyperbilirubinemia. They did not show the acute symptoms of kernicterus. Results of MRI imaging showed characteristic findings located at the posteromedial border of the globus pallidus. These MRI findings are consistent with a diagnosis of athetoid cerebral palsy. This shows that assessing the risk of kernicterus may not be completely possible in the newborn period, and awaits later imaging techniques for a definitive diagnosis.

A large study in which 140 infants were compared to 419 control newborn infants has recently been published (Newman, T., et al. 2006). In this prospective study, subjects were selected from over 106,000 live births from a Kaiser program.

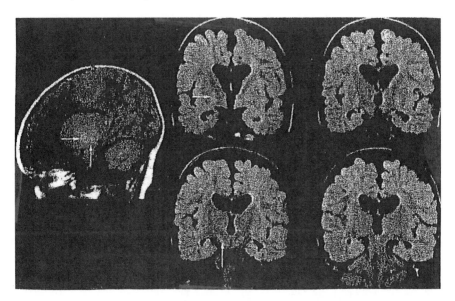

Fig. 1 MRI from infant with bilirubin encephalopathy showing injury to the sub-thalamic nuclei. High signal in the pallidum. Permission granted from American Academy of Pediatrics

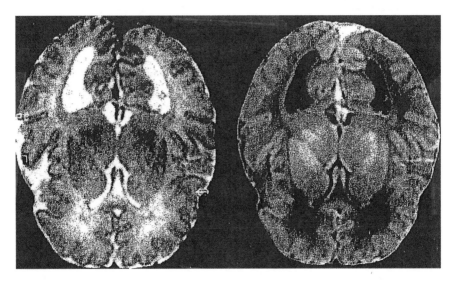

Fig. 2 MRI (T1 weighted) showing kernicteric high signal in the globus palidus, and hyperintense T2 discerning pallidum vrs putamen. Permission granted: American Academy of Pediatrics

Cases with hyperbilirubinemia exceeding 25 mg/100 ml within the first 30 days after birth and weighing at least 2,000 g were selected. Gestational age was greater than 35 weeks.

The study patients were examined for neurological development at about 5 years of age by both neurologists and psychologists. Child psychologists administered standard intelligence tests, as well as visual-motor tests. Neurological examinations were performed in which the examiner did not know to which group the subject belonged. Results were quantitated as to normal, or to several levels of disability. Questions were also given to the parents in which their concerns regarding their children's development were asked.

Results showed that most infants had bilirubin levels below 27 mg/100 ml; however, 10 infants' serum bilirubin levels exceeded 30 mg/100 ml. One infant had a bilirubin of slightly over 45 mg/100 ml. Most patients with hyperbilirubinemia showed a drop in the levels to below 20 mg/100 ml in less than 24 h. Four of the hyperbilirubinemic group tested positive for glucose-6-phosphate dehydrogenase deficiency, and only five newborns received exchange transfusions.

Results of outcome studies showed that there was no statistical differences in intelligence between the jaundiced group and the control group. The results of the visual motor test were also not significantly different between the two groups. Scores on neurological exams showed that in general, there was no difference in the jaundiced group. In fact, the control group showed a slightly higher incidence of "questionable" neurological exams than the jaundiced group. Results from the parental questionnaire were about the same between the two groups.

The authors comment that because most of their cases were only slightly higher than 25 mg/100 ml, the ability to detect "catastrophic" events was limited. The authors state, however, that their study is reassuring in that outcomes such as mild cognitive, behavioral, or motor impairment, did not occur in newborn infants with hyperbilirubinemia as high as 25 mg/100 ml. There were no newborn infants who on follow up had any signs or symptoms of kernicterus.

This was a selective study from the standpoint of the inability to evaluate all eligible subjects. The study does, however, indicate that exchange transfusions are probably not needed until bilirubin exceeds 25 mg/100 ml. Most infants in this study with bilirubin levels between 25 and 30 mg/100 ml, or lower, were treated with phototherapy with no adverse developmental or neurological effects.

A commentary on the above paper was published in the same issue (Watchko, J. 2006). In the commentary, most of the reservations regarding the study enumerated by the authors are repeated. These include the fact that most infants' serum bilirubin levels were only slightly over 25 mg/100 ml, and that phototherapy alone resulted in effective management in these cases. Only five infants received exchange transfusion. In addition, peak bilirubin levels were less than 6 h in 75% of jaundiced because of starting treatment quickly.

Only one subgroup had divergent results. That subgroup was one with nine newborns, all with a positive direct antiglobulin test for immune-mediated hemolytic disease. This subgroup had stastically significant lower IQ results as compared to controls. It has been previously suggested (Ozmert, E., et al. 1996) that hemolysis may act to increase the chances for deposition of bilirubin into brain tissue. This current study lends further credence to this concept.

Other factors may also influence the occurrence of kernicterus, including acidosis, low serum albumin, presence of substances which compete with bilirubin for albumin-binding sites, prematurity, etc. All these factors must be considered in choosing treatment for hyperbilirubinemia.

There is a voluntary kernicterus registry (Bhutani, V., et al. 2004), and of 116 infants in the registry, 76 had serum bilirubin levels greater than 35 mg/100 ml. The other side of that is that 40 infants had bilirubin levels less than 35 mg/100 ml. Estimates are that 1/10,000 infants have serum bilirubin levels in excess of 30 mg/100 ml, and since kernicterus is potentially preventable in a country with excellent health care, health care workers must be diligent as regards awareness of hyperbilirubinemia.

Without a doubt, cerebral palsy caused by any factor including high serum unconjugated bilirubin levels, can result in both behavioral and emotional problems not encountered in non-cerebral palsy children. These issues are best dealt with through counseling. Counseling is beneficial not only to patients, but also to parents. Manifestations of unpredictable variations of behavior include hyperactivity, impulsiveness, irritability, overreaction to stimuli, and difficulties with abstract thinking. When several of these symptoms occur together, they suggest brain damage (kernicterus/cerebral palsy), and a clear need for counseling.

In terms of counseling, several principles have evolved for good patient counselor relationships. First is to treat each client with respect regardless of age, ethnic group, etc. Each patient is different, and so the therapist should ask questions and respond positively to needs and concerns. The counselor should give the proper amount of information, but not so much that the client is overwhelmed. The information should be "customized" to fit the clients circumstances. Clients should be helped to remember instructions, and provided with plans which are suitable to the client and his/her family.

In the case of a child with cerebral palsy, guidance should be directed toward helping them with becoming independent and productive, and able to function in everyday life. For example, in physical therapy, a child might be helped to walk, or pick up an object on the floor. In behavioral therapy, a piece of candy might be placed in a box, and the patient instructed to remove it with the weak hand. This is a reward method of behavioral therapy. When aggressive behavioral is displayed, it is the role of the counselor or behavioral therapist to direct anger for release in new constructive ways.

A goal of a counselor will be to direct the patient toward as much self-sufficiency as is possible. Activities will vary greatly according to age and the degree of disability. Ultimately, a counselor will assume the role of coach, suggesting ways to improve, but always with a positive nondemeaning manor. The counselor may also serve as an occupational therapist in the sense that he/she may suggest to the client or the client's parents basic ways of feeding, dressing, and other everyday activities. As the patient grows older, counseling will be directed toward helping the patient maintain physical health, and reduce depression. Maintaining good relations and regularly seeing physicians and dentists is essential. Nutritional aspects should not be overlooked as many cerebral palsy patients need altered recommended daily allowances to correct for saliva loss and fatigue.

Open communication is essential as aging cerebral palsy patients often become withdrawn and incommunicative. The depression of both patient and family member due to cerebral palsy leads to feelings of helplessness and bitterness. These feelings can increase, and communication can in part, avert this condition. Treatment of cerebral palsy patients, and interactions with family members must be performed in a highly positive manor. Making sure that the family understands and appreciates the circumstances will help to minimize secondary behavioral problems. Firm constructive and positive guidance is a much better approach than permissiveness. It should be pointed out that there are drugs which may be useful. For example, amphetamines may act to prolong the attention span. Certain tranquilizing agents may serve to modify irritability and decrease impulsiveness.

In a recent paper (Cooley, W. 2004), some of the objectives in cerebral palsy therapies which counseling of both parents and patients can facilitate are discussed. One of these is the prevention of orthopedic complications of cerebral palsy. The counselor is in an ideal position to inquire of patients and parents as to the possible deterioration of locomotor skills. If this is occurring, then orthopedic consultation is imperative. If communication is a problem for the patient, counseling could suggest alternative means of communication, and serve to facilitate that process. In addition, the counselor is in a position to watch for normal growth and nutrition in a cerebral palsy child. This activity is positive, and will have a positive effect on the well-being, and help children realize full potential.

Another important role of a counselor is to provide an early response to the family support needs. Such quick response helps enhance resilience and helps equip families with the wherewithal to deal with the ever-changing condition of children with cerebral palsy. Prompt referral to the appropriate professional is key to correcting impending problems even before they occur if possible. This is an important role for a counselor to play in providing care for cerebral palsy children. The daily demands of caring for a cerebral palsy child at home may become too disruptive and demanding, thereby requiring placing the child in an outside facility equipped to handle disabled children. This is another setting in which counseling may play a key role.

Another paper (Berven, N. 2008) addresses the concept of rehabilitation counseling. Rehabilitation counseling is a method which acts to help disabled patients accomplish goals and become as independent as is possible. This type counseling is a subspecialty of counseling which focuses on the needs of disabled people in all aspects of their lives. Frequently, this focuses on long-term consequences of disabling aspects of illnesses or injuries (at birth or later) which exist even after other medical treatments and interventions have been exhausted.

Rehabilitation counselors will certainly believe in the dignity of disabled patients, and believe in full access to opportunities for participation in all aspects of community life. Disabled clients should be encouraged in a positive way to pursue goals which are within reach. In these counseling activities the patient should be viewed as a full partner and collaborator in discussions and decisions. This should all be done in a positive manner, emphasizing the patient's strengths.

Another consideration is that the patient, if possible, should be in an integrated situation. In the past, cerebral palsy patients and other disabled people were placed together in large facilities with staffs to provide care. They may have worked in low-pressure factory-like settings, but always surrounded by many other disabled people. Most of these facilities have closed, and cerebral palsy patients and other former residents are integrated into the community, and became part of the regular work force when possible.

In addition, the rehabilitation counselor practices counseling along the general counseling guidelines, but in the context of a partner with the cerebral palsy patient. The process is a collaborative one, with specific goals and needs formed in the minds of counselor and patient. These goals and needs include the successful completion of school in the case of cerebral palsy children. Many cerebral palsy children have normal intelligence, and 90% live into adulthood. Thus as much education as possible should be a key goal. Facilitating this is a critical need. The establishment of independence from others for help is also a prime consideration. This should be an agreed upon goal of all children with cerebral palsy.

Rehabilitation counseling is a process often needed at times of change, such as moving to a new area, or taking a new job. Needs such as identifying resources for assistance are situations likely to be encountered. The counselor must establish a strong relationship with the individual being counseled. The counseling practices employed will vary from one case to another, depending on the patient, and the nature of the disability. The complexity of needs of various patients often involve several different health care professionals, and obtaining and coordinating them often falls to the counselor. Following school, job placement and a focus on career goals are common tasks performed by counselors. Frequently, a counselor will visit a potential job site and meet with the immediate supervisor to explain the nature of the patient's disability. This advocacy role of a counselor may extend to the workplace in terms of making facilities accessible, and accommodating the needs of a cerebral palsy patient.

In summary, several key components are essential for successful counseling. These include a positive attitude in which the counselor emphasizes the abilities rather than the disabilities of the cerebral palsy patient. At the same time, the counselor and patient must be realistic. Most cerebral palsy is actually diagnosed in the second to third years, so emphasis on proper schooling should be the first and key goal. Later, socialization and integration into a regular classroom, if possible, is desirable. Because of the potential life span of cerebral palsy patients, long-term counseling will be needed as goals and aging dictate ever-changing challenges.

Neurological Sequelae from Jaundice

Numerous neurological sequelae result from exposure of the newborn infant to hyperbilirubinemia. These range from severe kernicterus and cerebral palsy to mild behavioral and/or intelligence quotient loss. Frequently, the findings in newborns exposed to high serum bilirubin levels depend in some measure on the sophistication of those examining the infants, or later, children who have recovered. Many manuscripts have been published describing studies aimed at elucidating the possible effects of bilirubin on neurodevelopment, with conflicting results.

In an early paper (Day, R., and Haines, M. 1954), IQs were measured in children, the majority of whom had received exchange transfusions for hyperbilirubinemia associated with erythroblastosis fetalis. This paper was a description of results after the institution of exchange transfusion as therapy for hyperbilirubinemia. The method was first described by Wallerstein (Wallerstein, H. 1946). Prior to that about 10% of erythroblastosis patients developed kernicterus and cerebral palsy. Untreated children who recovered from erythroblastosis and later were tested for IQ were about 11 points lower than comparable controls.

In the Day and Haines study, 136 children were tested, 68 were erythroblastosis-recovered patients, and 68 were controls. Of the 68 recovered erythroblastosis patients, 41 received replacement transfusion therapy, while the other 27 received a small (partial) transfusion, or no transfusion at all. Another group of patients served as controls who were matched in terms of socioeconomic criteria. IQ was assessed using the Stanford- Binet test. This test allows assessment of children aged 2 years of age and older.

Results showed that the group who had recovered from erythroblastosis had a mean IQ of 106.5, 6.1 points below the control groups. This difference was statistically significant at the 0.01 level of confidence. Differences between treated patients (transfusion) and not treated erythroblastosis patients were small, but just caught significance. The most jaundiced group of patients had average IQs 14.5 points lower than their siblings, while patients minimally jaundiced were only 2.5 points below controls.

The authors note that the differences in IQ between groups, although statistically significant, are small which is encouraging. These data also indicate that hyperbilirubinemia can produce small degrees of damage as well as major

D.W. McCandless, *Kernicterus*, Contemporary Clinical Neuroscience,
DOI 10.1007/978-1-4419-6555-4_21, © Springer Science+Business Media, LLC 2011

neurological disorders, and death. So, bilirubin damage is not necessarily an all or none event. The efficacy of exchange transfusion is affirmed by the finding that prior to this treatment, the IQ difference between erythroblastosis jaundiced newborns and controls was 11.8 points lower as compared to the 6.1 point spread in the present study.

A later study (Johnston, W., et al. 1967) examined the after effects of erythroblastosis fetalis and hyperbilirubinemia in terms of IQ, hearing, and psychological and neurological development 5 years after the disease was treated.

In this study, 129 cases of hyperbilirubinemia were studied. All had indirect bilirubin levels of at least 20 mg/100 ml, and 42 had indirect bilirubin levels over 25 mg/100 ml. Ninety-five percent received one or more exchange transfusions. Of these cases, three had evidence of mild athetosis, and seven had sensorineural hearing loss. There were no cases of hearing loss in the control group. There were no significant differences in IQs between the jaundiced and control groups.

The authors note that the question always exists as to whether "minor" hyperbilirubinemia, while not producing overt kernicterus and cerebral palsy, might be doing minor brain damage. This low level of damage might only be demonstrable at a later age.

In the present study, one infant did show overt signs of kernicterus including opisthotonus, shrill cry, and ocular signs. This patient had athetoid sequelae. While newborns, 11 cases were suspected of having kernicterus; however, on follow-up examination, two had athetoid movements. All seven with neurological sequelae had hearing loss. The loss was bilateral and involved higher frequencies. The IQ was measured as being the same as the patients siblings, and the same as the general population. The average serum bilirubin level in affected children was 30 mg/100 ml. The two patients with the highest indirect bilirubin levels of 30.8 and 29.7 mg/100 ml were both athetoid (see Fig. 1).

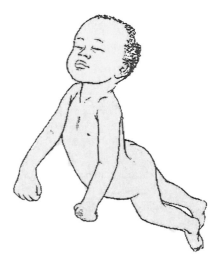

Fig. 1 Schematic figure showing opisthotonus posturing. Note neck extension, rotation of arms, and flexed hands

In another study (Holmes, G., et al. 1968), an attempt was made to look for objective evidence of minimal neurological sequelae such as hearing loss and slight incoordination. Subjects had unconjugated hyperbilirubinemia between 13 and 23.4 mg/100 ml. A mild hyperbilirubinemia group with levels up to 13 mg/100 ml was also examined. The "high" bilirubin group consisted of 34 children. Nine of these children had exchange transfusions and four received streptomycin.

Follow-up exams were performed at ages between 4.5 years and 7.5 years of age. The examinees were blind as to which group any individual belonged. Results showed no neurological abnormalities in any children. Athetoid and choreiform movements were specifically looked for. There were no statistical differences between groups as regards audiometric evaluations. This result indicates no relationship in the mild/moderate hyperbilirubinemic groups for hearing performance.

When examining motor maturation, a modified Oseretsky test was administered. After examining the data, and adjusting for age, the conclusion was that there was no relationship between mild/moderate hyperbilirubinemia and later motor development. The authors conclude that normal full-term infants with mild to moderate (5.5–23.4 mg/100 ml) were found to have no demonstrable evidence of hearing or motor involvement when thoroughly examined in follow ups to 7 years of age.

In another study (Hyman, C., et al. 1969), infants admitted to a hospital over a 6-year period with hemolytic disease or hyperbilirubinemia were studied. The jaundiced group consisted of 405 infants who in addition, were able to be followed up for a minimum of 4 years. None of the infants was actually born at the hospital where the study was done, but were referred from various other hospitals. Therapy was aimed at keeping indirect (unconjugated) bilirubin at or below 20 mg/100 ml for the first 5 days in term infants, and 7 days in premature infants. Three hundred and fifty one infants received exchange transfusions. All of this group also received penicillin plus streptomycin. Clinical observation included those evaluating feeding, neurological status, muscle tone, the cry, and the Moro reflex.

All had follow-up exams until 4 years or more. At this time a hearing test was performed. When central nervous system involvement was suspected, further evaluation was performed including EEG, and a full neurological/psychological work up.

Results showed that 346 children had no evidence of central nervous system dysfunction as indicated by history or examination. Conversely 59 children (15%) had some neurological involvement. This consisted of hearing loss, athetosis, seizures, behavioral problems, psychotic behavior, etc.

Neonatal complications other than prematurity and hyperbilirubinemia included respiratory problems, hemorrhagic problems, cardiac failure, and sepsis. All of these may act to alter important features such as albumin binding of bilirubin, and central nervous system function. Forty-three patients were found to have one or more of the above complications. Data show that the highest number of central nervous system abnormalities occurred in neonates who were exposed to high levels of unconjugated bilirubin.

The authors note that minimal brain dysfunction occurred at a frequency to indicate the possibility that there could be an etiological association with high levels of unconjugated bilirubin. Minimal brain dysfunction may have been more frequent than detected since it may not appear until the patient is placed in a stressful situation, such as school. This feature could be examined with further follow-up studies.

Confounding problems such as prematurity in patients might also influence the presence of central nervous system symptoms. When these patients were excluded, the low bilirubin group (less than 20 mg/100 ml) had an incidence of 9.5% brain damage as compared to 15% for those exposed to high bilirubin levels (over 29 mg/100 ml). The authors state that further studies are needed since the number of qualifying patients is small. A sample size several times that of the present study would be ideal.

This study, the authors note, suggests that the level of 20 mg/100 ml should remain as a major criteria for treatment. All infants with marked neurological abnormalities when newborns, showed central nervous system sequelae when examined later. Twenty-nine percent of newborn jaundiced patients suspected of central nervous system involvement as neonates showed brain damage by 4 years, whereas only 12% of newborns not suspected of neurological involvement had damage on follow up.

In another report of IQ following hyperbilirubinemia at infancy, measures were made at 6 years of age (Scheidt, P., et al. 1991). The patients studied consisted of over 200 who had low birth weights of less than 2,000 g. Bilirubin levels were kept at 20 mg/100 ml or less by use of exchange transfusion.

Results showed the rates of occurrence of cerebral palsy were similar between patients with high bilirubin levels as compared to patients with low bilirubin serum levels. Similarly, there was no difference in IQ measurements between the two groups. Further, IQ was not related to bilirubin levels, time of exposure, or length of time of exposure. Results showed that there were no measures of central nervous system function which were different between the two groups that could be discerned at the 6-year follow up. These data indicate that if hyperbilirubinemia is kept at or below the recommended levels for birth weight, etc., there are no measurable sequelae.

In another long-term study (Seidman, D., et al. 1991), a prospective study was performed on 1948 subjects at the age of 17 years of age who had hyperbilirubinemia at birth. This was a study with a large number of subjects, obtained through records at a military draft board. Data from birth records included a variety of information including bilirubin levels. At 17 years of age, results of physical examination, and IQ scores were obtained. The study took place in Israel where every male is inducted into the military.

Subjects were divided into three groups according to bilirubin levels as newborns. Mild hyperbilirubinemia was defined as those newborns who were not visually jaundiced, and/or whose serum bilirubin levels were not measured. Moderate hyperbilirubinemia was defined as serum levels of 5–8 mg/100 ml on day 1, 10–15 mg/100 ml on day 2, and 13–20 mg/100 ml, thereafter. Severe hyperbilirubinemia was defined as serum levels of over 8 mg/100 ml on day 1, over 15 mg/100 ml by day 2, and levels of 20 mg/100 ml or more on day 3.

Results showed that of 585 infants whose bilirubin levels were measured, 308 had moderate hyperbilirubinemia, and 144 had severe hyperbilirubinemia. The mild group, who did not have bilirubin measured probably contained a few patients who would have qualified as "moderate."

In terms of IQ scores collected at age 17 from these subjects, the mean score was statistically significantly lower in the severe neonatal hyperbilirubinemia group. However, after exclusion from the severe group those with low birth weight, preterm, gestational age, Apgar scores, birth order, etc., there was no significant decrease in IQ scores in this smaller group. The addition of phototherapy and exchange transfusion as independent variables did not influence the outcome. No effect of hyperbilirubinemia was discerned in terms of hearing/vision loss, or in neurological performance.

The authors state that these data support the concept that the risk is low for mental/physical impairment in term infants without hemolytic disease if serum bilirubin levels are maintained below 20 mg/100 ml. It should be emphasized, however, that in males in the severe group for hyperbilirubinemia, IQ scores were lower than in other groups (scores at or less than 85). Other studies have not been consistent as regards results of IQ measurement because of low subject numbers in various groups.

The present study is a clinical prospective one, not a controlled randomized experimental study. Therefore, interpretation of data is by necessity somewhat limited. Nevertheless, the authors note that these results suggest that severe neonatal hyperbilirubinemia may be associated with a risk for a lowering of IQ scores. This association between severe hyperbilirubinemia and IQ scores below 85 did not hold up when confounding factors such as birth weight were excluded from analysis. The authors conclude that severe neonatal hyperbilirubinemia should be of concern as a primary neurotoxic factor.

In another very similar study (Newman, T., and Klebanoff, M. 1993), a large number of newborn infants who were term and had birth weights at or greater than 2,500 g and had bilirubin levels documented at birth were included in this study. The study was a prospective cohort study coordinated in 12 US medical centers. The Wachsler Intellegence Scale was used to evaluate children at 7 years of age. In addition, a blinded neurological examination, and sensorineural hearing exam was also performed.

Results showed no association between IQ and bilirubin levels when comparing high bilirubin level children with low bilirubin level children. About 4.5% of children in the high bilirubin group (20 mg/100 ml or greater) showed abnormal neurological results, largely consisting of motor abnormalities. There was a stepwise increase in these sequelae as hyperbilirubinemia increased. There was no correlation between hyperbilirubinemia and sensorineural hearing loss.

The authors concede that there was no effect of high bilirubin levels on IQ scores, and only minor motor abnormalities. These findings suggest that bilirubin may lead to only relatively trivial consequences. It should be remembered, however, that hyperbilirubinemia was at a relatively low level when no consequences were evident, but high levels of serum bilirubin may lead to serious consequences.

A review of the literature shows adverse neurological outcomes in 37 patients with Crigler–Najjar syndrome type 1 have been reported. The authors note that the adverse outcomes could be grouped into four categories: (1) kernicterus, (2) cognitive decline, first noted in performance in school; this is accompanied by athetosis and chorea which implies damage to the basal ganglia, (3) cognitive decline with ataxia and tremor, suggesting cerebellar involvement, and (4) predominant cerebellar ataxia and tremor (Shevell, M., Majnemer, A., and Schiff, D. 1998).

The authors state that the undeniable toxicity of severe unconjugated hyperbilirubinemia especially in infants with hemolysis and prematurity, is well established. The pathological involvement in kernicterus is described in the literature as occurring in the basal ganglia, hippocampus, cerebellum, and some cranial nerve nuclei. Again, as in many other metabolic encephalopathies (i.e., thiamine deficiency, McCandless, D. 2010a, b), the lesions are punctate, and bilaterally symmetrical. Toxicity at the neurochemical level involves effects on cellular respiration, mitochondrial function, and energy metabolism, as well as other alterations. Entry into the brain is related to concentrations of bilirubin, binding of bilirubin to albumin, and levels of free unbound bilirubin.

The authors note that prior to phototherapy, the mortality rate reported in the literature was 23/28 infants. In addition, the decrease in morbidity of phototherapy is also significant. Phototherapy has permitted relatively intact survival of Crigler–Najjar syndrome type 1 patients beyond infancy. Also, heme oxygenase inhibitors (tin- mesoporphyrin) provide a therapeutic adjunct (Rubaltelli, F., et al. 1989). It is no surprise that neurological sequelae seen in kernicteric patients are irreversible even when the hyperbilirubinemia is reduced to non-toxic levels. Also of interest is that development of overt symptoms seems to be abrupt as opposed to gradual.

The authors state that in the absence of gene therapy, phototherapy and exchange transfusion remain the mainstem of treatment. Hepatic transplantation represents the only "cure" available. Plasmaphoresis is an emergency procedure for acute exacerbation of serum bilirubin levels. Phototherapy has resulted in no known adverse long-term outcomes, and is clearly an extensively used valuable treatment.

In a recent review of all aspects of jaundice in low birth weight infants, pathology, development of kernicterus, outcomes, etc., are carefully reviewed (Watchko, J., and Maisels, M. 2003). The authors point out that jaundice in preterm and full-term infants occur by similar circumstances, but that of preterm newborn infants is usually more prevalent, the course lasts longer, and levels of unconjugated bilirubin are higher than in full-term infants. The neuropathologic consequences of hyperbilirubinemia have long been known to include unconjugated bilirubin deposition in select brain areas, such as basal ganglia, hippocampus, cerebellum, and brainstem nuclei. It was also shown that premature infants could develop kernicterus at serum unconjugated levels than 20 mg/100 ml. The literature data on neurodevelopmental outcomes from hyperbilirubinemia in the neonatal period are mixed.

Much depends on treatment and abilities to discern subtle changes in central nervous system function.

The authors conclude commenting that neurological injury in hyperbilirubinemic preterm infants is a highly complex issue. The relative rarity of kernicterus in the USA is attributable to phototherapy and its aggressive use. Very few cases of preterm hyperbilirubinemia are beyond the reach of phototherapy treatment. Exchange transfusion remains an option in especially intractable cases.

In a study (Newman, T., Liljestrand, P., and Escobar, G. 2003), sequelae including psychometric testing and neurological evaluations at age 5 years were performed on a group of children who had hyperbilirubinemia exceeding 30 mg/100 ml as newborn infants. The patients were 11 who had total serum bilirubin levels equal or greater than 30 mg/100 ml during the first 30 days of life. All infants studied had birth weights equal to or above 2,900 g. Two received phototherapy during hospitalization, and 9 were discharged on the second day following birth. The actual range of bilirubin was 30.7–45.5 mg/100 ml.

These 11 infants were followed up between age 4 and 5 years. The neurological exam was deemed normal in 9 of the 11. In one, the exam was not done, and in another, the result was normal (questionable-possible mild ataxia). IQ tests were either not done or were normal. A parental evaluation questionnaire regarding the development of their offspring was either not done, or was judged "no concerns." One was in speech therapy, and another had a "bad temper."

The authors state that their study should not be an indicator that bilirubin levels over 30 mg/100 ml are safe. The low numbers of patients in this study, and the continued occurrence, although rare, of kernicterus suggest awareness must be maintained at a high level. Thus, aggressive therapy with phototherapy and exchange transfusion if necessary, are key to preventing kernicterus in hyperbilirubinemic newborn infants. Yet another paper (Jangaard, K., et al. 2008) examines outcomes of term and near term newborns with hyperbilirubinemia in excess of 19 mg/100 ml. The goal of the authors was to determine the incidence of kernicterus, developmental delay, cerebral palsy, and possible hearing loss in hyperbilirubinemic newborn infants as compared to controls who had lower bilirubin levels, or normal physiological jaundice. Data were prospectively collected from infants born at or after 35 weeks gestation in Nova Scotia from 1994 to 2000.

Cases of hemolysis, congenital or chromosomal abnormalities, or asphyxia were omitted from study. Follow-up times ranged from 2 to 9 years. A variety of variables such as multiple births, smoking, breast feeding, type of delivery, etc., were all tested for final outcomes in an adjusted model.

Results showed a total of 3,779 infants had bilirubin levels equal to or greater than 13.5 mg/100 ml. Three hundred and forty eight newborns had hyperbilirubinemia exceeding 19 mg/100 ml. In this study, hyperbilirubinemia was more common in breastfed newborns and in infants whose mothers were over 35 years old. Results also showed no cases of kernicterus, and the incidence of cerebral palsy was similar between jaundiced babies and controls. The risk for developmental delay was increased in the moderate hyperbilirubinemic group, but not in the severe group.

Hearing tests and vision were not judged to be altered by exposure to hyperbilirubinemia. There was an increase in attention deficit disorder among newborn infants who were exposed to hyperbilirubinemia greater than 19 mg/100 ml.

The authors comment that while there were no cases of kernicterus, only 7% of infants in the high bilirubin group had serum levels in the 23–30 mg/100 ml range, and less than 1% had unconjugated bilirubin levels over 30 mg/100 ml. Except for attention deficit disorder and developmental delay, no other neurological sequelae were diagnosed. Finally, there was an association between hyperbilirubinemia and autism. This possible link deserves closer study.

The usual reservations are cited by the authors of this study. They state that the population studied was predominately white and may not be representative of other populations. They state that some untested factors might have influenced the study, and that there may have been some infants who were actually jaundiced, but not recognized. If included as controls, they could have brought the groups closer together in terms of tested outcomes, in effect underestimating the hyperbilirubinemia group's outcomes. Some of the hyperbilirubinemic infants were lost to follow up, and there were more 2-year-old children examined than 4-year-old children. Since some of the sequelae of hyperbilirubinemia occurs well after 2 years of age, these sequelae may have been under represented.

The authors suggest that an association of hyperbilirubinemia to developmental delay, attention-deficit disorder, and autism are disconcerting. These disorders need and deserve more focused study to see if there is a causal relationship. There has long been a suggestion that more subtle neurological deficits resulted from hyperbilirubinemia. Neurological damage would be expected to be a gradual occurrence, "not an all or none" phenomenon. In fact, early entry of bilirubin into the brain produces more subtle signs and symptoms, characteristic of any metabolic encephalopathy. Even in bilirubin encephalopathy/kernicterus, changes in mutation can be reversed (and intracellular bilirubin levels decreased) by aggressive and early treatment. This represents significant hope for those with developing metabolic encephalopathy. Increasing awareness of health care workers will result in an improved outcome for patients with bilirubin encephalopathy.

Neurobehavioral Teratology

A question frequently asked relates to whether neonatal unconjugated hyperbilirubinemia might, at relatively low levels, have an adverse effect on mental acuity and behavior. The concept is that even when indirect bilirubin levels are in the 12–18 mg/100 ml range, some small amount of free unbound bilirubin could gain access to the brain, and have a deleterious effect. The issue of free unbound bilirubin entering cerebral tissue is discussed in the chapter in this volume entitled "Hyperbilirubinemia Revisited."

One difficulty in trying to study this possible phenomenon is that the damage may occur in the very first few postnatal days, and there is no predamage data/test results upon which to make a comparison. Therefore, studies compare test results between groups matched as best as possible for gestational age, sex, birth weights, socioeconomic background, etc. The variables are serum bilirubin values and related factors. These studies may be well done, or various contributing factors may be overlooked.

Studies of this type have evolved into and defined a new field termed neurobehavioral teratology. In this field, many experiments have centered on developmental alterations occurring in utero. But it must be remembered that development does not stop at birth, but that many organ systems continue to develop long after birth. In the case of the CNS, continued postnatal brain development includes: the blood–brain barrier, increasing numbers of neurons, increasing neural connections, and possibly the most important, myelination, which continues for many months/years.

In some ways, arguing against the concept of bilirubin altering mental acuity and behavior are neuropathological results. Unconjugated bilirubin clearly stains specific areas such as basal ganglia, cerebellum, hippocampus, and brainstem nuclei. Little or no staining is seen in the cerebral cortex. And if bilirubin were to damage the cerebral cortex at for example, levels of 15 mg/100 ml, then why would not such staining be obvious when levels reach 50 mg/100 ml. One consideration is that possible partial or complete deafness (from damage by bilirubin to the cochlear nucleus) may lead to altered perceptions, etc., which in turn would certainly influence intelligence and behavior.

In a landmark study (Butcher, R., Stutz, R., and Berry, H. 1971), the first published examination of behavioral abnormalities in Gunn rats was published. In these studies, homozygous jaundiced (jj) young animals were compared to heterozygous

(Jj) littermate controls on performance in three experiments. The first series of experiments was called an open field experiment. In this procedure, 60-day-old jaundiced and littermate control Gunn rats were placed in an open area 114 cm square, and divided into a checker board of smaller squares. Rats were introduced into the corner of the square, and allowed to explore freely for 5 min. The number of subsquares in which the rats placed all 4 feet was recorded.

In the second experiment, the rats were treated as in experiment one, except a "fear factor" was introduced. A loud bell was sounded for 1 s every minute for the 5-min time for exploration in the open field. An experiment to assess activity of jaundiced rats vs. littermate controls was performed in which a light beam bisected an activity cage measuring $33 \times 20 \times 25$ cm, and the rats were left in the box for 24 h. The number of times the animals crossed the light beam was recorded.

Finally, to test learning, jaundiced and heterozygous controls were tested for several consecutive days in a water-filled T maze. In these experiments, rats were acclimated to the water maze, then placed in the T maze, which allowed escape. The time for learning the escape route was measured for all rats. Exhaustion was prevented by removing the rats after 10 min if needed.

Results showed that in the first series of experiments, despite the reduced speed of the jaundiced Gunn rats, the overall numbers of squares entered by each group (jaundiced and heterozygous littermate controls) was not statistically different. While the numbers of squares entered was similar, the procedure between groups was different. The jaundiced animals seemed to move in a relentless "patrolling" of the field, whereas the littermate controls displayed a more normal method of exploration consisting of looking around, then moving to another area. The gait of the jaundiced Gunn rats was somewhat ataxic, but did not impede their ability to ambulate.

Similarly, in the second series of experiments, the number of squares entered between the two groups was similar. Thus, the fear factor did not alter the response of the jaundiced Gunn rats. The loud tone did reduce activity in littermate controls, who "froze" at hearing the buzzer. The act of thrusting a hand at the head of the rats also produced the "freeze" response in the heterozygotes, but not the jaundiced homozygote Gunn rats.

Results from experiment 3, activity for 24 h, showed that the jaundiced homozygotes had a higher activity rate as compared to the heterozygous littermate controls.

Results of the pretest swimming trials demonstrated that homozygous Gunn rats swam slower than littermate controls. In all cases, errors in the swimming maze by jaundiced Gunn rats exceeded those made by littermate controls. These results were statistically significant at the 0.01 level of confidence. The jaundiced Gunn rats failed to reach the goal in the swim maze in less than 10 min, whereas the littermate controls always completed the maze in less than 10 min.

The authors note that in spite of ataxia and locomotion deficits, the jaundiced Gunn rats had activity which equaled or exceeded that shown by littermate controls. The hyperactivity continued in the jaundiced Gunn rats in spite of noxious stimuli such as loud tones, or rapid hand approach. Previous studies (Johnson, L., et al.

1959) had shown that brain damage in jaundiced Gunn rats was a feature of the first month of life, when unconjugated hyperbilirubinemia was greatest.

The suggestion is made that there may be similarities between jaundiced Gunn rats and brain damaged children. Further, the learning deficits seen in the jaundiced rats are consistent with the clinical findings of kernicterus in children, many of whom are mentally retarded. Thus, the authors state, the Gunn rat model may serve as a means of evaluating various treatment paradigms.

A potential structural answer to the question of cerebral cortex involvement was provided by an electron microscopic study of Gunn rats (Jew, J., and Sandquist, D. 1979). In this study, neurons from several brain areas were examined including those of the: cerebral cortex, hippocampus, cochlear nucleus, locus coeruleus, and olfactory bulb. Structural alterations in neurons were similar in all areas examined. Thus, cerebral cortex neurons were reported to show characteristic mitochondrial and endoplasmic enlargement (swelling), and vacuolization. Also noted were glycogen deposits, myelin figures, and nerve terminal degradation.

The authors note that the Gunn rat is such an excellent model of hyperbilirubinemia that results might also apply to newborn human infants with elevated serum unconjugated bilirubin. The authors state that the involvement of neurons in the cerebral cortex, as well as areas such as the hippocampus, could explain symptoms of minimal cerebral dysfunction which have been described in children who had hyperbilirubinemia as newborns.

In another study of Gunn rats (Izquierdo, I., and Zand, R. 1978), the behavior of these rats in shuttle boxes stimulated by a buzzer/shock regime were observed. Various theme and variations to the buzzer/shock paradigm were studied. One such variation is a random delivery of buzzers and shocks to the Gunn rats (D test). Another is the DP test, a classical Pavlovian paradigm in which the buzzer/shock are paired. The DC test is one in which the interval between the two stimuli is varied.

Results showed that in experiments which require a pairing (P) factor, a situation which tends to enhance the rat's tendency to "freeze" results in impairment of behaviors such as shuttling. The non-jaundiced Gunn rat littermates showed this response to a greater extent than the jaundiced Gunn rats. It was also found that all Gunn rats, jaundiced or not, did not demonstrate a normal avoidance contingency (shuttle = no shock) when subjected to a varied interval between buzzer/shock.

The authors note that the findings of similar responses between non-jaundiced Gunn rats and jaundiced Gunn rats suggested the abnormal behavior is not related to bilirubin-induced brain damage. The suggestion is made that in breeding of Gunn rat colonies may have produced other genetic changes which might explain the similarities in behavior between littermate groups. Another explanation might be that even the heterozygous non-jaundiced littermates may have a partial decrease in the conjugating enzyme glucuronyl transferase resulting in a low level of hyperbilirubinemia. The possibility therefore exists that some unconjugated bilirubin does enter the brains of heterozygous Gunn rats, altering behavior.

Yet another paper (Escher-Graub, D., and Fricker, H. 1986), has examined the short-term behavioral complications in newborn infants who had hyperbilirubinemia at a low enough level that phototherapy was not indicated. The study involved a total of 76 newborn infants with hyperbilirubinemia levels of about 11.5–14.5 mg/100 ml. The control group of newborn infants consisted of 401 patients with serum bilirubin levels not over 5.75 mg/100 ml. Many criteria were compared between the two groups, including both mothers and infants, and almost all were similar between the two groups, including criteria such as Apgar scores.

The newborn infants were evaluated using the Brazelton Neonatal Behavioral Assessment Scale. Newborns were examined on day 4 or 5, and the examiners were blind to the topic of the study. All examiners were trained in the testing procedures.

Results showed that the two groups, mild–moderate jaundice and controls were statistically significantly different in almost one half of the items tested. In terms of habituation, jaundiced infants were more sleepy than controls, and were not as responsive to repeated noxious stimuli. As regards orientation, the jaundiced group had less capacity to process environmental events than did controls. For example, when alert, 70% of jaundiced vs. 57% of controls showed only brief periods of responsiveness to animate or inanimate stimuli. In other words, the jaundiced newborns were not as alert.

As regards motor performance, the jaundiced newborn infants were more floppy than controls when pulled to a sitting position. The jaundiced infants' defensive responses to a cloth being placed over the eyes was also judged to be more immature than the response of controls. The regulation of state was also different between the two groups in that the jaundiced newborns needed more outside stimulation to function. Jaundiced newborn infants were more difficult to console than control infants.

The authors state mild hyperbilirubinemia at a level low enough not to require phototherapy was associated with many behavioral problems as compared to control infants. Generally speaking, the behavior could be regarded as less mature in jaundiced infants ascompared to controls. The jaundiced infants were regarded as less responsive and more difficult with respect to interaction with caregivers. These developmental criteria, self-organization and interaction with caretakers, are deemed very important features of newborn behavior.

The authors point out that their findings of alterations of important behavioral criteria would make it difficult for the mother to efficiently interact with her infant. This in turn could modify the later attachment of mother and child.

The authors speculate that the data support the concept of a "subtle bilirubin encephalopathy." This might influence the initiation of phototherapy. The authors also say that an early hypothesis regarding bilirubin toxicity was that there is a primary impairment of some other area of brain function which produces impaired behavior, and at the same time reduces the infant's ability to metabolize unconjugated bilirubin. This supposed that the presence of hyperbilirubinemia and altered brain function were a curious coincidence (Levine, R. 1976). This hypothesis is no longer taken seriously. When this paper was published, the relative reversibility of the changes in behavior described above, was not known.

Another study of newborn jaundiced patients was conducted using the Brazelton Scale as described above. In this study (Mansi, G., et al. 2003), 28 jaundiced newborns with hyperbilirubinemia (range 13–20 mg/100 ml) were compared to a control group (bilirubin range 5.3–11 mg/100 ml). The newborn infants fulfilled the criteria of being full term, no drugs at delivery, Apgar score greater than 7 at 1 min, and 9 at 5 min, and no obvious malformations.

Results showed that the control group (28 subjects) performed better than the jaundiced group in social interactive categories. Visual and auditory items were statistically significantly different between the two groups. There were other differences in characteristics such as lability of state, self-quieting ability, and a greater incidence and frequency of tremors in the hyperbilirubinemic group.

The authors say that theirs is not the first report of results when mild/moderate hyperbilirubinemic infants are compared as regards behavioral results to those of control newborn infants. The Brazelton test is used in many contexts, including that of elevated serum bilirubin. The present study did not include any newborns in the jaundiced group who had received any phototherapy which might have confounded the results.

The authors state in the discussion that their results show, after stratification for sources of bias, that levels of elevated bilirubin claimed to be safe (below 20 mg/100 ml) are indeed associated with many statistically significant alterations in behavior as shown by the Brazelton test.

Many questions remain unanswered about this study. The authors say that the clinical significance of these findings is still not clear. Also not clear is whether these changes are permanent, or whether they may affect other behavioral traits such as socialization. Since more severe results of unconjugated hyperbilirubinemia, in this country, occur less and less, the detection of more subtle changes due to mild/moderate hyperbilirubinemia may be more difficult to study. Nevertheless, this paper presents more data indicating a possible negative behavioral outcome in cases of mild/moderate unconjugated hyperbilirubinemia.

There are many such papers in press from many different sites demonstrating this potential problem. There are also papers which are "negative" in that they do not show any changes in behavior in moderate hyperbilirubinemic newborn infants. Given the potential risks, the burden of proof remains with those who fail to show behavioral changes.

Another study (Hansen, T., et al. 1988) looked at the effects of unconjugated bilirubin on post synaptic potentials in rat transverse hippocampal slices. In this in vitro study, hippocampal slices were prepared, then exposed to bilirubin concentrations from 100 µmoles/l to 1 mmole/l. Bovine serum albumin served as a "stabilizer" for bilirubin.

Results showed that bilirubin had a depressive effect by reducing the slope of the field excitatory postsynaptic potential in the rat transverse hippocampal slices. There was a concomitant increase in the peak latency of the population spike. These changes were statistically significant, and were noted at a bilirubin concentration of 100 µmoles/l in as little as 10–30 min. With added bovine serum albumin, the changes reverted toward normal.

The nature of the bilirubin-induced changes indicates that bilirubin inhibits synaptic activation. This inhibition could be due to impaired neurotransmitter release or to some effect on the neuronal membrane. This is possibly related to the observation that jaundiced Gunn rats have a higher level of hippocampal no repinephrine than do litter mate non-jaundiced Gunn rats (Swenson, R., and Jew, J. 1982). There is also evidence that bilirubin has an effect on neuronal membranes, producing a depolarization effect (Mayor, F., et al. 1986). These data all point back to the observation that the behavior of jaundiced Gunn rats strongly resembles that seen in rats with experimental hippocampal lesions (Butcher, R., Stutz, R., and Berry, H. 1971, op. cit.).

The authors correctly point out that extrapolation of these in vitro rat data to human clinical situations must always be cautious. Also, the use of bovine serum albumin may have modulated bilirubin's effect differently than human albumin.

As regards albumin, a study was done examining bilirubin and albumin on psychoeducational outcomes (Hansen, R., Hughes, G., and Ahlfors, C. 1991). Seventy-four children 9–11 years old (who required ICU treatment as newborns for hyperbilirubinemia) were given four tests measuring psychoeducational outcomes including the Kaufman Mental Processing and Achievement Scales, the Beery Visual Motor Integration Test, and the Vineland Adaptive Behavior Scale.

Results showed that the bilirubin binding of albumin (based on albumin concentration) significantly correlated with the Kaufman Mental Processing Composite. Maximum bilirubin levels and cumulative exposure to bilirubin as newborn infants did not correlate with psychoeducational testing. The authors suggest that somehow the albumin binding value might be a useful criteria for determining clinical treatment.

Another study looked at behavior of newborn infants with mean bilirubin levels at day 3 of 7.2 mg/100 ml. These 73 newborns were administered the Brazelton Neonatal Behavioral Assessment Scale on the 60th and 80th hours of life. The full-term infants were matched for normal weights and health. Results showed an inverse relation between serum bilirubin levels and performance on the assessment scale as regards orientation, motor maturity, and habituation. The authors note that higher levels of physiological jaundice may produce measurable differences from those with lower levels of bilirubin.

In yet another study of neonatal behavior, 50 term neonates with moderate hyperbilirubinemia (14.3 mg/100 ml) were matched with 50 control infants with lower serum bilirubin levels (9.1 mg/100 ml), and behavior assessed (Paludetto, R., et al. 2002). The Brazelton Neonatal Behavioral Scale was administered to the neonates at an average age of 87 h, and again at 3 weeks. The infants in each group were matched for normal weight, gestational age, race, uneventful pregnancy, no drugs at birth, etc. Apgar scores were greater than 7 at 1 min, and equal to or greater than 9 at 5 min.

Results from this study are interesting. At the first examination (87 h of life, range 72–110 h), most of the 27 behavioral items were statistically significantly different in the moderate hyperbilirubinemia group as compared to the group with lower bilirubin levels. The moderate hyperbilirubinemic (14.3 mg/100 ml) group also had a negative correlation with the autonomic nervous system, and these newborn infants had more tremors. As the serum bilirubin dropped, the differences and symptoms decreased. When tested again at 3 weeks there were no significant differences between the low and moderate hyperbilirubinemic groups.

The authors state that even in the presence of moderate hyperbilirubinemia, significant changes occur in newborn infant's motor and sensory components of behavior. This study excluded other possible factors such as phototherapy, hypoxia, infections, etc. The authors also state that their results are generally in agreement with another study showing minor motor dysfunctions in newborn infants with moderate hyperbilirubinemia (Soorani-Lunsing, I., Woltil, H., and Hadders-Algra, M. 2001). Yet another study has shown that moderate hyperbilirubinemia may be associated with the greatest increase in serum bilirubin to brain transfer in, for example, the central auditory system (Roger, V., et al. 1996). The authors suggest that their study does not address the clinical relevance of their findings as regards issues such as feeding, bonding, etc.

In a recent study (Newman, T., et al. 2006), 140 hyperbilirubinemic newborn infants were compared to 419 randomly selected controls. Of the 140 hyperbilirubinemic newborn infants, 130 had bilirubin levels of 25 mg/100 ml or greater, and ten infants had levels over 30 mg/100 ml. Subjects were from a total of 106, 627 live births at the Kaiser Care Program from 1995–1998, with gestational age of at least 36 weeks and birth weights of at least 2, 000 g.

Formal neurologic development evaluations were performed at a mean age of 5.1 years of age by blinded examiners. Child psychologists administered the Wechsler Preschool and Primary Scale of Intelligence-Revised, and the Beery-Buktenica Developmental Test of Visual-Motor Integration, 4th edn.

Results were obtained from a formal evaluation, parental questionnaires, or outpatient visits in 94% of hyperbilirubinemic patients, and 89% of controls. In terms of outcomes, there were no significant differences between the hyperbilirubinemic and control groups as regards intelligence testing. Similarly, no differences were noted between jaundiced and control subjects as regards visual–motor integration. There was also no statistically significant differences between the two groups in terms of the neurological examinations.

The results of the parental questionnaire showed only minor differences in responses between the two groups. Out of 120 behaviors, 2 (impulsiveness and sleep problems) were reported significantly more often in the hyperbilirubinemic group. When the hyperbilirubinemic group was statistically reevaluated in terms of levels of elevated serum bilirubin, or duration of exposure to high bilirubin, the results remained the same.

The authors state that scant little evidence was found to suggest that hyperbilirubinemia at these levels showed any adverse neuro developmental effects. Most patients in this study had bilirubin levels only slightly above 25 mg/100 ml, so the study cannot make any predictions regarding the development of kernicterus at much higher levels of bilirubin.

While other reports of minor neurological dysfunction exist, the present study examined hyperbilirubinemic patients at a mean age of 5 years. At this time, minor neurological deficits might well have corrected themselves. The present study did find lower IQ levels in children with bilirubin levels of 25 mg/100 ml who also had

Table 1 This table collates various author's results from examination of intelligence and behavior in patients who showed clinical evidence of bilirubin encephalopathy/kernicterus. Note wide variation in results

Author	IQ	Behavior
Day, R., and Haines, M. (1954)	Decreased	Not done
Johnston, W., et al. (1967)	Unchanged	Not done
Holmes, G., et al. (1968)	–	Unchanged
Hyman, C., et al. (1969)	–	Behavioral problems
Scheidt, P., et al. (1991)	Unchanged	Unchanged
Seidman, D., et al. (1991)	Decreased	Not done
Newman, T., and Klebanoff, M. (1993)	Unchanged	Not done
Newman, T., et al. (2006)	Unchanged	Unchanged
Escher-Graub, D., and Fricker, H. (1986)	–	Behavioral problems
Mansi, G., et al. (2003)	–	Behavioral problems
Paludetto, R., et al. (2002)	–	Behavioral problems

positive direct antiglobin tests. The authors state these results must be viewed with caution because of the low numbers of such patients (9).

The authors suggest their results have important implications for managing jaundiced infants. They suggest that in terms of risk, exchange transfusions' benefits will exceed risks at a level of serum bilirubin greater than 25 mg/100 ml. These data provide reassurance to families with hyperbilirubinemic newborn infants that prompt treatment (phototherapy) will serve to lower bilirubin levels, and there will be no neurological sequelae (see Table 1).

Gene Therapy for Hyperbilirubinemia

The term "gene therapy" is used to designate a process in which viruses are used to carry specific genetic information into a cell, thereby transducing the cells' function. In\the case of the Crigler–Najjar syndrome, and the Gunn rat, the absence of the bilirubin-conjugating enzyme glucuronosyl transferase can be corrected in cells infected by the modified viral vectors. Overall the numbers of inborn errors of metabolism susceptible to treatment with gene therapy, as well as the potential treatment of other types of disorders will be highly significant.

While simply stated above, the actual performance of this feat is hardly simple. There are several potential pitfalls, and a lack of knowledge of these has resulted in the deaths of several patients upon whom these treatments have been attempted. Issues such as immune reaction, and uncontrolled replication (cancer) after transduction, need to be carefully addressed. And, of course, law, ethics, politics, etc., all have a vocal position on gene therapy.

A major method of gene therapy involves the use of a viral vector. Common to all viruses is the ability to bind to the host cells, and inject genetic material into the host cell, which in turn begins to replicate the virus, and the cycle repeats itself many times. Retroviruses insert reverse transcriptase into the host, and thereby using the host RNA as a map of action. This serves to incorporate the viral genes into those of the host. This process can be utilized to deposit correct and functional genes into the host, thereby correcting missing or defective host genes. In the case of Gunn rats, this methodology can replace defective or absent glucuronosyl transferase in hepatocytes. Following this, Gunn rats can conjugate bilirubin, correcting the hyperbilirubinemia. Viral vectors can carry information to the host cell in the form of RNA or DNA. Once in the cell, through various steps, the host ultimately will produce whatever proteins using whatever information was carried in.

In the case of retroviruses, as an example, the genetic material introduced to the host is in the form of RNA. The virus adds its RNA information into the nucleus along with reverse transcriptase and integrase, enzymes involved in the mechanics of this complicated process.

The introduced RNA molecule then produces a DNA molecule which now carries the viral information such as that of glucuronosyl transferase. The enzyme reverse transcriptase facilitates this process. This DNA molecule is then added to

D.W. McCandless, *Kernicterus*, Contemporary Clinical Neuroscience,
DOI 10.1007/978-1-4419-6555-4_23, © Springer Science+Business Media, LLC 2011

the genome of the host. Once inserted, the cell has been transduced. Problems such as where is the new DNA inserted are critical. Some problem results are described below.

In adeno-associated viral vectors, it is the DNA molecule which transfects host cells, and almost always inserts its genetic material on a specific site on chromosome 19. The adeno-associated viral vector is a low capacity transfector, and is hard to prepare for use. Trials looking at adeno-associated viral vectors for treating various muscle and eye diseases are underway. Advantages of adeno-associated viral vectors include a relative lack of pathogenicity. It can infect both dividing and non dividing cells, and is stable. It is therefore more predictable than other viral vectors. The adeno-associated viral vectors also have a low immunogenicity.

Several other viruses are under investigation for use for treating disease states. It should also be noted that several non-viral methods of gene therapy are being tried, or have been tried. Many of these have advantages, such as relatively easy production, and low or absent immunogenicity. Problems such as low levels of transfection have been overcome. Many laboratories are working diligently on these issues associated with gene therapy, and hope is high that problems will soon be resolved such that gene therapy could be a mainstay in treating a variety of heretofore devastating diseases.

The process of modifying the genome of viruses such that they cannot replicate renders them suitable for serving as a vector to deliver modified genetic information to a cell. This is the rationale for safely using a variety of viral vectors to carry sequences necessary to initiate the production of glucuronosyl transferase in, for example, Gunn rats. If Gunn rats, a nearly perfect model of Crigler–Najjar syndrome type 1, could be treated successfully then that would suggest maybe viral vectors could be used in human cases.

The overall idea is that this gene therapy approach may ultimately be the basis for cure of a host of genetic diseases which result from a single genetic mutation. There are several problems to be addressed, such as the virally induced immune response, and the possibility that there might be an uncontrollable replication of transfected cells, resulting in cancer. Indeed, some early human trials resulted in these complications, and several patients died from gene therapy attempts.

Retroviruses are commonly used for gene therapy because they are readily rendered unable to replicate. The retroviral vectors require that the target tissue cells be in a dividing cycle. This is not a major problem in liver since a partial hepatectomy stimulates hepatocyte replication. Lentiviruses are a class of retroviruses and have been used to treat hyperbilirubinemia in Gunn rats. Their main feature is that they are able to infect non-dividing cells, suggesting that a partial hepatectomy might not be necessary in Gunn rats. As it turns out, in the case of liver, dividing cell recipients greatly improves the efficacy of lentiviral vectors.

In general, features of lentiviral vectors which are beneficial include the ability to introduce a new gene into animal models of certain diseases, and also into humans. There is a mouse model of hemophilia which is cured by this method. As stated above, lentiviral vectors can infect both dividing and non-dividing cells. There is usually a stable result from lentiviral vectors, implying one single treatment may have a long-lasting result.

Adenoviruses transmit their information in the form of double-stranded DNA. The DNA molecule is "free standing" in the nucleus, but of course is not replicated when the cell divides. This means adenovirus vectors might need to be used in more than one treatment. An adenoviral vector system called Gendicine has been approved for some cancer treatments in China, but not in the USA.

Adenoviruses have been used to attempt to correct the glucuronosyl transferase defect in Gunn rats. These vectors do not have an envelope; therefore, they gain entry to a cell without envelope fusion. After entry into the host cell, the vector is transported to the nucleus, where entry enables viral gene expression to occur. In this way, new viral components are generated and can cause, in the case of Gunn rats, effective glucuronosyl transferase to form. Again, there are concerns about the safety of adenoviral vectors, since there has been at least one death associated with an adenoviral vector clinical trial.

Hyperbilirubinemia due to the Crigler–Najjar syndrome can be particularly devastating for several reasons. First, if there is a complete absence of the conjugating enzyme for bilirubin, glucuronosyl transferase, unconjugated hyperbilirubinemia is a constant feature, with little hope for any natural improvement. Second, treatment paradigms are possible in some cases, but may not have long-term effectiveness, and do have inherent risk. If the enzyme deficit is partial (Crigler–Najjar syndrome type 2), then phototherapy may serve to lower bilirubin levels, but daily treatment for several hours each day could be necessary. The enzyme defect in both type 1 and 2 is similar, except that it is "total" in type 1, and only partial in type 2. This results in a higher level of hyperbilirubinemia in type 1 as compared to type 2. In each case, the high levels of serum bilirubin result from the genetic alteration of one enzyme – glucuronosyl transferase.

The genetic defect in glucuronosyl transferase alone, suggests the possibility of treatment based on either gene therapy or enzyme replacement therapy. Both of these treatments have been successfully used in other metabolic encephalopathies. The availability of an excellent animal model of Crigler–Najjar syndrome (the Gunn rat) makes it possible to explore these new treatment modalities in vivo. The importance of this animal model cannot be overstated.

Gene therapy, as opposed to enzyme replacement therapy, has received the most attention as a possible treatment strategy for hyperbilirubinemia. In a paper (Askari, F., et al. 1996), adenoviral vectors were used to correct the glucuronosyl transferase deficiency in Gunn rats. The use of adenoviral vectors are viewed as an efficient mechanism to express recombinant genes in liver cells by gene transfer (Yang, Y., et al. 1994). The study was undertaken in order to determine if in vivo adenoviral gene transfer of human glucuronosyl transferase isoform into the livers of Gunn rats would serve to ameliorate the enzymatic deficiency.

Results showed that there were proviral sequences in Gunn rat livers infused with adenoviral vectors. Transfected Gunn rat liver showed hybridization 3 days after treatment, and this decreased over 3 weeks. Liver microsomes from transduced Gunn rats were isolated and reacted in vitro with unconjugated bilirubin in order to ascertain the activity of the conjugating enzyme. Results demonstrated the presence of bilirubin diglucuronides, monoglucuronides, and unconjugated bilirubin in the reaction media. Control liver preparations from liver transfected with galactosidase showed no effect on glucuronosyl transferase activity.

The authors note that these data, in an excellent model of Crigler–Najjar syndrome type 1, lend evidence that the enzymatic defect may be correctable by the expression of a single unconjugated bilirubin glucuronosyl transferase isoform. This represents a further verification of the importance and validity of the Gunn rat model, and the ability of this treatment to correct the enzyme deficiency. It also offers promise for treatment of the human form of Crigler–Najjar syndrome type 1.

Since serum unconjugated bilirubin levels were reduced in Gunn rat serum by 3 days after treatment, the outcome of the treatment was rapid. The treatment was effective in eliminating the occurrence of kernicterus in homozygous Gunn rats. These Gunn rat results suggest gene therapy may be beneficial to humans with Crigler–Najjar syndrome type 1.

While the Askari paper demonstrated a positive effect on Gunn rat liver glucuronosyl transferase activity, the changes were transient. In another series of experiments (Takahashi, M., et al. 1996), the long-term normilization of defective glucuronosyl transferase activity in Gunn rat liver was examined.

In this study, adenoviral vectors expressing bilirubin glucuronosyl transferase were injected into 1–3-day-old Gunn rats. On days 56 and 112, further injections were administered. Adenoviral vectors expressing galactosidase activities were injected to serve as a control.

Results showed that at 4 h and 24 h after injection of the vector, 90–95% of the glucuronosyl transferase DNA was present in the liver, with only traces present in other organs. Further, significant amounts of mono and diglucuronides were being excreted in bile 3.5 h after injection of the vector. By 6 h, all levels were indistinguishable from normal rats. Serum bilirubin levels examined 48 h after injection of vector into 1-day-old homozygous Gunn rats were not significantly different between experimental groups and controls. An additional indicator of efficacy is that the yellow colorization of the skin of homozygous Gunn rats disappeared shortly after vector treatment.

The authors observe that their data show unequivocal evidence for the conjugation of bilirubin in homozygous Gunn rats following vector administration. This was shown 3.5 h after treatment. Data presented showed that the loss over time of effectiveness resulted from a loss of gene content as opposed to some mechanism rendering the gene unable to function. The authors state that obstacles to human application of adenoviral gene therapy are the development of neutralizing antibodies.

Another paper (Nguyen, T., et al. 2002), examines the efficacy of using human immunodeficiency (HIV) derived vectors as a mechanism to deliver trans genes into primary human hepatocytes. This can occur in the absence of hepatocyte division. These methods include a minimal manipulation after harvest, and injection into a recipient within only a few hours after removal of a liver sample. These methodologies may offer viable methodology to enable the treatment of liver disease by transplanting genetically modified hepatocytes.

In a more recent study (Toietta, G., et al. 2005), elimination of unconjugated hyperbilirubinemia was achieved for a period of over 2 years (lifelong) in homozygous Gunn rats using a single injection of helper-dependent adenoviral vector.

Results from this study showed that helper-dependent adenoviral vector (HD-Ad) had a dramatic effect on unconjugated hyperbilirubinemia. There was a complete normalization of serum bilirubin levels after about 1 week following a single treatment with HD_Ad. Bilirubin levels dropped by as much as 90% as compared to non-treated homozygous rats. Going forward from 1 week, there was never again a statistical difference between the normal 1 week values of serum bilirubin and any subsequent bilirubin levels. Stated differently, the drop in serum bilirubin attributed to HD-Ad was stable for 2 years. Three doses were used – high, intermediate, and low, and the results applied to all three doses.

Phenobarbital administration to a group of rats treated with low-dose HD-Ad gene therapy showed no increase in rate or degree of hyperbilirubinemia. Two years after treatment with HD-Ad, bile from homozygous rats and saline injected homozygous controls were compared. Results showed mono and diglucuronides were present in the HD-Ad-treated group, whereas homozygous saline treated rats' bile did not have conjugated bilirubin present.

To determine if UGT 1A1 was present in excess in the liver after HD-Ad administration, bilirubin isomers were injected into treated homozygous Gunn rats 2 years after treatment. Results showed that the livers of these Gunn rats were able to conjugate these isomers, their conjugates being extracted from bile. This demonstrated that even 2 years after a single HD-Ad treatment, glucuronyl transferase was present and active in concentrations in excess of the required activity. This is also the case in normal non-jaundiced rats – there is excess conjugating enzyme. There was a positive correlation between the HD-Ad dose and conjugating capacity – the highest HD-Ad treatment-dosed animals had the highest conjugating capacity.

Measures of toxicity consisting of assessing serum alanine aminotransferase, and platelet levels were used. Results showed no hepatotoxicity as reflected in serum alanine aminotransferase activity. There was a decrease (50%) in platelets in the intermediate dose of HD-Ad rats. However, the drop in platelets was transient, and had returned to normal within 3 weeks of dosage.

The authors note that the HD-Ad method has advantages over other gene therapy methodologies. The correction of the glucuronosyl transferase deficiency can be achieved in Gunn rats using retroviral vectors, but a partial hepatectomy is required to induce hepatocyte regeneration. Partial correction of glucuronosyl transferase deficiency has also been achieved using in utero transfer of lentiviral vectors. A major advantage of Ad vectors is that they localize in the liver after systemic administration. Problems with toxicity and immunogenicity of viral proteins have been largely overcome.

The authors state that the significance of their study is that a complete "lifetime" correction (greater than 2 years) of the enzyme deficiency was achieved in a single treatment. This is relevant to human Crigler–Najjar syndrome because the Gunn rat model represents an ideal model of the human counterpart. In addition, the correction of serum hyperbilirubinemia remained constant for over 2 years, the approximate life span of rats.

There was a transient toxicity related to HD-Ad treatment reflected in a drop in platelets. This had disappeared within 3 weeks of HD-Ad administration, but the authors note that the long-term safety and efficacy need to be established before HD-Ad gene therapy is attempted in humans.

The previous paper makes mention of the fact that lentiviral vectors require hepatocyte proliferation in order to effectively serve as a mode for gene transfer. A definitive paper examining these criteria has been published (Ohashi, K., Park, F., and Kay, M. 2002). In this paper, experiments are described which examined the role of liver cell proliferation which enhances the efficacy of lentiviral gene transfer. Three methods of liver proliferation were employed: regeneration stimulated by a two third partial hepatectomy, hyperplasia induced by administration of dichloropyridyloxy benzyne, and lastly a combination of the two treatments listed above.

Results were assessed by using a lentiviral vector which expressed beta galactosidase. Gene transfer was measured by beta-D-galactoside staining in the liver. Results showed a 30-fold increase in galactoside staining in the first two experimental groups (partial hepatectomy, and dichloropyridyloy benzyne administration). Both of these treatments together produced an 80-fold increase in the efficacy of transduction as compared to control mice.

Other results from this study showed that administering the vector in several small doses greatly reduced any evidence of hepatotoxicity from the vector. Also, the biodistribution of vector was limited to the liver and spleen. No detectable beta galactoside staining was seen in any other organ. The authors state that the similarity in lentivirus vector-mediated gene transfer between the three mechanisms of producing hepatic damage suggests that this vector can be used for the potential treatment of different liver disorders. And yet, liver regeneration needs to be induced by some hepatotoxic treatment, or invasive surgical treatment.

Not all gene therapy treatment regimes involve viral vectors. In a recent study (Jia, Z., and Danko, I. 2005), repeated IV administration of plasmid DNA served to have a positive effect on hyperbilirubinemia in Gunn rats. The optimal gene delivery was determined to be via the tail vein. This mode of delivery resulted in high levels of gene expression in skeletal muscles of the hind limb. When plasmid Cl-Loc was injected, expression of luciferase was dose dependent, and was highest 7 days after injection. About 25–40% of muscle fibers were found to be beta galactosidase-positive 7 days after injection of plasmid Cl-Lac Z I.V.

HPLC analysis of bile from treated Gunn rats showed the presence of both monoglucuronides and diglucuronides. This conjugation to water-soluble bilirubin compounds occurred in skeletal muscle, and was similar to that seen in normal Wistar rats.

The total serum bilirubin levels dropped to their lowest at 7 days after IV injection, then gradually rose. Repeat injection of plasmid DNA once every month served to maintain total bilirubin at a low level for at least 4 weeks. A relative lack of toxicity of this procedure was indicated by a slight and transient rise in creatine kinase activity after infusion of plasmid DNA. Western blot analysis of Gunn rat skeletal muscle showed a persistence of UGT 1A1 for 4 weeks or more after venous administration. Further, there was no longer amplification of results following several cycles of gene delivery.

The authors note that IV administration proved successful with no adverse effects, especially at the injection site. They speculate that the treatment paradigm could be extended indefinitely by using different injection sites/muscle groups. This could give a Crigler–Najjar patient a longer time before the necessity for a liver transplant. The gradual diminution of efficacy after the first 7 days probably results from a loss of vector from the muscle group, and an immune elimination of these fibers which were transfected. The authors correctly note that further studies are warranted to determine more efficient ways to deliver gene therapy, and to try to gain more stability to the enzyme UGT 1A1 glucuronosyl transferase.

Oncoretroviral vectors have also been used in gene therapy, and one such example has been reported (Bellodi-Privato, M., et al. 2005). In this study, 2-day-old homozygous Gunn rats were injected with recombinant oncoretroviruses. Blood and bile were collected for analysis. Non-injected and beta galactosidase-injected homozygous Gunn rats served as controls.

Results showed that 6 weeks after gene transfer, Gunn rats treated with UDP-glucuronosyl transferase (UGT1) had serum bilirubin levels significantly lower than non-treated or beta galactosidase-treated homozygous controls. The values were less than 1.0 mg/100 ml. It was found that the correction of hyperbilirubinemia lasted in the normal range for up to 42 weeks after injection of UGT1 vectors. Control homozygous Gunn rats had bilirubin levels of around 6 mg/100 ml.

Some Gunn rats underwent bile duct cannulation at 13 weeks of age to determine bile constitution. Results showed that in treated homozygous Gunn rats, the bile had a predominance of bilirubin diglucuronides as compared to monoglucuronides (ratio about 2.5:1). In Wistar rats, the major component in bile is monoglucuronides.

At the time of bile analysis, livers were removed to determine the numbers of transduced cells. Using PCR, an estimate of transduced cells was between 5 and 10% of cells. Immunohistochemistry of the beta galactosidase-injected rats showed a frequency of about 2% of hepatocytes transformed. The concentration of retroviral vectors injected was 25 times greater in the UGT1 treated rats, an estimation of about 5% of hepatocytes would have been transduced. This value is in general agreement with the value obtained from PCR experiments. It is expected that retroviral vectors can integrate the transgene of the recipient genome, the actual proportion of transfected hepatocytes remains constant over time.

Examination of other organs as well as liver for a positive UGT1 signal was performed at 32 weeks after treatment. Results showed positive signal in various organs in some Gunn rats. Signals were found in heart, intestine, and skeletal muscle, as well as liver. The percent contribution was small, with heart contributing 8%.

The authors of this paper state that neonatal gene therapy using retroviral vectors yielded a complete resolution of hyperbilirubinemia, and prevented cerebral involvement in homozygous Gunn rats. This single treatment resulted in the excretion of water-soluble bilirubin diglucuronides and monoglucuronides in bile. The liver is the ideal target for gene therapy; however, other organs may participate. The authors further state that Crigler–Najjar type 1 is an ideal candidate for gene transfer. The use of retroviral vectors is especially efficacious in young animals since the liver is still developing. By the use of integrative vectors, permanent expression is assured in the infected cell's genome. One potential caution is the use of insertional mutogenesis. On the other hand, there are no reports of uncontrolled proliferation of target cells. The advantage for retroviral therapy in newborns is that tolerance to the transgene product is induced, thereby reducing and/or eliminating risk for an immune response to the UGT1 gene. Adult rats by comparison, do show a pronounced immune response to retroviral delivery of the UGT1 gene.

Another variation of using viral vectors for transduction of hepatocytes has been described (Nguyen, T., et al. 2006). The basic method is that hepatocytes are quickly transduced in suspension with lentiviral vectors and then immediately transplanted into the recipient. The acronym SLIT (suspension, lentiviral, immediately transplanted) has been created to describe the process. The present paper uses this technique to evaluate its effect on hyperbilirubinemia on the Gunn rat.

Lentiviral vectors which coded for UGT1 were prepared. Hepatocytes from homozygous Gunn rats were harvested and transduced. The transduced hepatocytes were immediately transplanted into recipient jaundiced homozygous Gunn rats. After this process, hyperbilirubinemia and bile pigments were assayed in order to determine the efficacy of the gene therapy treatment. The efficiency of the transplanted hepatocytes was analyzed by immunohistochemistry.

Results from this study showed that the hyperbilirubinemia decreased by about 30%. The decreased bilirubin levels remained constant for up to 240 days. Measurement of glucuronides showed their presence in bile, reaffirming conjugation was active. Hyperbilirubinemia in the control (green fluorescent protein) Gunn rats actually increased, and GFP-positive hepatocytes ranged from 0.5 to 1.0%.

The authors state that this experiment represents the first to show a long-term effect from lentiviral vectors. It therefore demonstrates the efficacy of the SLIT approach in the treatment of inborn errors of liver metabolism.

The liver is ideally suited for gene transfer due to its anatomic location and vascularity. The liver is the site of many metabolic abnormalities, many of which might be amenable to treatment using lentiviral gene transfer. Several gene therapy attempts are being explored, including for example, treating Wilson's disease (Merle, U., et al. 2006).

In this study, an animal model of Wilson's disease, the Long Evans cinnamon rat was used for study. HIV-derived lentiviral vectors were used because they can be administered in vivo and integrate into the genome of cells.

Results showed that lentiviral vector (HIV derived) served to permit long-lasting gene transfer in the rat model of Wilson's disease. The lentiviral vectors expressed ATP7B under the control of phosphoglycerokinase promotor. The efficacy of this treatment showed when lower liver copper content was seen in all treated animals as compared to untreated controls. There was a restoration of depleted holoceruloplasmin and its oxidase activity. This is an important parameter of assessment of the efficacy of the treatment. This was briefly reviewed because it shows a similar evolution of treatment of Wilson's disease as is seen in Crigler–Najjar syndrome. Many steps seem to be examined, such as long-term toxicity, uncontrolled hepatocyte growth, etc., before clinical trials are attempted.

In another Gunn rat/Crigler–Najjar syndrome study, experiments were conducted to examine effects of combined gene therapy to correct glucuronosyl transferase deficiencies in Gunn rats. UGT 1A1, UGT 1AG, and UGT 1A7 can all three be expressed and controlled to some extent depending on the dose of adenovirus. This type of manipulation is deemed essential because it provides a mechanism by which to assess the actual role of specific liver UGT. For example, UGT 1A1 may have some role in thyroid hormone glucuronidation, and this methodology can explore specific roles of UGT in toxicological studies (Miles, K., et al. 2006).

This study shows that a combination of rat UGT encoding adenovirus can be administered to homozygous Gunn rats which in turn express a range of liver enzyme profiles. The authors state that this experimental design may serve to allow for pharmacokinetic analysis of mycophenolic acid in Gunn rats. The technique is also an important new tool to examine characteristics of altered rates of hepatic glucuronidation, as well as after treatments.

Another paper examined the lifelong safety of gene therapy in the treatment of the hyperbilirubinemia of Crigler–Najjar syndrome type 1. In this study, homozygous newborn Gunn rats were injected on day 2 with lentiviral or oncoretroviral vectors which encoded human beta-uridine glucuronosyl transferase. This treatment resulted in a 95-week correction of the hyperbilirubinemia to normal levels. The complete normalization of the elevated serum bilirubin occurred with less than 5% of beta-UGT copy per haploid. Liver histology was normal following treatment, and no adverse immune responses were observed. This "protection" may be a result of delivery of the vector at a very young age (2 days). The authors state their data document a lifelong safety from adverse effects of viral vectors when administered in this fashion. These data bode well for future human viral vector gene therapy.

In a recent paper (Birraux, J., et al. 2009), a combination of methodologies involving transduction of hepatocytes using lentiviral vectors were used, and then transplanted into mice to assess long-term survival and efficiency.

In this study, hepatocytes were isolated from a 4-year-old Crigler–Najjar Syndrome type 1 patient. The cells were prepared for transduction using standardized methods. The lentiviral vector expressed human enhanced GFP, or UGT 1A1 protein. Transduction was performed on both fresh and previously pelleted and frozen hepatocytes. These hepatocytes were incubated with the lentiviral vectors, then assayed in cultured cells, or injected into mice to determine longevity.

Results showed that hepatocytes incubated with lentiviral vectors which expressed GFP maintained differentation and did not divide. After 5 days, results showed that more than 90% of the hepatocytes from the Crigler–Najjar type 1 patient were transduced. Increasing dosages of vector did not change this result.

Therefore, this shows that these hepatocytes from a Crigler–Najjar patient are amenable to alteration of expression of glucuronyl transferase enzyme activity in a straightforward transductional procedure. The demonstration of hepatocyte enzyme activity was shown by Western blot analysis using anti-UGT 1A1 in culture.

This study also examined the long-term capacity of the lentiviral vector-transduced Crigler–Najjar type 1 hepatocytes to remain viable. This was accomplished by injecting the appropiate lentiviral vector transduced hepatocytes into the inferior splenic pole of adult immunodeficient non-obese diabetic mice. Transduced hepatocytes represented from 0.5 to 1.0% of the recipient liver. Both GFP-transduced and UGT 1A1-transduced mice were injected, and expression of GFP was still present after 26 weeks.

In terms of human UGT 1A1 expression, real time PCR data showed a similar response of UGT 1A1-transduced hepatocytes as compared to the response of GFP-transduced hepatocytes. Thus, the UGT 1A1 expression was also still present after 26 weeks.

The authors note that orthotopic liver transplantation can be an efficient method of treatment in Crigler–Najjar syndrome type 1. But access to livers for transplant in cases such as Crigler–Najjar type 1 is not always available. The idea of using autologous hepatocytes which have been transduced (ex vivo therapy) is an attractive approach which avoids several problems. This approach has been tried for treatment of inherited hypercholesterolemia without great success because of the low percentage of transduced hepatocytes. The authors state that in their study, these problems were avoided by refinements in methodology which allowed a high percentage of hepatocytes to be transduced.

The authors note that their study shows that: (1) liver cells from a Crigler–Najjar syndrome type 1 patient can be transduced to express UGT 1A1 protein and (2) the ability of these transduced hepatocytes to engraft into liver, and (3) that the host liver (mice) were partially hepatectomized to enhance engraftment, and finally, (4) this procedure would be similar to a human clinical setting where the Crigler–Najjar liver would be the source of hepatocytes. This would in fact be a partial hepatectomy, and the capacity of the lentiviral vector to restore UGT 1A1 activity has been shown in the Gunn rat (Nguyen, T., et al. 2005, 2007).

Another huge advantage of this for Crigler–Najjar syndrome patients is that in normal individuals, the glucuronosyl transferase enzyme is about 20 times what is required to conjugate bilirubin. Therefore, 5% success in hepatocyte engraftment is enough to lower serum bilirubin dramatically. The fact that Crigler–Najjar syndrome type 2 patients have only about 10% of glucuronosyl transferase activity and do not as a rule develop brain damage, argues in favor of this concept.

Finally these authors believe that this report is the first to demonstrate that liver cells from a patient with Crigler–Najjar syndrome type 1 (or any other inherited inborn error of metabolism) can be transduced by lentiviral vectors. Secondly, this report shows that rapidly transduced hepatocytes were able to express UGT 1A1 enzyme for a significant period of time. The authors state that this could now be tried in primates as a precursor study before clinical trials in human Crigler–Najjar syndrome type 1 patients.

Epilogue: Comments and Future Directions

The study of unconjugated hyperbilirubinemia, and resultant kernicterus is a highly interesting and important endeavor. This area of investigation has been an active one for well over a 100 years, and occupied the minds of many outstanding scientists.

The incidence and cost in terms of suffering and dollars from hyperbilirubinemia worldwide is enormous. Glucose-6-phosphate dehydrogenase deficiency (G-6-PDH) is estimated by WHO to have a worldwide incidence of at least 400 million cases. The vast majority of these people are not even aware they have the disease. Hyperbilirubinemia is a consistent feature of this disorder, and newborn infants may develop symptoms. Malaria has a very similar geographic distribution as G-6-PDH deficiency.

The number of newborn infants with G-6-PDH varies from region to region, and among different ethnic groups. But given the number of worldwide cases, it is easy to see that thousands of newborns annually suffer from hyperbilirubinemia. The treatment (phototherapy) in Third World countries is minimal. The net result is that many die, or are brain damaged by the toxic effects of bilirubin. The sequelae from kernicterus run the gamut from mild nearly imperceptible symptoms to severe cerebral palsy.

Phototherapy can effectively reduce serum unconjugated bilirubin to levels which are not considered to be toxic to brain. Yet these phototherapy units cost several thousand dollars each, and are not available in most hospitals in for example, in sub-Saharan Africa where there are many cases of G-6-PDH deficiency. As an example of the status of medicine in Third World countries, several years ago there was only one neurologist in Zimbabwe. He has since immigrated to the USA Monies for health care in the majority of Third World countries is almost non-existent.

On the other hand, it has been shown that simple white fluorescent light has a beneficial effect in lowering serum unconjugated bilirubin levels. These lights can be constructed for just a couple 100 dollars, and along with educational materials distributed to each hospital in endemic areas, a significant level of treatment could be provided. The financial cost of this would be trivial, yet could save many lives, and reduce crippling sequelae of untreated unconjugated hyperbilirubinemia.

D.W. McCandless, *Kernicterus*, Contemporary Clinical Neuroscience, DOI 10.1007/978-1-4419-6555-4_24, © Springer Science+Business Media, LLC 2011

It should be relatively easy to find a foundation or philanthropic organization which would be amenable to financing such a project. Just a few 100 units, some spare parts, and educational brochures would be sufficient for Africa.

Another area which has seen increased interest relates to the concept that bilirubin can enter brain at measured serum levels which may vary considerably. The idea being advanced (see the chapter "Hyperbilirubinemia Revisited," this volume) is that the critical serum bilirubin level of importance as regards brain damage is the unbound, unconjugated bilirubin level. Unbound in this context refers to that moiety not bound to albumin.

Conjugated bilirubin is water soluble, and so, does not enter brain. Unconjugated (lipid soluble), but bound to albumin also does not gain entrance to brain. It is the unbound (to albumin) unconjugated bilirubin which is eligible to enter brain and stain cerebral nuclei. The unbound unconjugated bilirubin moiety may have various serum levels in any individual case depending on a variety of known and unknown factors.

In order to properly assess the risk of kernicterus, the unbound unconjugated bilirubin moiety should be measured. Unfortunately, the assay method currently in use is not easy, and beyond the capabilities of many clinical laboratories' abilities. Apparently, a somewhat simplified technique is being developed by a company specializing in such "assay kits." The current bottom line is that many hospitals only measure total bilirubin in newborn infants. The supposition is that in newborns, the total serum bilirubin levels are almost all indirect reacting bilirubin. Even this is not always true. In addition, it has been shown that kernicterus can result from indirect serum levels as low as 15 mg/100 ml, a level not always even recommended for phototherapy. Cases such as this would be missed, with resultant possible brain damage. All of the above not withstanding, the fact is that the compliance of pediatricians as regards recommendations on phototherapy are quite low (see the chapter "Phototherapy Treatment," this volume).

There needs to be developed a simple method to assess the unbound, unconjugated bilirubin levels in newborn infants. There needs to be an increased awareness on the part of health care workers as to the risks of newborn hyperbilirubinemia. Early discharge from hospitals of newborn infants (as a money saver) must be coupled with an educational session with parents. In some instances, follow-up visits in the home could be appropriate. This should be a judgment call based on an appraisal of whether the parents are capable of responding positively to their infant's rising bilirubin.

In another somewhat controversial area, various investigators have examined the issue of whether slight staining of brain areas might produce subtle alterations in mentation and/or behavior. One clear rationale for this possibility is that bilirubin encephalopathy would not be expected to be an "all or none" phenomenon. When the point is reached that unbound unconjugated bilirubin might cross the blood–brain barrier, it would be slow and in small quantities. Only as the serum levels increased, would the amount of bilirubin entering brain increase.

The landmark study was done (Butcher, R., Stutz, R., and Berry, H. 1971) in which Gunn rats were examined for minor alterations in brain function. Results showed that learning deficits were present in homozygous Gunn rats, and that they were similar in nature to the deficits seen in children who had high levels of unconjugated bilirubin as infants. An electron microscopic study shortly thereafter showed that neurons in the cerebral cortex of Gunn rats were damaged in a characteristic way: mitochondria were swollen and vacuolated.

Many studies have been performed on children who were jaundiced as newborns to determine if any alterations in intellect or behavior could be detected. Overall, results have been equivocal. Some studies show no demonstrable differences between previously jaundiced children and those not jaundiced. Other studies do demonstrate a difference in that previously jaundiced children have lower IQs , and/or behavioral deficits. Several factors may serve to explain such differences in study results. These include actual bilirubin levels, and how they were determined, length of elevated jaundice,, and confounding factors such as length of gestation, birth weight, hemolysis, etc.

It is certainly intuitive that some minor subtle permanent structural damage caused by bilirubin could act to produce change in behavior and mentation. Many studies, starting with the work of Butcher and colleagues, support this concept. When studies result in negative data regarding these mental changes, the burden of proof rests with the investigators. Studies must be carefully done and sources of error eliminated wherever possible.

In keeping with the concept of possible subtle changes are the findings in the early 1980s of alterations in hearing in jaundiced newborns as reflected by hearing tests. This was also shown by deviations from normal in evoked brainstem auditory responses recorded from the eighth cranial nerve (see the chapter "Auditory Brainstem Responses," this volume).

These studies showed rather conclusively that there was a slowing in the frequency of evoked auditory brainstem responses in children who had been jaundiced as newborns. There was also a diminution in the amplitude of the evoked waves. These are highly significant sequelae of neonatal hyperbilirubinemia. This is demonstrated in that a partial to complete deafness has a profound effect on language acquisition, social interaction, and learning abilities. Similar changes in auditory brainstem responses have been conclusively demonstrated in homozygous Gunn rats as compared to non-jaundiced littermate controls.

Obviously the worst neurological sequelae of bilirubin encephalopathy/kernicterus is cerebral palsy. In this country, the widespread use of phototherapy has all but eliminated this severe level of brain damage in jaundiced newborn infants. Perhaps only a few dozen bilirubin-induced cases of cerebral palsy occur each year. Before phototherapy, these cases, numbering in the hundreds, accounted for as many as 10% of all cerebral palsy patients (Winter, S., et al. 2002). This is a completely different story in underdeveloped countries endemic for G-6-PDH deficiency and malaria. Thousands of kernicterus cases occur annually in Third World countries with little knowledge of or wherewithal for of treatment. As stated above, this is indefensible.

The occurrence of the genetic disorder, the Crigler–Najjar syndrome continues at a fixed frequency. This disorder appears in two varieties: type 1 and type 2.

Type 2 has been defined as a partial deficiency of the conjugating enzyme glucuronyl transferase. Type 2 is responsive to phenobarbital treatment, whereas type 1 is not responsive to phenobarbital, because the conjugating enzyme is absent. In addition to phototherapy, liver transplantation, hepatocyte transplantation, and exchange transfusions have all been successfully administered (see the chapter "Crigler–Najjar Syndrome," this volume).

As regards the Crigler–Najjar Syndrome, the Gunn rat model of neonatal jaundice is one of the best animal models of a human disease state that exists. The deficiency of the conjugating enzyme is the sole cause of hyperbilirubinemia and kernicterus in Gunn rats, and exactly reflects the Crigler–Najjar syndrome. The availability of this model has certainly facilitated translation in this disorder.

The Gunn rat has been used for experiments which are difficult in humans, such as electron microscopy of brain tissue. These studies in the Gunn rat are consistent with the concept, advanced many years ago, that the primary cerebral defect in kernicteric Gunn rat brain is related to the mitochondrian and energy metabolism (see chapter on the Gunn rat model). Other experimental studies which can be conducted in Gunn rats which would be impossible in humans include the direct measurement of labile energy metabolites in brain.

In fact, several studies in Gunn rats have been performed looking for the primary biochemical defect in the brain of symptomatic animals. Results from most studies, both biochemical and structural, are consistent with much earlier in vitro data suggesting that bilirubin is an uncoupler of oxidative phosphorylation. These studies include direct energy metabolite measurement in vivo, and several electron microscopic studies.

Further studies in newborn infants with hyperbilirubinemia confirming Gunn rat data would be useful, but probably not practical. Enhanced imaging studies could increase knowledge of cerebral mechanisms in human hyperbilirubinemia.

It is important to remember that many energy and other metabolites are labile, and efforts are necessary to measure them under optimal conditions (see Fig. 1), represents this technical problem. If control and experimental metabolite values are

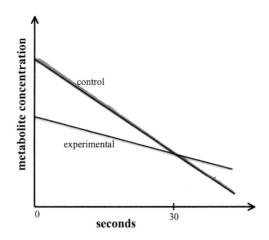

Fig. 1 This figure shows how results from measuring labile metabolites can vary depending on the speed of arresting the decay by freezing. Thus, if 30 s are required before the freezing front reaches a key brain area, divergent rates of decay of metabolites combined with different "0" time values may lead to artifactual results showing no difference

initially different, but also have different decay levels, then a suboptimal fixation method could yield similar and spurious final results. This situation does occur, especially if a decrease in initial values are the result of a decrease in metabolite utilization.

The exact cause of the highly localized brain lesions in metabolic encephalopathy has always been a subject of speculation. In the case of hyperbilirubinemia, the lipid solubility of unconjugated bilirubin could lead to a simple explanation: varying lipid content between stained and unstained regions. In a study (Chavko, M., and Nemoto, E. 1992), differences in rat brain lipids between hippocampus and cerebral cortex were measured in an ischemia study. Several classes of lipids were measured including cerebrosides. Both hydroxy fatty acid (HFA), and non-hydroxy fatty acid (NFA) were measured.

Results from this study (ischemia) included control (nonischemic) regional brain levels HFA and NFA. These control results, as compared to the cerebral cortex, the levels in the hippocampus were double those of the cortex. The levels in the basal ganglia were about 1.5 times the levels in the cerebral cortex (see Table 1). These data lend credence to the concept that a higher levels of normal concentrations of HFA and NFA in regions which are pigmented in hyperbilirubinemic newborns could explain selective vulnerability.

Table 1 Concentrations of hydroxy fatty acids and nonhydroxy fatty acids in specific brain regions. Note the twofold differences between cerebral cortex and hippocampus

Cortex		Hippocampus		Basal ganglia	
HFA	NFA	HFA	NFA	HFA	NFA
3.0 ± 0.4	1.8 ± 0.4	6.1 ± 1.4	3.6 ± 0.9	4.5 ± 0.4	2.3 ± 0.6

Adapted from Chavko, M., and Nemoto, E. (1992)

Additional support for this concept is derived from a paper on absence seizures which examines in part, the steroid sensitivity of the hippocampus (Tolmacheva, E., and van Luijtelaar, G. 2007). Some data supports the concept that limbic structures (hippocampus) might have an effect on spike–wave discharges through progesterone action. Indeed, progesterone increases in the hippocampus-reduced spike–wave discharges, whereas that was not the case in the cerebral cortex. These data are suggestive that hormonal modulation of hippocampal excitability may play a role in absence seizures. From the standpoint of hyperbilirubinemia, this suggests that the lipid content of the hippocampus may be high as compared to other brain regions, thereby serving to permit lipid-soluble hormones (and bilirubin) access to the newborn infant hippocampus.

Yet another paper examines the effects of kainic acid on lipid composition (lipid/protein ratios) in the hippocampus in status epilepticus (Dal-Pizzol, F., et al. 2007). Results from this study showed that the lipid/protein ratio increased with time following kainic acid administration in Wistar rats' hippocampus, whereas the cerebral cortex was unaffected. These data once again show a selective vulnerability of the hippocampus based on its lipid characteristics.

The possibility of gene therapy for the Crigler–Najjar type 1 syndrome has been examined for several years (for review, see Nguyen, T., and Ferry, P. 2007). In a Gunn rat study (animal model of Crigler–Najjar type 1), lentiviral or oncoretroviral vectors encoding glucuronyl transferase were administered when the homozygous rats were 2 days old. This treatment resulted in a lowering of hyperbilirubunemia to normal levels. The normalization of serum bilirubin levels lasted as long as 95 weeks. Conjugated bilirubin appeared in bile, demonstrating that conjugation was proceeding. Liver histologic examination of the treated Gunn rats showed no abnormalities. This study provides evidence for the efficacy of the treatment in Gunn rats (Nguyen, T., et al. 2007). A very recent review examines Gunn rat studies, and the possibility of use in humans (Miranda, P. and Bosma, P. 2009). Future gene therapy for many inborn errors of metabolism hold great promise for treatment.

Appendix – Sources

The following sites represent potential sources of information, equipment, animals, etc. These sites are only suggestions, and those interested should seek other venders as well.

(1) Help site: P.I.C.K. Parents of infants and children with kernicterus

> One West Superior St.,
> Suite 2400,
> Chicago, Il., 60610
> 312-274-9695
> info@pickonline.org

(2) Help site: Cerebral Palsy 4 my child

> 3955 Orchard Hill Place,
> Suite 365
> Novi Mi., 48375
> 1-800-469-2445
> www.cerebralpalsy.org
> UCP National
> 1660 L St., N.W.
> Suite 700,
> Washington, D.C. 20036
> 1-800-872-5827

(3) Bilirubin meter:

> Respironics International
> Immeuble Hermes
> 20 Rue Jacques Daguerre
> Ruell Maimalson
> Cedex, 92565, France

(4) Blue light phototherapy units:

> Narang Medical Limited
> Narang Tower
> 46 Community Center,
> Naraina Ph-1

D.W. McCandless, *Kernicterus*, Contemporary Clinical Neuroscience,
DOI 10.1007/978-1-4419-6555-4, © Springer Science+Business Media, LLC 2011

New Delhi, India
e-mail: net@narang.com

(5) Bilirubin blankets and phototherapy units:

G.E. Health Care
www.gehealthcare.com/perinatal/products/phototherapy.html

(6) Gunn rats:

National Bioresource Project
Institute of Laboratory Animals,
Kyoto University, School of Medicine
e-mail: serikawa@anim.med.kyoto-u.ac.jp
www.nbrp.jp/center

(7) Jaundiced mice:

The Jackson Laboratory
600 Main Street
Bar Harbor, Maine, 04609
www.jax.org/imrcom/nih_utility/aq8167_j.biol.chem2008.pdf
Nguyen, N., et. al. (2008) Disruption of the UGT1 locus in mice resembles human
Crigler–Najjar type 1 disease. J. Biol. Chem. 283:7901-7911

(8) ABR equipment:

Natus Medical Incorp.
1501 Industrial Rd.,
San Carlos, Ca., 94070
650-802-0400
www.natus.com

References

Abro, A., et al. (2009) Jaundice with hepatic dysfunction in p. falciparum malaria. J. coll. Phys. Surg. Pak. 19:363–366

Agrawal, V., et al. (1998) Brainstem auditory evoked responses in newborns with hyperbilirubinemia. Ind. Ped. 34:513–518

Ahboucha, S., and Butterworth, R. (2007) The neurosteroid system: an emerging therapeutic target for hepatic encephalopathy. Metab. Brain Dis. 22:291–308

Ahdab-Barmada, M., and Moossy, J. (1984) Theneuropathology of kernicterus in the premature neonate: diagnostic problems. J. Neuropath. Expt. Neurol. 43:45–56

Ahlfors, C. (2000) Measurement of plasma unbound unconjugated bilirubin. Anal. Biochem. 279:130–135

Ahlfors, C. (2010) Predicting bilirubin neurotoxicity in jaundiced newborns. Curr. Opin Pediatr. 22 doi:10.1097/mop.obo13e328336e628

Ahlfors, C., and Parker, A. (2008) Unbound bilirubin concentration is associated with abnormal auditory brainstem response for jaundiced newborns. Pediatr. 121:976–978

Ahlfors, C., and Shapiro, S. (2001) Auditory brainstem responses and unbound bilirubin in jaundiced Gunn rat pups. Biol. Neonate 80:158–162

Ahlfors, C., et al. (1986) Changes in the auditory brainstem response associated with intravenous infusion of unconjugated bilirubin into infant rhesus monkey. Ped. Res. 20:511–515

Akobeng, A. (2005) Neonatal jaundice. Am. Fam. Physician. 71:21–23

Al-Rimawi, H., et al. (1999) Effect of desferrioxamine in acute haemolytic anaemia of glucose-6-phosphate dehydrogenase deficiency. Acta Haematol. 101:145–148

Altman, K. (1959) Some enzymologic aspects of the human erythrocyte. Am. J. Med. 23:936–951

Amato, M., et al. (1988) Effectiveness of single versus double volume exchange transfusion in newborn infants with ABO hemolytic disease. Helvetica Paed. Acta 43:177–186

American Academy of Pediatrics (2004) Management of hyperbilirubinemia in the newborn infant 35 or more weeks of gestation. Pediatr. N114:297–316

Amin, S., et al. (2001) Bilirubin and serial auditory brainstem responses in premature infants. Pediatr. 107:664–670

Anderson, G. (2004) Pharmacogenetics and enzyme induction/inhibition properties of antiepileptic drugs. Neurology 63:S24–S29

Arias, I., and Gartner, L. (1970) Breast milk jaundice. BMJ 4:177

Arias, I., et al. (1963a) Effect of several drugs and chemicals on hepatic glucuronide formation in newborn rats. Proc. Soc. Exp. Biol. Med. 123:1037–1045

Arias, I., et al. (1963b) Proc of the 55th meeting of the Amer. Soc. Clin. Invest

Arias, I., et al. (1964a) Prolonged neonatal unconjugated hyperbilirubinemia associated with breasr feeding and a steroid that inhibits glucuronide formation. J. Clin. Invest. 43:2037–2047

Arias, I., et al. (1964b) Unconjugated hyperbilirubinemia. J. Vclin. Invest. 43:1249–1250

Arias, I., et al. (1969) Chronic non hemolytic unconjugated hyperbilirubinemia with glucuronyl trandferase deficiency. Am. J. Med. 47:395–404

Arias, I., Schorr, J., and Fraad, L. (1959) Congenital hypertrophic pyloric stenosis with jaundice. Pediat. 24:338–342

Arkwright, J.A, (1910) Icterus gravis of the newborn. Edin. Med. J. 12:156–162

Askari, F., et al. (1996) Complet correction of hyperbilirubinemia in the Gunn rat model of Crigler-Najjar syndrome type 1 following transient in vivo adenovirus mediated expression of human bilirubin UDP-glucuronsyl transferase. Gene Ther. 3:381–388

Atkinson, L., et al. (2003) Phototherapy use in jaundiced newborns in a large managed care organization: how do clinicians adhere to the guideline? Pediatr. 111:e555–e561

Auclair, C., et al. (1976) Bilirubin and paranitrophenol glucuronyl transferase activities and ultrastructural aspect of the liver in patients with chronic hemolytic anemias. Biomedicine 25:61–72

Auden, G. (1905) A series of fatal cases of jaundice in the newborn occuring in successive pregnancies. St. Bart. Hosp. Rev. 41:135–144

Bag, S., et al. (1994) Compliocated falciparum malaria. Ind. Pediatr. 31:821–825

Balistreri, W. (1997) Bile acid therapy in pediatric hepatobiliary disease: the role of ursodeoxy-cholic acid. J. Pediatr. Gastro. Nutrit. 24:573–589

Batty, H., and Millhouse, O. (1976) Ultrastructure of the Gunn rat substantia nigra Acta Neuropathol. 34:7–19

Baumes, J. (1806) Traite d L'Lictere infans des naissance. McQuignon, Paris

Bayon, J., et al. (2001) Pitfalls in preparation of unconjugated bilirubin by biosynthetic labeling from precursor H-5-aminolevulinic acid in the dog. J. Lab. Clin. Med. 138:313–321

Behrman, R., and Hibbard, E. (1963) Bilirubin: acute effects in newborn Rhesus monkeys. Science 144:545–546

Behrman,R., and Hsia, D. (1969) Summary of a symposium on phototherapy for hyperbilirubine-mia. J. Pediatr. 75:718–726

Belanger, M., and Butterworth, R. (2005) Acute liver failure: A critical appraisal of available animal models. Met. Brain Dis. 20:409–424

Bellodi-Privato, M., et al. (2005) Successful gene therapy of the Gunn rat by in vivo neonatal hepatic gene transfer using murine oncoretroviral vectors. Hepatol. 42:431–438

Berk, P. (1970) Constitutional hepatic dysfunction (Gilbert's syndrome) Am. J. Med. 49:296–305

Berk, P., and Blaschke, T. (1972) Detection of Gilbert's syndrome in patients with hemolysis. Ann. Inter. Med. 77:527–540

Berk, P., Wolkoff, A., and Berlin, N. (1975) Inborn errors of bilirubin metabolism. Med. Clin. North Am. 59(4):803–816

Berven, N. (2008) Rehabilitation counseling. In Encyclopedia of counseling, vol. 2. Tinsley, H., and Lease, S., eds. Sage, Los Angeles, CA

Beutler, E. (2006) Disorders of red cells resulting from enzyme abnormalities. In Williams hematology. Lichtman, M., et al. eds. McGraw Hill, New York, NY, pp 603–631

Beutler, E., and Mitchel, M. (1968) Special modifications of the fluosercent screening method for glucose-6-phosphate dehydrogenase deficiency. Blood 32:816–818

Beutler, E., Duparc, S., and the G6PD deficiency working group (2007) Glucose-6-phosphate dehydrogenase deficiency and antimalarial drug development. Am. J. Trop. Med. Hyg. 77:779–789

Bevis, D. (1953) The composition of liquor amnii in haemolytic disease of the newborn. J. Obstet. Gynaec. Brit. Emp. 60:244–250

Bhasin, M., and Walker, H. (2001) Genetics af castes and tribes of India. Kamala-Raj Enterprises, Delhi

Bhatia, V., Batra, Y., and Acharya, S. (2004) Prophylactic phenytoin does not improve cerebral edema or survival in acute liver failure-a controled clinical trial. J. Hepatol. 41:89–96

Bhutani, V., et al. (2004) Kernicterus:epidemiological strategies for its prevention through systms based approaches. J. Perinatol. 24:650–662

Bienzle, U., Effiong, C., and Luzzatto, L. (2008) Erythrocyte glucose – phosphate dehydrogenase deficiency and neonatal jaundice. Acta Paediatrica 65:701–703

Billing, B. (1963) Bilirubin Metabolism. Postgrad Med. 39:176–182

Billing, B. (1982) Bilirubin metabolism. in Diseases of the liver. Schiff, L., and Schiff, E. eds. Lippincott, Philadelphia, PA

Billing, B., Cole, P., and Lathe, G. (1956) Studies on the directly reacting bilirubin. Biochem. J. 63:6–9

Birraux, J., et al. (2009) A step toward liver gene therapy: efficient correction of the genetic defect of hepatocytes isolated from a patient with Crigler-Najjar syndrome type 1 with lentiviral vectors. Transplantation 87:1006–1012

Black, M., and Billing, B. (1969) Hepatic bilirubin UDP glucuronyl transferase activity in liver disease and Gilberts syndrome. N. Engl. J. Med. 280:1266–1271

Black, M., and Billing, B. (1971) Hepatic bilirubin glucuronyl transferase in liver disease and Gilbert's syndrome. N. Engl. J. Med. 280:1266–1272

Black, M., et al. (1973) Hepatic bilirubin UDP glucuronyl transferase activity and cytochrome P450 content in a surgical population and the effects of preoperative drug thrapy. J. Lab. Clin. Med. 81:704–713

Black, M., Fevery, J., and Parker, D. (1974) Effect of phenobarbitone on plasma bilirubin clearance in patients with unconjugated hyperbilirubinemia. Clin. Sci. Mol. Med. 46:1–17

Blackfan, K. Diamond, L., and Leister, C. (1944) Atlas of blood in children. The Commonwealth Fund, Oxford University Press, London

Blanc, W., (1961) Kernicterus in Gunn's strain of rats. In Kernicterus, Sass-Kortsak, A. ed. University of Toronto Press, Toronto, ON, pp 150–152

Blaschke, T., et al. (1974) Crigler–Najjar syndrome: an unusual case with development of nerve damage at age 18. Ped. Res. 8:573–590

Bloomer, J., and Sharp, H. (1984) The liver in Crigler-Najjar syndrome, protoporphyria, and other metabolic disorders. Hepat. 4: 18s–21s

Bloomer, J., et al. (1971) Studies on the mechanism of fasting hyperbilirubinemia. Gastroent. 61:479–486

Bonnet, J. (1995) Is jaundice an early manifestation of Gilbert's syndrome in hypertrophic pyloric stenosis? Pediat. Surg. 10:551–552

Bosma, P., et al. (1995) The genetic basis of the reduced expression of bilirubin UDP glucuronyl-transferase 1 in Gilbert's syndrome. N. Engl. J. Med. 333:1171–1175

Brazie, J., Ibbott, F., and Bowes, W. (1966) Identification of the pigment in amniotic fluid of erythrocytes as bilirubin. J. Pediat. 69:354–358

Brewster, D., Kwiatkowski, D., and White, N. (1990) Neurological squealae of cerebral malaria in children. Lancet 336:1039–1043

Brown, A., and Zuelzer, W. (1957) Soc. Pediatr. Res., 27th annual meeting, Carmel, CA

Brown, W., Grodsky, G., and Carbone, J. (1964) Intracellular distribution of tritiated bilirubin during hepatic uptake and excretion. Am. J. Physiol. 207:1237–1241

Burchard, G., et al. (1997) Exchange blood transfusion in severe falciparum malaria. Trop. Med. Int. Health 2:733–740

Burke, B., et al. (2009) Trends in hospitalization for neonatal jaundice and kernicterus in the United States, 1988–2005. Pediatr. 123:524–532

Butcher, R., Stutz, R., and Berry, H. (1971) Behavioral abnormalities in rats with neonatal jaundice. Am. J. Ment. Def. 75:755–759

Cabana, M., et al. (1999) Why don't physicians follow clinical practice guidelines? JAMA. 282:1458–1465

Calmus, Y., et al. (1990) Hepatic expression of class I and class II major histocompatability complex molecules in primayt biliary. Hepatology 11:12–15

Cappadoro, M., et al. (1998) Early phagocytosis of glucose-6-phosphate dehydrogenase deficient erythrocytes parasitized by Plasmodium falciparum may explain malaria protection. Blood 92:2527–2534

Carmichael, S., and Abrams, B. (1997) A critical review of the relationship between gestational weight gain and preterm delivery. Obstet. Gynecol. 89:865–873

Carson, P., et al. (1956) Enzymatic deficiency in primaquine sensitive erythrocytes. Science 124:484–485

Cauli, O., et al. (2005) Altered modulation of motor activity by group 1 metabotropic glutamate receptors in the nucleus accumbens in hyperammonemic rats. Metab. Brain Dis. 20:347–358

Chalmers, T. (1960) Pathogenesis and treatment of hepatic failure. N. Engl. J. Med. 263:23–30

Chan, T., Chan, W., and Weed, R. (1982) Erythrocyte hemighosts:a hallmark of severe oxidative injury in vivo. Br. J. Haematol. 50:575–582

Chan, T., et al. (1971) The survival of G6PD deficient erythrocytes in patients with typhoid fever on choloramphemicol therapy. J. Lab. Clin. Med. 77:177–184

Chang, F.-Y., et al. (2009) Unconjugated bilirubin exposure impairs hippocampal long term synaptic plasticity. PLoS ONE 4(6):e5876

Chaves-Carballo, E., Harris, L., and Lynn, H. (1968) Jaundice Associated with pyloric stenosis and neonatal small bowel obstructions. Clin. Pediat. 4:198–202

Chavko, M., and Nemoto, E. (1992) Regional differences in rat brain lipids during global ischemia. Stroke 23:1000–1004

Chenard-Neu, M., et al. (1996) Auxiliary liver transplantation: regeneration of the native liver and outcome in 30 patients with fulminant hepatic failure. Hepatology 23:1119–1127

Chin, K., Taylor, M., and Perlman, M. (1985) Improvement in audiyory and visual evoked potentials in jaundiced preterm infants after exchange transfusion. Arch. Dis. Child. 60:714–717

Chiu, H., Brittingham, J., and Laskin, D. (2002) Differential induction of heme oxygenase in macrophages and hepatocytes during acetaminophen induced hepatotoxicity. Toxicol. Appl. Pharm. 181:106–115

Claireaux, A., Cole, P., and Lathe, G.(1953) Icterus of the brain in the newborn. Lancet 4: 1226–1230

Claireaux, A. (1961) Pathology of human kernicterus. In Sass-Kortsak, A. ed. Kernicterus. University of Toronto press, Toronto, ON

Clegg, L., and Lindup, W. (1984) J. Pharm. Pharmacol. Phenobarbitol in hyperbilirubinemia. 36:776–779

Cohen, A., et al. (1985) Effects of phenobarbitol on bilirubin metabolism and its response to phototherapy in the jaundiced Gunn rat. Hepatology 5:310–316

Cole, S., et al. (1992) Over expression of a transporter gene in a multidrug resistant human lung cancer cell line. Science 258:1650–1654

Condie, D. (1853) A practical treatise on the diseases of children. Blanchard and Lea. Philadelphia, pp 123–167

Conney, A. (1965) Enzyme induction and drug toxicity. In Drugs and Enzymes, Brodie, B. and Gillette, J. eds., Pergamon Press, Oxford, 4:277–290

Conney, A., and Burns, J. (1962) Factors influencing drug metabolism. In Advances in pharmacology. Garattini S. ed. Academic, New York, NY, p 31

Cooley, W. (2004) Providing a primary care medical home for children and youth with cerebral palsy. Pediatr. 114:1106–1113

Cooper, W. et al. (2002) Very early exposure to erythromycin and infantile hypertrophic pyloric stenosis. Arch. Pediat. Adol. Med. 156:647–650

Cowger, M., Igo, R., and Labbe, R.F. (1965) The mechanism of bilirubin toxicity studied with purified respiratory enzyme and tissue culture systems. Biochem. J. 4:2763–2770

Cremer, R., Perryman, P., and Richards, D. (1958) Influence of light on the hyperbilirubinemia of infants. Lancet 1:1094–1097

Crigler, J., and Gold, N. (1969) Effect of sodium phenobarbital on bilirubin metabolism in an infant with congenital nonhemolytic unconjugated hyperbilirubinemia and kernicterus. J. Clin. Invest. 48:42–55

Crigler, J., and Najjar, V. (1952) Congenital familial nonhemolytic jaundice with kernicterus. Pediat. 10:169–180

Csoma, Z., Kemeny, L., and Olah, J. (2008) Letter to the editor: phototherapy for neonatal jaundice. N. Engl. J. Med. 358:2522–2525

Dal-Pizzol, F., et al. (2007) Changes in lipid composition in hippocampus early and late after status epilepticus induced by kainic acid in wister rats. Metab. Brain Dis. 22:25–29

Dancis, J., et al. (1958) Hemoglobin metabolism in the premature infant. J. pediat. 54:748–755

Day, R. (1954) Inhibition of brain respiration in vitro by bilirubin. Am. J. Dis. Child. 88:504–509

Day, R. (1954) Inhibition of brain respiration in vitro by bilirubin: reversal of inhibition by various means. Am. J. Dis. Child. 88:504–510

Day, R., and Haines, M. (1954) Intelligence quotients of children recovered from erythroblastosis fetalis since the introduction of exchange transfusion. Pediatr. 13:333–338

De Carvalho, M., et al. (2007) Intensified phototherapy using daylight fluorescent lamps. Acta Pediat. 88:768–771

de Lacerda, G. (2008) Treating seizures in renal and hepatic failure. J. Epil. Clin. Neurophys. 14:1–8

de Vree, J., et al. (1998) Mutations in the MDR3 gene cause progressive familial intrahepatic cholestasis. Proc. Natl. Acad. Sci. 98:282–287

Delage, Y., et al. (1977) Rotor's syndrome: evidence for an impairment of hepatic uptake and storage of cholephilic organic anions. Digestion 15:228–234

DeLeon, A., Gartner, L., and Arias, I. (1967) The effect of phenobarbitol on hyperbilirubinemia in glucuronyl transferase deficient rats. J. Lab. Clin. Med. 70:273–280

Diamond, I., and Schmid, R. (1966a) Experimental bilirubin encephalopathy. The mode of entry ob bilirubin into the central nervous system. J. Clin. Invest. 45:678–689

Diamond, I., and Schmid, R. (1966b) Oxidative phosphorylation in experimental bilirubin encephalopathy. Science 155:1288–1289

Diamond, I., and Schmid, R. (1968) Neonatal hyperbilirubinemia and kernicterus. Arch. Neurol. 18:699–702

Diamond, L., Blackfan, K., and Baty, J. (1932) Erythroblastosis fetalis and its association with universal edema of the fetus, icterus gravis neonatorum, and anemia of the newborn. J. Pediatr. 1:269–301

Dublin, W. (1951) Neurological lesions in erythroblastosis fetalis in relation to nuclear dafness. Am. J. Clin. Path. 21:935–939

Dutton, G., and Storey, I. (1954) Uridine compounds in glucuronic acid metabolism. Biochem. J. 57:275–283

Eleftheriadis, N., et al. (2003) Status epilepticus as a manifestation of hepatic encephalopathy. Acya Neurol. Scand. 107:142–144

Emad, S., et al. (2008) Congenital malaria. Pak. J. Med. Sci. 24:765–767

Ernster, L., Herlin, L., and Zetterstrom, R. (1957) Experimental studies on the pathogenesis of kernicterus. Pediatr. 20:647–659

Esbjorner, E., et al. (1991) The serum reserve albumin concentration for monoacetyldiaminodiphenyl sulphone and auditory evoked responses during neonatal hyperbilirubinemia. Acta Periatr. Scand. 80:406–412

Escher-Graub, D., and Fricker, H. (1986) Jaundice and behavioral organization in the full term neonate. Helv. Paediat. Acta. 41:425–435

Escobar, G., et al. (1999) Rehospitalization in the first two weeks after discharge from the neonatal intensive care unit. Pediatr. 104:e2

Farivar, M., et al. (1976) Effect of insulin and glucagon on fulminant murine hepatitis. N. Engl. J. Med. 295:1517–1519

Farrell, D., et al. (1973) Feline GM3 gangliosidosis. J. Neuropath. Epxt. Neurol. 32:1–18

Felsher, B., et al. (1973) Hepatic bilirubin glucuronidation in Gilbert's syndrome. J. Lab. Clin. Med. 81:829–836

Fenerty, C., and Lindup, W. (1989) Brain uptake of L-tryptophan and diazepam: The role of Plasma protein binding. J. Neurochem. 53:416–422

Fernandes, A., et al. (2006) Inflammatory signalling pathways involved in astroglial activation by unconjugated bilirubin. J Neurochem. 96:1667–1679

Fevery, J., et al. (1977) Unconjugated bilirubin and an increased proportion of bilirubin mono-conjugates in the bile of patients with Gilbert's syndrome and Crigler-Najjar disease. J. Clin. Invest. 60:970–982

Field, J. (1949) Blood examination and prognosis in acute falciparum malaria. Trans. R. Soc. Med. Hyg. 43:33–48

Fischer, H., and Pleininger, H. (1942) Synthese des biliverdins and bilirubins, der biliverdine XIII alpha and III alpha, sowie der vinylneoxanthosaure. Hoppe Seylers Z. Physiol. Chem. 274: 231–240

Fischer, P. (2003) Malaria and newborn. J. Trop. Pediatr. 49:132–134

Fouts, J., and Hart, L. (1965) Hepatic drug metabolism during the perinatal period. Ann. N. Y. Acad. Sci. 123:245–250

Fox, I., et al. (1998) Treatment of the Crigler-Najjar syndrome type 1 with hepatocyte transplantation. N. Engl. J. Med. 338:1422–1427

Funato, M., et al. (1994) Vigintiphobia, unbound bilirubin, and auditory brainstem responses. Pediatr. 93:50–53

Garray, E., Owen, C., and Flock, E. (1966) Formation of bilirubin in normal, damaged, and Gunn rat livers. J. Lab. Clin. Med. 67:817–829

Gardos, G., and Straub, F. (1957) Uber die rolle der ATP in der K-permeabilitat der menschlichen roten blutkorperchen. Acta physiol. Hung. 12:1–12

Gartner, L., and Herschel, M. (2001) Jaundice and breastfeeding. Pediatr. Clin. North Am. 48: 389–399

Gartner, L., et al. (1970) Kernicterus: high incidence in premature infants with low serum bilirubin concentrations. Pediatr. 45:906–917

Gartner, L., et al. (1977) Development of bilirubin transport and metabolism in the newborn rhesus monkey. J. Pediatr. 90:513–531

Genc, S., et al. (2003) Bilirubin is cytotoxic to rat oligodendrocytes in vitro. Brain Res. 985: 135–141

Gennuso, F., et al. (2004) Bilirubin protects astrocytes from its own toxicity by inducing up regulation and translocation of multidrug resistance associated protein 1. Proc. Natl. Acad. Sci. 101:2470–2475

Gibbs, W., Gray, R., and Lowry, M. (2008) Glucose-6-phosphate dehydrogenase deficiency and neonatal jaundice in Jamaica. Brit. J. Haematol. 43:263–274

Gilbert, A. (1907) Les trois cholemies congenitales. Bull. Mem. Soc. Med. Hop. Paris. 24: 1203–1208

Gilbert, W., Nesbitt, T., and Danielsen, B., (2003) The cost of prematurity: Quantification by gestational age and birth weight. Obstet. Gynecol. 102:488–492

Gimson, A. (1996) Fulminant and late onset hepatic failure Brit. J. Anaesth. 77:90–98

Goldenberg, R., and Rouse, D. (1998) Prevention of premature birth. N. Engl. J. Med. 339:313–320

Goldenberg, R., et al. (1995) The effect of zinc supplimentation on pregnancy outcome. JAMA 274:463–468

Goldenberg, R., et al. (2008) Epidemiology and causes of preterm birth. Lancet 371:75–84

Goldenberg, R., Hauth, J., and Andrews, W. (2000) Intrauterine infection and preterm delivery. N. Engl. J. Med. 342:1500–1507

Goldstein, G., and Lester, R. (1964) Reduction of biliverdin to bilirubin in vivo. Proc. Soc. Expt. Biol. Med. 117:681–683

Gollan, J., et al. (1975) Prolonged survival in three brothers with severe type 2 Crigler-Najjar syndrome: ultrastructeral and metabolic studies. Gastroent. 68:1543–1555

Gonatas, G., and Gonatas, J. (1965) Ultrastructural and biochemical observations on a case of systemic late infantile lipidoses. J. Neuropath. Exp. Neurol. 24:318–340

Gonzalez-Quiroga, G., et al. (1994) Frequency and origin of G-6-PD deficiency among icteric newborns in the metropolitan area of Monterrey, Nuevo Leon, Mexico. Gene Geog. 8:157–164

Goresky, C., et al. (1978) Definition of a conjugation dysfunction in Gilbert's syndrome. Clin. Sci. Mol. Med. 55:63–69

Gourley, G. (2000) Breastfeeding, diet, and neonatal hyperbilirubinemia. Pediatr. In review e25–e31

Gourley, G. (2009) Breast-feeding, jaundice, and kernicterus. Semin. neonat. 7:135–141

Grant, A. (1989) Cervical cerclage to prolong pregnancy. In Effective care in pregnancy and childbirth., Chalmers, I., Enkin, M., and Keirse, M., eds. Oxford University Press, Oxford

Graziani, L. (1992) Neurodevelopment of preterm infants: neonatal neurosonographic and serum bilirubin studies. Pediatr. 89:229–234

Grodsky, G., and Carbone, J. (1957) The synthesis of bilirubin glucuronide by tissue homogenates. J. Biol. Chem. 226:449–458

Grojean, S., et al. (2001) Bilirubin exerts additional toxic in hypoxic cultured neurons from the developing rat brain by the recruitment of glutamate neurotoxicity. Pediatr. Res. 49: 507–513

Grojean, S., Vert, P., and Daval, J. (2009) Combined effects of bilirubin and hypoxia on cultured neurons from the developing rat forebrain. Semin. Perinatol. 26:416–424

Grunebaum, E., et al. (1991) Breast milk jaundice. Euro. J. Ped. 150:267–270

Guindo, A., et al. (2007) X-linked G6PD deficiency protects hemizygous males but not heterozygous females against severe malaria. Plos Med. 4: e66 doi: 10.1371/journal.pmed.0040066

Guinta, F., and Roth, P. (1969) Effect of environmental illumination on prevention of hyperbilirubinemia of prematurity. Pediatr. 43:191–201

Guldutuna, S., et al. (1995) Crigler-Najjar syndrome type 2: new observation of possible autosomal recessive inheritance. Digest. Dis. Sci. 40:28–32

Gunn, C. 1938 Hereditary Acholuric Jaundice in a New Mutant Strain of Rats. J. Hered. 29: 137–139

Gupta, A. and Mann, S. (1998) Is auditory brainstem response a bilirubin toxicity marker? Am. J. Otolaryngol. 19:232–236

Guthrie, L. (1914) Icterus gravis neonatorum. Proc. R. Soc. Med. 7:767–771

Hahm, J., et al. (1992) Ionization and self association of unconjugated bilirubin determined by rapid solvent partition from chloroform with further studies of bilirubin solubility. J. Lipid Res. 33:1123–1137

Hanefeld, F., and Natzschka, J. (1971) Histochemical studies in infant Gunn rayts with kernicterus. Neuropadiatrie 2:428–438

Hansen, R., Hughes, G., and Ahlfors, C. (1991) Neonatal bilirubin exposure and psychoeducational outcome. J. Dev. Behav. Pediatr. 12:287–293

Hansen, T., (2002) Kernicterus: an international perspective. Semin. Neonatol. 7:103–109

Hansen, T., Allen, J., and Tommarello, S. (1999) Oxidation of bilirubin in the brain-further characterization of a potentially protective mechanism. Mol. Genet. Metab. 68: 404–409

Hansen, T., and Cashore, W. (1995) Rates of bilirubin clearance from rat brain regions. Biol. Neonate. 68:135–140

Hansen, T., and Tommarello, S. (1998) Effect of phenobarbital on bilirubin metabolism in rat brain. Biol. Neonate. 73:106–111

Hansen, T., Bratlid, D., and Walaas, S. (1988) Bilirubin decreases phosphorylation in intact synaptosomes from rat cerebral cortex. Pediatr. Res. 23:219–223

Hansen, T., et al. (1988) Short term exposure to bilirubin reduces synaptic activation in rat hippocampal slices. Pediatr. Res. 23:453–456

Hansen, T., et al. (2009) Reversibility of acute intermediate phase bilirubin encephalopathy. ACTA Paediatr. 98:1689–1694

Hansen, T., Mathiesen, S., and Walaas, S. (1996) Bilirubin has widespread inhibitory effects on protein phosphorylation. Pediatr. Res. 39:1072–1077

Hargreaves, T., and Piper, R. (1971) Breast milk jaundice. Arch. Dis. Child. 46:195–198

Hauth, J., et al. (1995) Reduced incidence of preterm delivery with Metronidazole and Erythromycin in women with bacterial vaginosis. N. Engl. J. Med. 333: 1732–1736

Hillier, S., et al. (1993) The normal vaginal flora, H_2O_2 producing lactobacilli, and bacterial vaginosis in pregnant women. Clin. Infect. Dis. 16(suppl4): s273–s281

Hillier, S., et al. (1995) Association between bacterial vaginosis and preterm delivery of a low-birth-weight infant. N. Engl. J. Med. 333:1737–1742

Hollingsworth, J. (1955) Lifespan of fetal erythrocytes. J. Lab. Clin. Med. 45:469–473

Holmes, G., et al. (1968) Neonatal bilirubinemia in production of long term deficits. Am. J. Dis. Child. 116:37–43

Honein, M., et al. (1999) Infantile hypertrophic pyloric stenosis after pertussis prophylaxis with erythromycin. Lancet 354:2101–2105

Hoofnagle, J., et al. (1995) Fulminant hepatic failure: summary of a workshop. Hepatology 21:240–252

Hopkins, P., et al. (1996) Higher serum bilirubin isassociated with decreased risk for familial coronary artery disease. Arterioscler. Thromb. Vasc. Biol. 16:250–255

Hsia, D., et al. (1952) Erytyhroblastosis fetalis: studies of serum bilirubin in relation to kernicterus. N. Engl. J. Med. 247:668–673

Huang, A., et al. (2009) Differential risk for early breastfeeding jaundice in a multi-ethnic Asian cohort. Ann. Acad. Med. Sing. 38:217–224

Hyman, C., et al. (1969) CNS abnormalities after neonatal hemolytic disease or hyperbilirubinemia. Am. J. Dis. Child. 117:395–405

Iannucci, T., Tomich, P., and Gianopoulos, J. (1996) Etiology and outcome of extremely low birth weight infants. Am. J. Obstet. Gynecol. 174:1896–1902

Isselbacher, K., and McCarthy, E. (1959) Studies on bilirubin sulfate and other nonglucuronide conjugates of bilirubin. J. Clin. Invest. 38:645–651

Ives, N., et al. (1988) The effects of bilirubin on brain energy metabolism during normoxia and hypoxia: an in vitro study using NMR spectroscopy. Pediatr. Res. 23:569–573

Izquierdo, I., and Zand, R. (1978) Behavioral observations in Gunn rats. Psychopharm. 57:155–161

Jackson, J. (1997) Adverse effects associated with exchange transfusion in healthy and ill newborns. Pediatr. 99: e7–e22

Jangaard, K., et al. (2008) Outcomes in a population of healthy term and near term infants with serum bilirubin levels of 325 micromolar who were born in Nova Scotia, Canada, between 1994 and 2000. Pediatr. 122:119–124

Jangaard, K., Vincer, M., and Allen, A. (2007) A randomized trial of aggressive versus conservative phototherapy for hyperbilirubinemia in infants weighing less than 1500 g: short and long term outcomes. Pediatr. Child Health 12:853–858

Jansen, P., and Muller, M. (2000) The molecular genetics of familial intrahepatic cholestasis. GUT 47:1–5

Jansen, P., et al. (1989) Auxillary liver transplantation in jaundiced rats with UDP-glucuronyl transferase deficiency and defective hepatobiliary transport. J. Hepat. 8:192–200

Javitt, J. (1966) Cholestasis in rats induced by taurolithocholate. Nature 210:599–604

Jedlitschky, G., et al. (1997) ATP dependent transport of bilirubin glucuronides by the multidrug resistance protein MRP1 and its hepatocyte canalicular isoform MRP2. Biochem. J. 327: 305–310

Jew, J., and Sandquist, D. (1979) CNS changes in hyperbilirubinemia. Arch. Neurol. 36:149–154

Jew, J., and Williams, T. (1977) Ultrastructeral aspects of bilirubin encephalopathy in cochlear nuclei of the Gunn rat. J. Anat. 124:599–616

Jia, Z., and Danko, I. (2005) Long term correction of hyperbilirubinemia in the Gunn rat by repeated intravenous delivery of naked plasmid DNA into muscle. Mol. Ther. 12:860–866

Jiang, Z., et al. (2007) Changes in brainstem auditory evoked response latencies in term neonates with hyperbilirubinemia. Ped. Neurol. 37:35–41

Jirsova, V., Jirsa, M., and Janovsky, M. (1958) Importance of the quantitative determination of direct and indirect bilirubin in hemolytic diseasr of the newborn. Acta Pediat. 47:179–186

Jezequel, A., et al. (1980) In Familial hyperbilirubinemia. Okolicsanyi, L. ed., Wiley, New York, NY, P. 69

Johnson, J., et al. (1975) Efficacy of 17 hydroxyprogesterone caproate in the prevention of premature labor. N. Engl. J. Med. 293:675–680

Johnson, L., et al. (1959) Kernicterus in rats with an inherited deficiency of glucuronyl transferase. Am. J. Dis. Child. 95:591–608

Johnston, W., et al. (1967) Erythroblastosis fetalis and hyperbilirubinemia. Pediatrics, 39:88–92

Joseph, R., et al. (1999) Mass newborn screening for glucose-6-phosphate dehydrogenase deficiency in Singapore. Sotheast Asian J. Trop. Med. Public Health 30(Suppl2):70–71

Kaplan, E. (1961) The life span of erythrocytes in the newborn. In Kernicterus. Sass-Kortsak, ed. University of Toronto press, Toronto, ON

Kaplan, M., and Hammerman, C. (2005) Bilirubin and the genome: the hereditary basis of unconjugated neonatal hyperbilirubinemia. Curr. Pharmacogenomics 3:21–42

Kaplan, M., et al. (1997) Gilbert syndrome and glucose-6-phosphate dehydrogenase deficiency: a dose dependent genetic interaction crucial to neonatal hyperbilirubinemia. Proc. Natl. Acad. Sci. 94:12128–12132

Kaplan, M., et al. (2002) Hemolysis and bilirubin conjugation in association with UDP glucuronosyltransferase 1A1 promoter polymorphism. Hepatology 35:905–911

Kaplan, M., Hammerman, C., and Maisels, J. (2003) Bilirubin genetics for the nongeneticist: Heredity defects of neonatal bilirubin conjugation. Pediatr. 111:886–893

Kapoor, R. (2008) Letter to the editor: phototherapy for neonatal jaundice. N. Engl. J. Med. 358:2522–2525

Karp, W., (1979) Biochemical alterations in neonatal hyperbilirubinemia and bilirubin encephalopathy: a review. Pediatr. 64:361–368

Karplus, M., et al. (1988) The effects of brain bilirubin deposition on auditory brain stem evoked responses in rats. Early Hum. Dev. 16:185–194

Karrer, F., et al. (1984) A reproducible large animal model of acute hepatic failure. Curr. Surg. 41:464–467

Katoh, R., Kashiwamata, S., and Niwa, F. (1975) Studies on cellular toxicity of bilirubin: effect on the carbohydrate metabolism in the young rat brain. Brain Res. 83:81–90

Katoh-Semba, R., and Kashiwamata, S. (1984) Rates of protein synthesis and degradation in Gunn rat cerebellum with bilirubin induced cerebellar hypoplasia. Neurochem. Pathol. 2: 31–37

Kawahara, H., et al. (2005) Medical treatment of infantile hypertrophic pyloric stenosis. J. Pediat. Surg. 40:1848–1851

Kelly, J., et al. (1992) An improved model of acetaminophen induced fulminant hepatic failure in dogs. Hepatology 15:329–335

Khalil, A. (2000) Byler disease progressive familial intrahepatic cholestasis Hepatology 32: 1337–1341

Kim, H., et al. (1980) Lack of predictive indices in kernicterus. Pediatrics 66:852–858

Kniffin, C., and Wright, M. (2002) Crigler-Najjar syndrome type 2. OMIM #606785, Johns Hopkins University

Ko, J., et al. (2007) Neonatal intrahepatic cholestasis caused by citrin deficiency in Korean infants. J. Korean Med. Sci. 22:952–956

Kochar, D. (2006) The changing spectrum of severs falciparum malaria. J. Vector Borne Dis. 43:104–108

Kochar, D., et al. (2003) Malarial hepatopathy. J. Assoc. Physic. India 51:1069–1072

Koosha, A., and Rafizadeh, B. (2007) Evaluation of neonatal indirect hyperbilirubinemia at Zanjan Province of Iran in 2001–2003: prevalence of glucose-6-phosphate dehydrogenase deficiency. Singapore Med. J. 48:424–428

Krahenbuhl, S., et al. (1995) Toxicity of bile acids on the electron transport chain of isolated rat liver mitochondria. Hepatology 22:607–612

Kren, B., et al. (1999) Correction of the UDP glucuronosyltransferase gene defect in the Gunn rat model of Crigler-Najjar syndrome type 1 with a chimeric oligonucleotide. P. Natl. Acad. Sci. 96:10349–10354

Labrune, P. (2001) Crigler-Najjar syndrome. Orphanet

Labrune, P., et al. (1994) Genetic heterogeneity of Crigler-Najjar syndrome type 1: a study of 14 cases. Hum. Genet. 94:693–697

Labrune, P., Myara, A., and Huguet, P. (1989) Jaundice with hypertrophic pyloric stenosis: a possible early manifestation of Gilbert syndrome. J. Pediat. 115:93–95

Lanone, S., et al. (2005) Bilirubin decreases nos2 expression via inhibition of NAD(P)H oxidase: implications for protection against endotoxic shock in rats. FASEB J. 19:1890–1892

Lathe, G., and Walker, M. (1958) The synthesis of bilirubin glucuronide in animal and human liver. Biochem. 70:705–712

Lathe, G. (1972) The degradation of haem by mammals and its excretion as conjugated bilirubin. Essays Biochem. 8:107–148

Lee, C., et al. (1995) Postnatal maturation of the blood brain barrier for unbound bilirubin in newborn piglets. Brain Res. 689:233–238

Leitich, H., et al. (1999) Cervical length and dilation of the internal cervical os detected by vaginal ultrasonography as markers for preterm birth. Am. J. Obstet. Gynecol. 181:1465–1472

Lenhardt, M., McArtor, R., and Bryant, B. (1984) Effects of neonatal hyperbiliruninemia on the brainstem electric response. J. Pediatr. 104:281–284

Leong, A. (2007) Is there a need for neonatal screening of glucose-6-phosphate dehydrogenase deficiency in Canada? McGill J. Med. 10:31–34

Levine, G., et al. (1973) Jaundice, liver structure, and congenital pyloric stenosis. Arch. Pathol. 95:267–270

Levine, R. (1976) Bilirubin: worked out years ago? Pediatr. 64:380–385

Li, Q, et al. (1998) Gene therapy with bilirubin-UDP glucuronosyltransferase in the Gunn rat model of Crigler-Najjar syndrome type 1. Hum. Gene Ther. 9:497–505

Lightner, D., Wooldridge, T., and McDonagh, A. (1979) Configurational isomerization of bilirubin and the mechanism of jaundice phototherapy. Biochem. Biophys. Res. Comm. 86: 235–243

Lisker, R., Cordova, M., and Zarate, G. (1969) Studies on several genetic hematological traits of the Mexican population. Am. J. Phys. Anthrop. 30:349–354

Liu, Y., et al. (2003) Bilirubin as a potent antioxident supresses experimental autoimmune encephalomyelitis. J. Neuroimmunol. 139:27–35

Llansola, M., et al. (2007) NMDA receptors in hyperammonemia and hepatic encephalopathy. Metab. Brain Dis. 22:321–336

Lu, G., et al. (2001) Vaginal fetal fibronectin levels and spontaneous preterm birth in symptomatic women. Obstet Gynecol. 97:225–228

Lucey, J., Ferreiro, M., and Hewitt, J. (1968) Prevention of hyperbilirubinemia of prematurity by phototherapy. Pediatr. 41:1047–1054

Luzzatto, L., Usanga, E., and Reddy, S. (1969) Glucose-6-phosphate dehydrogenase deficienct red cells. Science 164:839–842

Main, D., et al. (1985) Can preterm deliveries be prevented? Am. J. Obstet. Gynecol. 151:892–898

Maisels, M., and McDonagh, A. (2008) Phototherapy for neonatal jaundice. N. Engl. J. Med. 358:920–928

Maisels, M., and McDonagh, A. Letter to the editor: phototherapy for neonatal. jaundice 358: 2522–2525

Maisels, M., et al. (2004) Evaluation of a new transcutaneous bilirubinometer. Pediatr. 113: 1638–1645

Mansi, G., et al. (2003) Safe hyperbilirubinemia is associated with altered neonatal behavior. Neonatol. 83:19–21

Mao, Q., Deeley, R., and Cole, S. (2000) Functional reconstitution of substrate transport by purified multidrug resistance protein MRP1 in phospholipid vesicles. J. Biol. Chem. 275:34166–34172

Marsh, K., et al. (1995) Indicators of life threatening malaria in African children. N. Engl. J. Med. 332:1399–1404

Martin, J., and Siebenthal, B. (1955) Jaundice due to hypertrophic pyloric stenosis. J. Pediat. 47:95–103

Maruo, Y., et al. (2000) Prolonged unconjugated hyperbilirubinemia associated with breast milk. Pediatr. 106:e59

Maurer, H., et al. (1968) Reduction in concentration of total serum bilirubin in offspring of women treated with phenobarbitone during pregnancy. Lancet 2:122–124

Mayor, F., et al. (1986) Effect of bilirubin on the membrane potential of rat brain synaptosomes. J. Neurochem. 47:363–369

McCaffrey, R., and Awny, A. (1970) Glucose-6-phosphate dehydrogenase deficiency in Egypt. Blood 36:793–796

McCandless, D. (2010a) Thiamine deficiency in mammals. In Thiamine deficiency and associated clinical disorders. McCandless, D. ed. Springer, New York, NY

McCandless, D. (2010b) Bilirubin encephalopathy. In Thiamine deficiency and associated clinical disorders. McCandless, D., ed. Springer, New York, NY

McCandless, D., and Abel, M. (1980) The effect of unconjugated bilirubin on regional cerebellar energy metabolism. Neurobeh. Toxicol. 2:81–84

McCandless, D., et al. (1979) Metabolite levels in brain following experimental seizures: The effect of electroshock and phenytoin in cerebellar layers. J. Neurochem. 32:743–754

McCandless, J., and McCandless, D. (2009) Bilirubin encephalopathy. In Metabolic Encephalopathy, McCandless, D. ed. Springer, New York, NY

McCoy, D. (1967) Spontaneous rupture of the liver in a case of sclerodermia. J. Irish Med. Assocn. 60:474–484

McDonagh, A., and Maisels, M. (2006) Bilirubin unbound: déjà vu all over again. Pediatr. 117:523–525

McDonagh, A., et al. (2008) Photoisomers: Obfuscating factors in clinical peroxidase measurements of unbound bilirubin? Pediatr. 123:67–76

McDonagh, A., Palma, L., and Lightner, D. (1980) Blue light and bilirubin excretion. Science 208:145–151

McDonald, J., et al. (1998) Role of glutamate recepter mediated excitotoxicity in bilirubin induced brain injury in the Gunn rat model. Exp. Neurol. 150:21–29

Medina, M., et al. (1997) Molecular genetics of glucose-6-phosphate dehydrogenase deficiency in Mexico. Blood Cells Mol. Dis. 15:88–94

Menendez, C. (1995) Malaria during pregnancy. Parasit. Today 11:178–183

Menken, M., and Weinbach, E. (1967) Oxidative phosphorylation and respiratory control of brain mitochondria isolated from kernicteric rats. J. Neurochem. 14:189–193

Menken, M., et al. (1966) Kernicterus. Arch. Neurol. 15:68–73

Menken, M., Waggoner, J., and Berlin, N. (1966) The influence of bilirubin on oxidative phosphorylation and related reactions in brain and liver mitochondria: effect of protein binding. J. Neurochem. 13:1241–1248

Merle, U., et al. (2006) Lentiviral gene transfer ameliorates disease progression in Long Evans cinnamon rats: An animal model for Wilson disease. Scand. J. Gastroent. 41:974–982

Miles, K., et al. (2006) Adenovirus mediated gene therapy to restore expression and functionality of multiple UDP glucuronosyl transferase 1A enzymes in Gunn rat liver. J. Pharm. Exp. Ther. 318:1240–1247

Miller, C., and Reed, H. (1958) The relation of serum concentrations of bilirubin to respiratory function of premature infants. Pediat. 19:362–369

Milligan, T. (2006) Single vrs double volume exchange transfusion in jaundiced newborn infants. Cochrane Database Syst Rev. Wiley 1–63

Miranda, P., and Bosma, P. (2009) Towards liver-directed gene therapy for Crigler-Najjar syndrome. Curr. Gene ther. 2:72–82

Mishra, S., Mahapatra, S., and Mohanty, S. (2003) Jaundice in falciparum malaria. JIACM 4: 12–13

Mitchell, L., and Risch, N. (1993) The genetics of infantile hypertrophic pyloric stenosis. A reanalysis. AJDC 147:1203–1211

Mohammadzadeh, A., et al. (2009) Prevalence of glucose-6-phosphate dehydrogenase deficiency in neonates of Northeast of Iran. J. Chinese Clin. Med. 44:448–451

Mohanty, A., and Schiff, E. (2009) The dilemma of idiopathic fulminant hepatic failure. Gastroent. Hepatol. 5:48–52

Mohapatra, M. (2006) The natural history of complicated falciparum malaria. J. Assoc. Physic. India 54:848–853

Molyneux, M., et al. (1989) Clinical features and prognostiv indicators in pediatric cerebral malaria. Q. J. Med. 71:441–459

Monaghan, G., et al. (1996) Genetic variation in bilirubin UDP-glucuronosyltransferase gene promotor in Gilbert's syndrome. Lancet 347:578–581

Monaghan, G., et al. (1999) Gilbert's syndrome is a contributary factor in prolonged unconjugated hyperbilirubinemia of the newborn. J. Pediatr. 134:441–446

Moore, P., and Karp, W. (1980) Further observations on the effect of bilirubin encephalopathy on the Purkinje cell population in Gunn rats. Epxt. Neurol. 69:408–413

Mordmuller, B., and Kremsner, P. (1998) Hyperparasitemia and blood exchange transfusion of children with falciparum malaria. Clin. Infect. Dis. 26:850–852

Morphis, L., et al. (1982) Bilirubin induced modulation of cerebral protein phosphorylation in neonatal rabbits. Science 218:156–158

Mukerjee, P., Ostrow, J., and Tiribelli, C. (2002) Low solubility of unconjugated bilirubin in dimethylsulfoxide water systems: implications for pKa determinations. BMC Biochem. 3:17–26

Muller, N., et al. (1991) Coincidence of schizophrenia and hyperbilirubinemia. Pharmacopsych. 24:225–228

Munch-petersen, A., et al. (1953) UDP dehydrogenase. Nature 172:1036–1038

Murphy, J., et al. (1981) Pregnanediols and breast milk jaundice. Arch. Dis. Child. 56:474–476

Mustafa, M., Cowger, M., and King, T. (1969) Effects of bilirubin on mitochondria reactions. J. Biol. Chem. 244:6403–6414

Nakamura, H., et al. (1992) Determination of serum unbound bilirubin for prediction of kernicterus in low birth weight infants. Acta Pediatr. Jpn. 34:642–647

Nakata, D., Zakim, D., and Vessey, D. (1976) Defective function of a microsomal UDP-glucuronyltransferase in Gunn rats. Proc. Natl. Acad. Sci. 73:289–292

Narian, J. (2008) Malaria in the south east Asia region. Ind. J. Med. Res. 128:1–3

Neuzil, J., and Stocker, R. (1994) Free and albumin bound bilirubin are efficient co-antioxidants for alpha tocopheral. J. Biol. Chem. 269:16712–16719

Newman, T., and Klebanoff, M. (1993) Neonatal hyperbilirubinemia and longterm outcome: another look at the collaborative perinatal project. Pediatr. 92:651–657

Newman, T., et al. (2004) 5-year neurodevelopmental outcome of newborns with total serum bilirubin levels 25 mg/dl. Pediatr. Res. 55:461A

Newman, T., et al. (2006) Outcomes among newborns with total serum bilirubin levels of 25 mg per deciliter or more. N. Engl. J. Med. 354:1889–1900

Newman, T., et al. (2009) Numbers needed to treat with phototherapy according to American Academy of Pediatrics Guidelines. Pediatr. 123:1352–1359

Newman, T., Liljestrand, P., and Escobar, G. (2003) Infants with bilirubin levels of 30 mg/dl or more in a large managed care organization. Pediatr. 111:1303–1311

Newton, C., et al. (1996) Epileptic seizures and malaria in Kenyan children. Trans. R. Soc. Trp. Med. Hyg. 90:152–155

Newton, C., et al. (1997) Intracranial hypertension in Africans with cerebral malaria. Arch. Dis. Child. 76:219–226

Newton, C., Taylor, T., and Whitten, R. (1998) Pathophysology of fatal falciparum malaria in African children. Am. J. Trop. Med. Hyg. 58:673–683

Newton, C., Hien, T., and White, N. (2000) Cerebral malaria. J. Neurol. Neurosurg. Psychiat. 69:433–441

Nguyen, T., and Ferry, P. (2007) Gene therapy for liver enzyme deficiencies: what have we learned from models for Crigler-Najjar and tyrosinemia? Expert. Rev. Gastroent. Hepatol. 1: 155–171

Nguyen, T., et al. (2002) Highly efficient lintiviral vector mediated transduction of nondividing fully reimplantable primary hepatocytes. Mol. Ther. 6:199–209

Nguyen, T., et al. (2005) Therapeutic lentivirus mediated neonatal in vivo gene therapy in hyperbilirubinemic Gunn rats. Mol. Ther. 12:852–860

Nguyen, T., et al. (2006) Ex vivo lentivirus transduction and immediate transplantation of uncultered hepatocytes for treating hyperbilirubinemic Gunn rat. Transplantation 82: 794–803

Nguyen, T., et al. (2007) Critical assessment of lifelong phenotype correction in hyperbilirubinemic Gunn rats after retroviral mediated gene transfer. Gene Ther. 17:1270–1277

O'Callaghan, J., and Miller, D. (1985) Cerebellar hypoplasia in the Gunn rat is associated with quantitive changes in neurotypic and gliotypic proteins. Pharmacol. 234:522–533

Odell, G. (1959) Studies in kernicterus. 1. The protein binding of bilirubin. J. Clin. Invest. 38: 823–833

Odell, G. (1966) The distribution of bilirubin between albumin and mitochondria. J. Pediatr. 68:164–180

Odell, G., (1959) Studies in kernicterus. J. Clin. Invest. 38:823–839

Odell, G., Natzschka, J., and Storey, G. (1967) Bilirubin nephropathy in the Gunn strain of rat. Am. J. Physiol. 212:931–938

Ohashi, K., Park, F., and Kay, M. (2002) Role of hepatocyte direct hyperplasia in lentivirus mediated liver transduction in vivo. Human Gene Ther. 13:653–663

Okolicsany, L., et al. (1978) An evaluation of bilirubin kinetics with respect to the diagnosis of Gilbert's syndrome. Clin. Sci. Mol. Med. 54:539–547

Okuda, H., et al. (1989) Dose related effects of phenobarbital on hepatic glutathione-S-transferase activity and ligand levels in the rat. Drug Metab. Dispos. 17:677–682

Okumura, A., et al. (2001) Preterm infants with athetoid cerebral palsy: kernicterus? Arch. Dis. Child. Fetal Neonatal Ed. 84:136–137

Okumura, A., et al. (2009) Kernicterus in preterm infants. Pediatr. 123: e1052–e1058

Onishi, S., et al. (1979) The separation of configurational isomers of bilirubin by high pressure liquid chromatography and the mechanism of jaundice phototherapy. Biochem. Biophys. Res. Comm. 90:890–896

Onishi, S., Kawade, N., and Itoh, S. (1979) Postnatal development of UDP-glucuronyl transferase activity. Biochem. J. 184:705–707

Orrenius, S., Ericsson, J., and Ernster, L. (1965) Phenobarbitol induced synthesis of the microsomal drug metabolizing enzyme system and its relationship to the proliferation of endoplasmic membranes. J. Cell Biol. 25:627–638

Orth, J. (1875) Ueber das vorkommen von bilirubinkrystallen bei neugeborenen kinder. Virchow's Arch. Path. Anat. 63:447–472

Ostrow, J., and Mukerjee, P. (2007) Revalidation and rationale for high pKa values of unconjugated bilirubin. BMC biochem. 8:7–19

Ostrow, J. (1967) Photo oxidative derivatives of 14C bilirubin and their excretion by the Gunn rat. In Bilirubin metabolism. Bouchier, I., and Billing, B., eds. Blackwell, Oxford, pp 117–127

Ostrow, J. (1986) Bile pigments and jaundice. Ostrow, J., ed. Marcell Dekker, New York, NY

Ostrow, J. and Tiribelli, C. (2003) Bilirubin, a curse and a boon. GUT 52:1668–1670

Ostrow, J. Pascolo, L., and Tiribelli, C. (2003) Reassessment of the unbound concentrations of unconjugated bilirubin in relation to neurotoxicity in vitro. Pediatr. Res. 54:98–104

Ostrow, J., and Tiribelli, C. (2003) Reassessment of the unbound concentrations of unconjugated bilirubin in relation to neurotoxicity in vitro. Pediatr. Res. 54: 98–104

Ostrow, J., et al. (2003) New concepts in bilirubin encephalopathy. Europ. J. Clin. Med. 33: 988–997

Ostrow, J., Hammaker, L., and Schmid, R. (1961) The preparation of crystalline bilirubin. J. Clin. Invest. 60:1442–1452

Ostrow, J., Wennberg, R., and Tiribelli, C. (2008) Letter to the editor: phototherapy for neonatal jaundice. New Engl. J. Med. 358:2522–2525

Ostrow, J.D. (1971) Photocatabolism of labeled bilirubin in the congenitally jaundiced Gunn rat. J. Clin. Invest. 50:707–718

Owa, J., and Osinaike, A. (1998) Neonatal morbidity and mortality in Nigeria. Indian J. Pediatr. 65:441–449

Owens, D., and Sherlock, S. (1973) Diagnosis of Gilbert's syndrome: role of reduced caloric test. BMJ 3:559–567

Ozmert, E., et al. (1996) Long term follow up of indirect hyperbilirubinemia in full term Turkish infants. Acta Pediatr. 85:1440–1444

Paludetto, R., et al. (2002) Moderate hyperbilirubinemia induces a transient alteration of neonatal behavior. Pediatr. 110:e50 doi: 10.1542/peds.110.4.e50

Panizon, F. (1960) Erythrocyte enzyme deficiency in unexplained kernicterus. Lancet 2:1093

Panosian, C. (1998) Editorial response:exchange blood transfusion in severe falciparum malaria. Clin. Inf. Dis. 26:853–854

Park, W., Chang, Y., and Lee, M. (2002) Effect of 7-nitroindazole on bilirubin induced changes in brain cell membrane function and energy metabolism in newborn piglets. Biol. Neonate 82:61–65

Pascolo, L., et al. (1998) ATP dependent transport of unconjugated bilirubin by rat liver canalicular plasma membrane vesicles. Biochem. J. 331:99–103

Passonneau, J., and Lowry, O. (1993) Enzymatic analysis. Humana press, Totowa, NJ

Peeters, T. (1993) Erythromycin and other macrolides as prokinetic agents. Gastroent. 105: 1886–1899

Peled, N., et al. (1992) Gastric outlet obstruction induced by prostaglandin therapy in neonates. N. Engl. J. Med. 327:505–510

Perlman, M., et al. (1983) Auditory nerve brainstem evoked responses in hyperbilirubinemic neonates. Pediat. 72:658–664

Peters, A., and Van Noorden, C. (2009) Glucose-6-phosphate dehydrogenase deficiency and malaria: cytochemical detection of heterozygous G6PD deficiency in women. J. Histo. Cytochem. doi: 10.1369/jhc.2009.953828

Petit, F., et al. (2006) Crigler Najjar type 2 syndrome may result from several types and combinations of mutations in the UGTA1A gene. Clin. Genet. 69:525–527

Pfannenstiel, H.J. (1908) Ueber den habituellen ikterus gravis der neugeborenen. Munch Med. Woch. 55:2169–2176

Phillips, M., and Latham, P. (1982) Electron microscopy of human liver disease. In Diseases of the liver. Schiff, L., and Schiff, E. eds. Lippencott, Phidelphia, PA

Powell, L., et al. (1967) The assessment of red cell survival in idiopathic unconjugated hyperbilirubinemia (Gilbert's syndrome) by the use of radioactive diisopropylfluorophosphate. Aust. Ann. Med. 16:221–232

Quastel, J., and Bickis, I. (1959) Metabolism of normal tissues and neoplasms in vitro. Nature 183:281–284

Rageb, A., El-Alfi, O., and Abboud, M. (1966) Incidence of glucose-6-phosphate dehydrogenase deficiency in Egypt. Am. J. Hum. Genet. 18:21–28

Ramchandran, S., and Pereeria, M. (1976) Jaundice and hepatomegaly in falciparum malaria. Trop. Med. Hyg. 79:207–210

Ramos, A., Silverberg, M., and Stern, L. (1966) Pregnanediol and neonatal hyperbilirubinemia. Amer. J. Dis. Child. 111:353–356

Rao, K., and Norenberg, M. (2007) Aquaporin-4 in hepatic encephalopathy. Metab. Brain Dis. 22:265–276

Rapoport, S., and Nieradt, C. (1951) Glycerate-2-3-diphosphatase. J. Biol. Chem. 189: 683–691

Rapoport, S., and Luebering, J. (1950) The formation of 2,3-diphosphoglycerate in rabbit erythrocytes. J. Biol. Chem. 183:507–520

Rela, M., et al. (1999) Auxillary partial orthotopic liver transplantation for Crigler-Najjar syndrome type 1. Ann. Surg. 229:565–569

Reyes, H., et al. (1969) Organic anion-binding protein in rat liver: drug induction and its physiologic consequence. Proc. Natl. Acad. Sci. 64:168–173

Reynolds, S., and Blass, J. (1976) A possible mechanism for selective cerebellar damage in partial PDH deficiency. Neurology 26:625–628

Rice, A., and Shapiro, S. (2008) A new animal model of hemolytic hyperbilirubinemia-induced bilirubin encephalopathy (kernicterus). Pediatr. Res. 64:265–269

Rigato, I., et al. (2004) The human multidrug resistance associated protein MRP1 mediates ATP dependent transport of unconjugated bilirubin. Biochem. J. 383:335–341

Rigato, I., Ostrow, J., and Tiribelli, C. (2005) Bilirubin and the risk of common non-hepatic diseases. Trends mol. Med. 11:277–283

Roger, V., et al. (1996) A good model of acute hepatic failure: 95% hepatectomy. Chirurgie 121:470–473

Rolleston, H.D. (1910) Case of recurring jaundice in four successive pregnancies with fatal jaundice in three successive infants. BMJ 12:864–865

Rollins, M., et al. (1989) Pyloric stenosis: congenital or acquired? Arch. Dis. Child. 64:138–139

Romero-Gomez, M. (2005) Role of phosphate activated glutaminase in the pathogenesis of hepatic encephalopathy. Metab. Brain Dis. 20:319–326

Rose, C., and Felipo, V. (2005) Limited capacity for ammonia removal by brain in chronic liver failure: potential role of nitric oxide. Metab. Brain Dis. 20:275–284

Roseth, S., et al. (1998) Bilirubin inhibits transport of neurotransmitters in synaptic vesicles. Pediat. Res. 44:312–316

Roth, E., et al. (1983) Glocose-6-phosphate dehydrogenase deficiency inhibits in vitro growth of Plasmodium falciparum. Proc. Natl. Acad. Sci. USA 80:298–299

Rozdilsky, B. (1961) Experimental studies on the toxicity of bilirubin. In Kernicterus. Sass-Kortsak, A. ed. University of Toronto press, Toronto, ON

Rubaltelli, F., et al. (1989) Tin-protoporphyrin in the management of children with Crigler-Najjar disease. Pediatr. 84:728–731

Rubboli, G., et al. (1997) A neurophysiological study in children and adolescents with Crigler-Najjar Syndrome type 1. Neuropediatrics 28:281–286

Saheki, T., et al. (2004) Adult onset type 2 citrullinemia and idiopathic neonatal hepatitus caused by citrin deficiency. Mol. Genet. Metab. 81:s20–s26

Santana, M., et al. (2009) Glucose-6-phosphate dehydrogenase deficiency in an endemic area for malaria in Manaus: a cross sectional survey in the Brazilian Amazon. Plos one $: e5259doi: 10.1371/journal.pone.0005259

Santhi, A., and Sachdeva, M. (2004) Glucose-6-phosphate dehydrogenase deficiency among the Jats and Brahmins of Sampla (Haryana). Anthropologist 6:291–292

Sass, D., and Shakil, A. (2005) Fulminant hepatic failure. Liver Transpl. 11:594–605

Sawasaki, Y., Yamada, N., and Nakajima, H. (1976) Studies on kernicterus. Proc. Japan Acad. 52:41–44

Sawasaki, Y., Yamada, N., and Nakajima, H. (1973) Change in brain bilirubin level in Gunn rat. Proc. Japan Acad. 49:84–88

Scheidt, P., et al. (1991) Intelligence at 6 years in relation to neonatal bilirubin levell. Pediatr. 87:797–805

Schenker, S. (1963) Disposition of bilirubin in the fetus and the newborn. Ann. N. Y. Acad. Sci. 111:303–305

Schenker, S., Dawber, N., and Schmid, R. (1964) Bilirubin metabolism in the fetus. J. Clin. Invest. 43:32–39

Schenker, S., et al. (1964) Bilirubin metabolism in the fetus. J. Clin. Invest. 43:32–40

Schenker, S., et al. (1967) Studies on the intracerebral toxicity of ammonia. J. Clin. Med. 46:838–848

Schenker, S., McCandless, D., and Wittgenstein, E. (1966) Studies in vivo of the effect of unconjugated bilirubin on hepatic phosphorylation and respiration. GUT 7:409–415

Schenker, S., McCandless, D., and Zollman, P. (1966) Studies of Cellular toxicity of unconjugated bilirubin in kernicteric brain. J. Clin. Med. 45:1213–1220

Schmid, R. (1956) Direct-reacting bilirubin, bilirubin glucuronide, in serum, bile, and urine. Science 124:76–77

Schmid, R., et al. (1958) Congenital jaundice in rats due to a defect in glucuronide formation. J. Clin. Invest. 37:1123–1130

Schmid, R., et al. (1966) Enhanced formation of rapidly labelled bilirubin by phenobabitol: Hepatic microsomal cytochromes as possible sources. Biochem. Biophysiol. Res. Comm. 24: 319–327

Schmorl, C. (1904) Zur kenntnis des ikterus neonatorium. Ver. Dtsch. Path. Ges. 6:109–115

Schoenfield, L., et al. (1963) Studies of chronic idiopathic jaundice (Dubin-Johnson syndrome) Gastroent. 44:101–108

Schroter, W. (1980) Successful long term phenobarbital therapy of hyperbilirubinemia in congenital hemolytic anemia due to glucose phosphate isomerase deficiency. Eur. J. Pediatr. 135:41–43

Schutta, H., Johnson, L., and Neville, H. (1970) Mitochondril abnormalities in bilirubin encephalopathy. J. Neuropath. Exp. Neurol. 29:296–310

Seidman, D., et al. (1991) Neonatal hyperbilirubinemia and physical and cognitive performance at 17 years of age. Pediatr. 88:828–833

Seppen, J., et al. (1996) Bilirubin glucuronidation by intact Gunn rat fibroblasts expressing bilirubin UDP-glucuronosyltransferase. Biochem. J. 314:477–483

Shah, S., et al. (2009) Malarial hepatology in falciparum malaria. J. coll. Phys. Surg. Pak. 19: 367–370

Shankaran, S., and Poland, R. (1977) The displacement of bilirubin from albumin by furosemide. J. Pediatr. 90:642–646

Shapiro, S., and Conlee, J. (1991) Brainstem auditory evoked potentials correlate with morphological changes in Gunn rat pups. Hear. Res. 57:16–22

Shapiro, S. (1988) Acute brainstem auditory evoked potential abnormalities in jaundiced Gunn rats given sulfonamide. Ped. Res. 23:306–310

Shapiro, S. (1993) Reversible brainstem auditory evoked potential abnormalities in jaundiced Gunn rats given sulfonamide. Ped. Res. 34:629–633

Shapiro, S. (2003) Bilirubin toxicity in the developing nervour system. Pediatr. Neurol. 29:410–421

Shet, R. (2004) Metabolic concerns associated with antiepileptic medications. Neurology 63:24–29

Shevell, M., Majnemer, A., and Schiff, D. (1998) Neurologic perspectives of Crigler-Najjar syndrome type 1. J. Child. Neurol. 13:265–269

Shigematsu, Y., et al. (2002) Newborn mass screening and selective screening using electrospray tandem mass spectrometry in Japan. J. Chrom B Analyt. Tech. Biomed. Life Sci. 776:39–48

Schutta, H., Johnson, L., and Neville, H. (1970) Mitochondrial abnormalities in bilirubin encephalopathy. J. Neuropath. Exp. Neurol. 29:907–925

Silberberg, D., and Schutta, H. (1967) The effects of inconjugated bilirubin and related pigments on cultures of rat cerebellum. J. Neuropath. Exp. Neurol. 26:572–583

Sirota, L., et al. (1988) Breast milk jaundice in preterm infants. Clin. Ped. 4:195–197

Smetana, H., Edlow, J., and Glunz, P. (1965) Neonatal jaundice. Arch. Path. 80:553–574

Smith, C., et al. (2004) Auditory brainstem response detects early bilirubin neurotoxicity at low indirect bilirubin values. J. Perinatol. 24:730–732

Snow, R., et. al. (2005) The global distribution of clinical episodes of p. falciparum malaria. Nature 434:214–217

Soorani-Lunsing, I., Woltil, H., and Hadders-Algra, M. (2001) Are moderate degrees of hyperbilirubinemia in healthy term neonates really safe for the brain? Pediatr. Res. 50:701–705

Soylu, A., et al. (2008) Intrahepatic cholestasis in subclinical and overt hyperthyroidism: two case reports. J. Med. Case Rep. 2:116–123

Springer, S., and Annibale, D. (2008) Kernicterus: treatment and medication. eMedicine972576

Srivastava, A., et al. (2009) Falciparum malaria with acute liver failure. Trop. Gastrol. 3: 172–174

Stanley, I., et al. (2004) An evidence based review of important issues Concerning neonatal hyperbilirubinemia, Pediatr. 114:130–153

Stern, L., et al. (1970) Effect of phenobarbitol on hyperbilirubinemia and glucuronide formation in newborns. Arch. Pediatr. Adol. Med. 120:26–31

Stocker, R., Glazer, A., and Ames, B. (1987) Antioxident activity of albumin bound bilirubin. Proc. Natl. Acad. Sci. 84:5918–5922

Stoll, M., Zenone, E., and Ostrow, J. (1981) Excretion of administered and endogenous photobilirubins in the bile of the jaundiced Gunn rat. J. Clin. Invest. 68:134–141

Storey, I., and Dutton, G. (1955) Uridine compounds in glucuronic acid metabolism. Biochem. J. 59:279–288

Strominger, J., et al. (1957) Enzymatic formation of uridine diphosphoglucuronic acid. J. Biol. Chem. 224:79–90

Stumpf, D., Eguren, L., and Parks, J. (1985) Bilirubin increases mitochondrial inner membrane conductance. Biochem. Med. 34:226–229

Sugama, S., Soeda, A., and Eto, Y. (2001) Magnetic resonance imaging in three children with kernicterus. Pediatr. Neurol. 25:328–331

Swenson, R., and Jew, J. (1982) Learning deficits and brain monoamines in rats with congenital hyperbilirubinemia. Exp. Neurol. 76:447–456

Takahashi, M., et al. (1996) Long term correction of bilirubin UDP glucuronosyl transferase deficiency in Gunn rats by administration of a recombinant adenovirus during the neonatal period. J. Biol. Chem. 271:26536–26542

Talafant, E. (1956) Properties and conjgation of the bile pigment giving a direct diazo reaction. Nature 178:312–314

Tenhunen, R., et al. (1968) The enzymatic conversion of heme to bilirubin by microsomal heme oxygenase. Proc. Natl Acad Sci. 61:748–755

Taubes, G. (2002) Gene therapy: the strange case of chimeraplasty. Science 298:2116–2120

Thong, Y., and Rencis, V. (1977) Bilirubin inhibits hexose-monophosphate shunt activity of phagocytosing neutrophils. Acta Paediatr. Scand. 66:757–759

Tiribelli, C., and Ostrow, J.D. (2005) The molecular basis of bilirubin encephalopathy and toxicity: Report of an EASL single topic conference, Trieste, Italy, 1–2 October, 2004. J. Hepatol. 43:156–166

Tobias, J. (2002) Ursodeoxycholic acid in the treatment of cholestasis and hyperbilirubinemia in pediatric intensive care unit patients. South. Med. J. online Nov 1, 2002, a09723387

Toietta, G., et al. (2005) Lifelong elimination of hyperbilirubinemia in the Gunn rat with a single injection of helper-dependent adenoviral vector. Proc. Natl. Acad. Sci. 102:984–989

Tolmacheva, E., and van Luijtelaar, G. (2007) Absence seizures are reduced by the enhancement of GABA-ergic inhibition in the hippocampus in WAG/Rij rats. Neurosci. Lett. doi: 10.1016/j.neulet.2007.01.038

Trioche, P., et al. (1999) Jaundice with hypertrophic pyloric stenosis as an early manifestation of Gilbert syndrome. Arch. Dis. Child. 81:301–303

Tripathy, R., et al. (2007) Clinical manifestations and predictors of severe malaria in Indian children. Pediatr. 120:454–460

Trolle, D. (1968) Decrease of total serum bilirubin concentration in newborn infants after phenobarbitone treatment. Lancet 2:705–708

van Nievwkerk, C., et al. (1996) Progressive familial intrahepatic cholestasis Gastroent 111: 165–171

Vanderwinden, J., et al. (1992) Nitric oxide synthase activity in infantile hypertrophic pyloric stenosis. N. Engl. J. Med. 327:511–515

Vaughan, V., Allen, F., and Diamond, L.K. (1950) Erythroblastosis fetlis: IV. Further observations on kernicterus. Pediatr. 6:706–716

Vaz, A., et al. (2010) Bilirubin selectively inhibits cytochrome c oxidase activity and induces apoptosis in immature cortical neurons. J. Neorochem. 112:56–65

Vazquez, J., et al. (1988) Interaction of bilirubin with the synaptosomal plasma membrane. J. Biol. Chem. 263:1255–1265

Vietti, T. (1961) The in vivo crystallization of bilirubin in hyperbilirubinaemic infants. In Kernicterus. Sass-Kortsak, A. ed. University of Toronto press, Toronto, ON

Virchow, R. (1847) Die pathologischen pigmente. Arch. Anat. Physiol. Klin. Med. 64:741–772

Vogt, M., and Basford, R. (1968) The effect of bilirubin on the energy metabolism of brain mitochondria. J. Neurochem. 15:1313–1320

Walker, O., et al. (1992) Prognostic risk factors and post mortem findings in cerebral malaria. Trans. R. Soc. Trop. Med. Hyg. 86:491–493

Wallerstein, H. (1946) Treatment of severe erythroblastosis by simultaneous removal and replacement of blood of newborn infant. Science 103:583–587

Wang, X., and Anderson, R. (1994) Hepatocyte transplantation : a potential treatment for acute liver failure. Scand. J. Gastroent. 30:193–200

Warrell, D., Molyneux, M., and Beales, P. (1990) Severe and complicated malaria. Trans R. Soc. Trop. Med. Hyg. 84:1–65

Watchko, J. (2006) Neonatal hyperbilirubinemia-What are the risks? N. Engl. J. Med. 354:1947–1949

Watchko, J., and Maisels, M. (2003) Jaundice in low birth weight infants: pathobiology and outcome. Arch. Dis Child. 88:455–458

Waters, W., and Bowen, W. (1955) Bilirubin encephalopathy: studies related to cellular respiration. Am. J, Dis. Child. 90:603–614

Waters, W. (1961) The protective action of albumin in bilirubin toxicity in newborn puppies in kernicterus. Sass Korsak ed., Toronto p 219–221

Weisiger, R., et al. (2001) Affinity of human serum albumin for bilirubin varies with albumin concentration and buffer composition. J. Biol. Chem. 276:29953–29960

Weiss, J., et al. (1983) Albumin binding and hyperbilirubinemia. N. Engl. J. Med. 309:147–150

Wennberg, R. (2000) The blood brain barrier and bilirubin encephalopathy. Cell. Mol. Neurobiol. 20:97–109

Wennberg, R., et al. (1991) Bilirubin induced changes in brain energy metabolism after osmotic opening of the blood brain barrier. Pediatr. Res. 30:473–478

Wennberg, R., et al. (1982) Abnormal auditory brainstem response in a newborn infant with hyperbilirubinemia: improvement with exchange transfusion. J. Pediatr. 100:624–626

Wennberg, R., et al. (1993) Brainstem bilirubin toxicity in the newborn primate may be promoted and reversed by modulating PCO2. Ped. Res. 34:6–9

Wennberg, R., et al. (2006) Toward understandingm kernicterus:A challenge to improve the management of jaundiced newborns. Pediatr. 117:474–485

Wennberg, R., Rasmussen, L., and Ahlfors, C. (1977) Displacement of bilirubin from human albumin by three diuretics. J. Pediatr. 90:647–650

White, N. (1996) The treatment of malaria. N. Engl. J. Med. 335:800–806

White, N., and Ho, M. (1992) The pathophysiologt of malaria. Adv. Parasitol. 31:83–173

White, N., et al. (1988) Single dose phenobaritone prevents convulsions in cerebral malaria. Lancet ii: 64–66

WHO Severe and complicated malaria: A report of the WHO malaria action programme. Trans. R. Soc. Med. Hyg. 80:1–50

WHO working group. (1989) Glucose-6-phosphate dehydrogenase deficiency. Bull. WHO 67:601–611

Wilairatana, P., Looareesuwan, S., and Charoenlarp, P. (1994) Liver profile changes and complications in jaundiced patients wioth falciparum malaria. Trop. Med. Parasit. 45:298–302

Winter, S., et al. (2002) Trends in the prevalence of cerebral palsy in a population based study. Pediatr. 110:1220–1225

Wong, Y., and Wood, B. (1971) Breast milk jaundice and oral contraceptives. BMJ 5:403–404

Woolley, M., et al. (1974) Jaundice, hypertrophic pyloric stenosis, and hepatic glucuronyl transferase. J. Pediat. Surgery 9:359–363

Wright, G., et al. (2007) Brain cytokine flux in acute liver failure and its relationship with intracranial hypertension. Metab. Brain Dis. 22:375–388

Wu, T., et al. (1994) Unconjugated bilirubin inhibits the oxidation of human low density lipoprotein better than Trolox. Life Sci. 54:477–481

Yamamoto, K., et al. (1998) Analysis of bilirubin uridine 5-prime-diphosphate glucuronosyltransferase gene mutations in seven patients with Crigler-Najjar syndrome type 2. J. Human. Genet. 43:111–114

Yang, Y., et al. (1994) Inactivation of E2a in recombinant adenoviruses improves the prospect for gene therapy. Natl. Genet. 7:362–369

Yen, J., et al. (2003) Idiopathic hypertrophic pyloric stenosis in identical twins. Chang Gung Med. J. 26:933–936

Yeung, C. (1992) Neonatal jaundice in Chinese. H. K. J. Paediatr. 9:233–250

Yeung, C. and Yu, V. (1971) Phenobarbitone enhancement of bromosulphalein clearance in neonatal hyperbilirubinemia. Pediatr. 48:556–561

Yilmaz, Y., et al. (2001) Prognostic value of auditory brainstem responses for neurologic outcome in patients with neonatal indirect hyperbilirubinemia. J. Child. Neurol. 16:772–775

Yin, J., Miller, M., and Wennberg, R. (1991) Induction of hepatic bilirubin metabolizing enzymes by the traditional Chinese medicine yin zhi huang. Dev. Pharmacol. Ther. 16:176–184

Yoshida, A. (1973) Hemolytic anemia and G6PD deficiency. Science 179:532–537

Young, F., and Cheah, S. (1977) An invitro study of the effect of free fatty acids on the bilirubin serum albumin complex. Res. Comm. Chem. Path. Pharm. 17:679–684

Youngs, J., and Cornatzer, W. (1963) Effect of oxidative phosphorylation inhibiters on synthesis of liver mitochondrial phospholipids. Proc. Soc, Expt. Biol. Med. 112: 308–311

Yu, M., et al. (1992) Association between glucose-6-phosphate dehydrogenase deficiency and neonatal jaundice: interaction with multiple risk factors. Int. J. Epidem. 21:947–952

Zetterstrom, R. (1961) Introduction. In Kernicterus. Sass-Kortsak, A., ed. University of Toronto press, Toronto, ON

Zetterstrom, R., and Ernster, L. (1956) Bilirubin, an uncoupler of oxidative phosphorylation in isolated mitochondria. Nature 178:1335–1338

Zieve, L. (1966) Pathogenesis of hepatic coma. Arch. Int. Med. 118:211–220

Zimmerman, H. (1986) Effects of alcohol on other hepatotoxins. Alcohol. Clin. Exp. Res. 10:3–15

Zuelzer, W., and Mudgett, R. (1950) Kernicterus: Etiologic study based on an analysis of 55 cases. Pediat. 6:452–474

Index